AF707412

2021 Volume 62

# Glass Technology

## European Journal of Glass Science and Technology

## Part A

**Society of Glass Technology**

9 Churchill Way, Chapeltown,

Sheffield S35 2PY, UK

The objects of the Society of Glass Technology are to encourage and advance the study of the history, art, science, design, manufacture, after treatment, distribution and end use of glass of any and every kind. These aims are furthered by meetings, publications, the maintenance of a library and the promotion of association with other interested persons and organisations.




| Website | www.sgt.org |
|---|---|
| Electronic journals | www.ingentaconnect.com/content/sgt/ |

ISBN 978-0-900682-91-9

# Contents


1 **News**
6 **Conferences and events**
9 Dissolution of sodium silicate glasses for the production of water glass – part IV: characterisation of reaction layer and glass surface
M. Dathe, W. Lebek, E. Koslowski, M. Bron, T. Bräuniger, I. Stoessel & H. Roggendorf
24 Some secrets of Renaissance Venetian opaque red glass revealed by analyses and glassmaking treatises
Mario Bandiera, Marco Verità, Sandro Zecchin & Marcia Vilarigues
34 The effect of the colemanite on the melting properties of rock glass wool
Qingwei Wang, Jie Gao, Jin Liu & Lida Luo
41 Effect of temperature change on sulphuric acid–water glass slurry
Zuochun Li, Guihe Wang & Feng Huang
45 How the United Nations International Year of Glass 2022 arrived and what happens now
Alicia Duran & John Parker
46 UN developmental goals and glass
51 **News**
60 Glass blowing becomes expression of Spanish culture
61 **Conferences and events**
64 Impossible Objects enters joint development agreement with Owens Corning
65 Experimental and numerical study on abrasive water jet machining of laminated glass
Ajitanshu Vedrtnam, Dheeraj Gunwant, Shashkant Chaturvedi, Dheeraj Sagar & S. J. Pawar
77 'Glass in our Lives' is a boon says the youth
79 Recycling of lead glass from landfill
81 "24-hour turnaround" for Mersen UK
83 **News**
92 **Conferences and events**
96 Thermal, optical and mechanical properties of new glass compositions containing fly ash
Clarissa Justino de Lima, Fred Veer, Hongzhi Zhang, Fernando França de Mendonça Filho, Oguzhan Copuroglu & Rob Nijsse
109 International Year of Glass events
111 Using near infrared thermal imaging to optimise production and quality from oxy-gas borosilicate furnace
112 World first as 100% hydrogen fired at Pilkington UK, St Helens
114 UK glass industry net zero strategy summary
116 SGT Furnace Solutions training seminars
117 **News**
127 **Conferences and events**
131 **Book review:** The First Thousand Years of Glass-Making in the Ancient Near East: Compositional Analyses of Late Bronze Age and Iron Age Glasses
132 A BRIEF HISTORY OF GLASS AND GLASSMAKING
Ian Donald
141 International Year of Glass
143 Rebuilding the UK telecommunications industry
145 SGT Furnace Solutions training seminars
146 **News**
153 Morley Glass triples post-consumer glass recycling capacity
154 Fountain – the first highly durable anti-reflective coated glass offered by AGCULTURE
156 **Conferences and events**
160 Effect of natural convection and internal radiation on glass conditioning
M. R. Hyre & K. Paul
168 Water in natural glasses
V. Bykov & V. Anfilogov
173 International Year of Glass
175 The Upcycled Glass Company of South Devon
177 Container glass collection for recycling in Europe
179 World's first all-oxide all-solid-state sodium (Na) ion secondary battery
181 **News**
195 **Conferences and events**
198 Four testing and evaluation methods of the spreading effects on E-glass woven fabrics
Xi Wang, Tianyong Zheng & Xiangchun Ning

**Annual contents**
**Author index**

# GLASS TECHNOLOGY

## Author Index


Anfilogov, V. 168

Bandiera, M. 24
Bräuniger, T. 9
Bron, M. 9
Bykov, V. 168

Chaturvedi, S. 65
Copuroglu O. 96

Dathe, M. 9
de Lima, C. J. 96
de Mendonça Filho, F. F. 96
Donald, I. 132
Duran, A. 45

Feng Huang 41

Guihe Wang 41
Gunwant, D. 65

Hongzhi Zhang 96
Hyre, M. R. 160

Jie Gao 34
Jin Liu 34

Koslowski, E. 9

Lebek, W. 9
Lida Luo 34

Nijsse, R. 96

Parker, J. M. 45
Paul, K. 160
Pawar, S. J. 65

Qingwei Wang 34

Roggendorf, H. 9

Sagar, D. 65
Stoessel, I. 9

Tianyong Zheng 198

Vedrtnam, A. 65
Veer, F. 96
Verità, M. 24
Vilarigues, M. 24

Xi Wang 198
Xiangchun Ning 198

Zecchin, S. 24
Zuochun Li 41

Speaking in front of a UK parliamentary committee, **British Glass** has raised concerns that the inclusion of glass beverage packaging in the governments proposed Deposit Return Scheme (DRS) would lead to an increase in plastic packaging and limit the sector's ability to recycle glass back into bottles and jars.

Oral evidence was provided to the Environmental Audit Committee's inquiry 'Next steps for deposit return schemes'.

Speaking after giving evidence to the Committee, British Glass' Technical Director Dr Nick Kirk said, "We welcome the Environmental Audit Committee's inquiry at the pivotal moment for deciding how we collect, manage and recycle our waste and resources for generations to come.

"British Glass does not oppose the introduction of a deposit return scheme but put simply it is the wrong solution for glass beverage packaging. It would split glass food and beverage packaging into two waste streams, to the detriment of both.

"It risks increasing emissions in the glass sector and reducing the amount of recycled material available to be used again in the manufacture glass bottles and jars. As international evidence has shown, it could well lead to more, not less, plastic packaging on the market.

"We already have a solution to improving glass recycling, and it's at our doorsteps. Instead of including glass packaging in a damaging and expensive DRS, we want to see more glass collected through improved household collections under extended producer responsibility, enhanced by consistent collections. Only this will create a truly circular economy for glass packaging."



**Verescence** has acquired a majority stake in **Pacificglas**, the Korean glass packaging to the local cosmetics sector and a long-term partner with **Amorepacific**.

For over 120 years, Verescence has been a privileged partner to the biggest perfume and cosmetics brands due to its extensive know-how and expertise. Verescence produces 500 million bottles per year in its three glass production sites and its four decoration sites in Europe and North America. In 2019, the company employed 2300 people worldwide and achieved sales revenue of €309 million.

Thomas Riou, CEO of Verescence said, "We are delighted with the acquisition of a majority stake in Pacificglas and the long-term partnership with Amorepacific. This is a major step in our Strategic Plan, Verescence 2022. Amorepacific will become a top customer of Verescence group and with Pacificglas, Verescence will reinforce its leadership in high-end Perfumery & Cosmetic glass packaging. From this strong base Verescence plans to become a pan-Asian leader and will invest in technology and capacity to meet growing demand of high-end glass."

Glass-manufacturing firm **View** is set to install smart windows in the new 350,000 ft$^2$ Terminal 5 expansion at Chicago's O'Hare International Airport (ORD) in the US. As part of an $8.5 billion project, the expansion of the terminal will upgrade the airport.

By using artificial intelligence (AI), View Smart Windows control the amount of natural light in the terminal and reduce heat and glare. Furthermore, these smart windows decrease energy consumption by minimising the need for cooling.

The new terminal is designed by the architecture companies **HOK** and **Muller2** in the form of an aircraft wing and consist of floor-to-ceiling windows, offering expansive runway view and increased natural light.

**O-I Glass** is investing approximately $75 million in an expansion at its Zipaquirá, Colombia facility. Upon completion by the end of 2022, the project is expected to add nearly 2% of capacity to the company's Americas segment and produce about 500 million bottles annually.

"Our customers recognize the strong brand building characteristics and superior sustainability profile that make glass the preferred packaging substrate. In particular, strong market growth and affinity for glass is fueling significant demand for glass packaging in the Andean market. Building on our highly successful expansion at the Zipaquirá facility in 2019, we are excited to add a fourth furnace with an attractive return on investment to support market growth. As a result, the facility will be one of O-I's largest and most cost effective plants in the global network," said Andres Lopez, O-I CEO.

Investment in a transport tester and new vertical load machine has enabled **Glass Technology Services** to add to their large range of glass performance tests.

The transport tester, an **Ohaus Orbital Shaker**, simulates the transportation process to assess performance during distribution and highlight any potential issues before they arise. New tests in the machine will assist customers to evaluate the robustness of printed and decorated glass packaging components by placing filled bottles or finished packs into a fully assembled case.

Testing is carried out using a continuous vibration motion – producing speeds of up to 500 revolutions per minute mimicking random vibration caused by the transporting of glass containers. This is used to assess the performance of a glass container with its interior packing material in terms of its ruggedness and the protection that it provides the contents when subjected to transportation.

To test smaller and more delicate glasses and glass containers the vertical load machine has been upgraded to an **Instron** 34TM30 which has the capability of testing up to 30 kN forces and with the addition of a calibrated 500 N load cell. This provides Glass Technology Services the opportunity to testing much smaller samples like those from cosmetic and pharmaceutical markets along with an increased potential for bespoke testing such as ring on ring and three-point bend. Three-point bend testing helps evaluate the reaction of the glass in realistic loading conditions by applying a flexural force. For ring on ring testing a disk or plate shaped sample is loaded by concentric rings to produce homogeneous biaxial stress – this testing is typically used for flat glass applications.

***NSG Group** recently announced that its windshield Head Up Display (HUD) technology will be featured in **General Motors'** all-new full-size SUVs: the 2021 Chevrolet Tahoe and Suburban, 2021 GMC Yukon/Yukon XL, and 2021 Cadillac Escalade/Escalade ESV models.*

*This new 15-inch multi-colour HUD will be among the many exclusive features making these models the best-selling SUVs in the United States.*

*NSG's unique Advanced Press Bending technology, with best-in-class surface control, enables the production of complex windshields with HUD, an increasingly popular option among new vehicles.*

**International Cookware**, licensee of the Pyrex® trademark, is taking over **Duralex International** confirming its leadership position in glass kitchenware in France and Europe.

Since the acquisition of a majority stake in International Cookware by the pan-European investment fund **Kartesia** in April 2020, the group has adopted an ambitious development and growth strategy in the glass sector, reflected today in this takeover.

This takeover confirms a coherent and complementary approach adopted by the two iconic brands in French glass-making expertise, creating a powerhouse with global ambitions, based on strong synergies between the two companies and their respective industrial sites. These are two world-class brands, using two different types of glass – Pyrex uses borosilicate glass to make kitchenware with resistance to thermal shock of up to 220°C, while Duralex® makes highly tempered soda–lime–silica glass for durable and stylish tableware. The two have extremely complementary and convergent expertise, product ranges, brand images and even geographic locations (Châteauroux in Indre for Pyrex and Chapelle-Saint-Mesmin in Loiret for Duralex, both in the Centre Val de Loire region).

**NSG Group** has successfully developed a new antibacterial and antiviral coating for glass substrates, based on its proprietary sol-gel technology.

Copper contained in the newly developed coating, reacts with water, oxygen and other substances in air to generate active oxygen ($H_2O_2$, OH, etc), which in turn inactivates bacteria and viruses by destroying their envelopes and decomposing lipids, protein and gene materials.

Laboratory evaluation conducted by a public institution based on international standards has confirmed a high virus inactivation effect of 99·99% or more without the need for light activation.

In addition to the excellent antibacterial and antiviral performance with effectiveness in darker places, the coating's other properties such as high durability and transparency make it suitable for a wide range of glass applications which may be directly contacted by people. In addition to mobile information devices such as smart phones and tablets, it can be used for the operating panels and buttons of various equipment such as cash registers, ATMs, medical equipment, elevators and household appliances. The Group believes that the technology can contribute to making home and public places requiring stringent hygiene, such as hospitals, care facilities and restaurants, safer and more comfortable.

**SCHOTT**'s pharmaceutical packaging business unit has delivered enough vials to provide more than 1 billion doses of COVID-19 vaccines, marking a massive contribution to the global fight against COVID-19. The company remains well on track to deliver vials for more than 2 billion vaccine doses through 2021.

The vials were delivered to projects around the world, with a focus on the US, Europe, and China. Approximately 90% of approved vaccines rely on SCHOTT vials.

"The reaction of the pharmaceutical community to COVID-19 is a testament to the power of scientific progress," said Dr Frank Heinricht, CEO, "In just under a year, all previous records for vaccine development have been shattered by not just one, but several research groups. The entire industry is successfully working together to ensure an adequate supply. We're also working with our government partners to evaluate ways to improve the supply chain and expand production capacity."

SCHOTT's global manufacturing footprint includes five sites dedicated to the manufacture of type-1 borosilicate glass tubes and another 16 plants converting the tubes into vials.

New York Governor Andrew M. Cuomo has launched a new nation-leading collaboration with the State College of Ceramics at **Alfred University** to bolster markets for recycled glass and improve the quality of glass available for recovery throughout New York State. The **Center for Glass Innovation** was first announced in the Governor's 2020 State of the State Address and will play an instrumental role in helping recycling markets and municipal recycling programs find new ways to produce and recycle glass.

"The College of Ceramics at Alfred University is playing a critical role in enhancing recycling markets and the overall economic revival of the Southern Tier," Governor Cuomo said, "This first-ever Center for Glass Innovation builds on our efforts partnering the best academic institutions with industry experts to develop strategies to make New York a cleaner and greener state. The investment will also help to ensure our students have the education and skills they need for new opportunities and jobs in the green economy."

Recycling markets continue to experience volatility. As a result, recycling operations are struggling to find suitable markets for material, impacting local solid waste recycling efforts. Glass is the heaviest component of the municipal waste stream and is costly to process. The State Department of Environmental Conservation (DEC) is working with key stakeholders, including municipalities, to strategize on how to bolster new markets to build capacity in the state and the northeast region to address these challenges.

As a major focus of this initiative, the New York State College of Ceramics (NYSCC) will form the Center for Glass Innovation as a research resource for glass producers in New York State and, ultimately, nationally. The Center will create space for basic and applied research, user facilities, and experimental glass tanks for applied, industrial-scale research, with an emphasis on creating higher value end markets for curbside collected glass. This will be the first center of its kind in the United States where glass companies will be able to test small batches of new glass compositions in a pilot production environment.

SUNY's statutory unit at Alfred University, the NYSCC, will receive nearly $1.7 million for this initiative through the State's Environmental Protection Fund (EPF) as part of a three-year agreement. NYSCC has a world-renowned reputation in the field of glass science and engineering and is the only institution in the country to offer a doctoral degree in glass science. Since NYSCC's glass science and engineering degree was established in 1932, thousands of graduates have found successful positions at the forefront of materials discovery, glass processing, and manufacturing. NYSCC's location in the Southern Tier of New York State is in excellent geographic proximity to important facilities such as: **Corning**; **Guardian Glass**; **Anchor Glass**; and **Tesla/SolarCity**. Other collaborators include **Monofrax**; **St. Gobain**; **Owens-Illinois**; and the **Glass Manufacturer Industry Council**.

Lumson has increased investment in processes that optimize resources to achieve the most ethically and economically sustainable solutions. The main target is to reduce, reuse, and recycle with ideas and initiatives aimed at decreasing the impact of the production process on the ecosystem, which are integral and fundamental parts of the daily operations. Following the latest market requests, Lumson will offer, starting from Q1/2021, a new sustainable solution: PCR glass. The brand new portfolio will guarantee the same high-quality level of the standard glass, and will be customizable with all the decorations and accessories that Lumson already offers.

**HORN Glass Industries** has taken over all assets of **EUROX Sauerstoff Mess-Systeme** due to an assets deal. HORN become the 100% shareholder of EUROX.

Technical engineer, Dr.-Ing. Lars Biennek with his team is responsible for the continuation and further development of the oxygen measurement systems, which will be carried under the name EUROX® Measurement-Systems in the future.

HORN takes on the current orders including the rental and service contracts as well as the leasing contracts.

A game-changing network of hydrogen technology clusters has been unveiled for Australia. **Orora** has joined the initiative as a founding member of the South Australian Hub-to-Hub (SA-H2H) Hydrogen Technology Cluster, led by **EfficientSee** and **Mumford Commercial Consulting**.

**National Energy Resources Australia** (NERA) recently announced the launch of 13 regional clusters across Australia to form a national network for Australian hydrogen technology, expertise and innovation. NERA is investing AUD $1.85M to establish the clusters across all states and territories that will facilitate the development of new capability in hydrogen focused technologies for Australia.

Simon Bromell, group general manager of Orora Beverage, said the cluster is a great opportunity for Orora to join a team that is helping move Australia towards a more sustainable energy future.

"Orora is committed to becoming a leading sustainable packaging solutions provider, with sustainability core to Orora's approach for reducing its impact on society and the environment. As a sponsor and founding member of the South Australian cluster, Orora has the opportunity to help shape the future direction of hydrogen-focussed technology development in the region. This is an exciting initiative for our business and for the community of South Australia and reinforces Orora's ongoing commitment to ensuring the sustainability of our operations."

Orora's state-of-the-art glass manufacturing facility in South Australia is one of the largest glass plants in the southern hemisphere.

The SA-H2H Hydrogen Technology Cluster will promote supply chain innovation, manufacturing, and commercialisation of hydrogen focused technologies in South Australia.

**Bowmer + Kirkland** has been appointed as stage 1 contractor for the glass innovation facility following an open procurement process which was managed through the North West Construction Hub. Developed for **Glass Futures**, the proposed 160,000 sq ft Centre of Excellence will be the world's first openly accessible, multi-disciplinary experimental glass innovation facility for research and development trials.

The facility will be used to deliver industry and Government-backed R&D projects aimed at reducing the carbon footprint of glass production.

The glass innovation facility is being funded through a mix of public and private investment. A £15 million **UKRI** grant has been secured for the installation of a globally unique, experimental furnace and state of the art infrastructure capable of producing 30 tonnes of glass per day and will enable the glass industry to implement change in a safe experimental space.

**Network Space** will secure private sector investment to support the construction costs alongside a £9 million grant from the **Liverpool City Region Combined Authority**.

Glass sector companies will also contribute a further £20 million in resource, time and equipment to support the project.

In addition to creating 80 new permanent jobs, the project will lead to apprenticeships, training and general up-skilling within the glass sector.

Glass Futures Ltd, chief executive, Richard Katz, stated: "The planning submission is a great step forward in creating a facility that will deliver innovative technologies to make glass net-zero.

"The Centre of Excellence will help us work collaboratively with our members, the supply chain and academia to achieve that aim."

The new **Cristalerías Toro** plant in the commune of Maipú will have a totally renewable and emission-free electricity supply after the Chilean company and the firm specializing in clean energy **Statkraft** agreed an annual supply for 43·8 GWh/year. The contract started at the end of 2020 when operation began. The agreement will have the I-REC certification, which will allow Cristalería Toro to accredit the match of the energy consumed in a certain time, with the energy effectively injected into the National Electric System (SEN), from renewable sources.

"We are providing clean energy to companies like Cristalerías Toro with a clear sustainable approach, since in addition to using a supply without emissions, they are implementing technology to take care of the environment. Our energy has an added value for customers, through the certification that proves the renewable sources," highlights María Teresa González, general manager of Statkraft Chile. For his part, the general manager of Cristalerías Toro, Francisco Ruiz, highlights: "For Cristalerías Toro this project has meant putting itself at the forefront of technology for the manufacture of glass containers and is part of our policy of producing high quality products in an environmentally friendly way. That is why the high percentage of recycled glass used in the manufacture of the containers and this agreement with Statkraft, allows us to offer all our customers a much more sustainable container."

Cristalerías Toro is a Chilean family business with more than 60 years of experience in the production and commercialization of glass containers. Statkraft is owned by the Norwegian State, has declared the challenge of curbing climate change urgent and, for this, renewable energies are the solution to the problem. The company has been in Chile for six years and is working on the construction of a hydroelectric project in the area of 52 MW of installed power, and on a wind project of more than 100 MW of power in the central area.

# CONFERENCE DIARY

**Spring/Summer**
Association for the History of Glass Study Day: Glass in the North, Vindolanda and Newcastle/Sunderland, UK.
*Web www.historyofglass.org.uk/study_days.html*

**29–31 March**
Experimental Archaeology: the Past in the Future. 12th Experimental Archaeology Conference, #EAC12, virtual conference.
*Email info@exarc.net Web exarc.net/meetings/eac12*

**~~17–19 April~~ (event cancelled)**
Deco '21, The Society of Glass and Ceramic Decorating Products annual conference, Pittsburgh, PA, USA.
*Myra Warne, Society of Glass and Ceramic Decorating Products Web www.sgcd.org*

**19–22 April (online only)**
SPIE Optics+Optoelectronics.
*SPIE, PO Box 10, Bellingham WA 98227-0010 USA Email MeetingInfo@SPIE.org Web spie.org*

**25–30 April (online only)**
International Congress on Ceramics (ICC8), Busan, South Korea.
*The Korean Ceramic Society, Room 403, Meolijae Building, 76, Bangbae-ro, Seocho-gu, Seoul, Republic of Korea. Email secretary@icc8.org Web www.icc8.org*

**6–9 May**
31st China International Glass Industrial Technical Exhibition, Shanghai, China.
*Wendy Yao, Beijing Zhonggui Exhibition, Sanlihe Road 11, Haidian District, Beijing 100831, China. Email ceramsoc@chinaglass-expo.com Web www.chinaglass-expo.com*

**10–12 May (online only)**
94th Annual Meeting of Deutsche Glastechnische Gesellschaft, Glastechnische Tagung, Aachen, Germany.
*Annette Doms, Deutsche Glastechnische Gesellschaft, Siemenstrasse 45, D-63071 Offenbach, Germany. Email dgg@hvg-dgg.de Web www.hvg-dgg.de*

**12–14 May**
XXXIV ATIV Int. Conf., Where Glass Science, Art and Technology Meet Together, Parma, Italy.
*ATIV Secretariat, MV Congressi spa Via Marchesi 26 d 43126 Parma, Italy. ativ@ativ-online.it Web www.ativ.eu*

**20–22 May** (online only)
GAS 2021, Glass Art Society conference.
*Glass Art Society, 2208 NW Market Street, Suite 200, Seattle, Washington 98107, USA. Email info@glassart.org Web www.glassart.org*

**23–26 May**
VITROGEOWASTES-II. Vitrification and Geopolimerization of Wastes for Inertizing and/or Recycling, Baeza, Spain.
*Prof. J. Ma. Rincón, Museo Nac. Ciencias Naturales- CSIC, c/José Gutiérrez Abascal 2, 28006-Madrid, Spain. Email rinconjma@mncn.csic.es Web vitrogeowastes.com/*

**29 May–2 June**
Ninth Colloquium of the French Association for the Archaeology of Glass, Nantes, France. [29–30 May experimental glass making, 31 May–2 June colloquium.]
*L'AFAV, Association Française pour l'Archéologie du Verre. Web afaverre.fr/Afaverre/*

**6–10 June** (hybrid, online and in person)
Optical Fiber Communications Conference and Exhibition (OFC), San Francisco, California USA.
*The Optical Society, 2010 Massachusetts Ave, NW, Washington, DC 20036 USA. Email custserv@osa.org Web www.ofcconference.org*

**9–10 June**
Furnace Solutions 15 and Training Day, online.
*Christine Brown, Society of Glass Technology, 9 Churchill Way, Chapeltown, Sheffield S35 2PY, UK. Email christine@sgt.org Web www.furnacesolutions.co.uk*

**13–17 June** (hybrid, online or in person)
14th International Conference on Solid State Chemistry (SSC 2021), Trenčín, Slovak Republic.
*Prof. Dušan Galusek, Alexander Dubček University of Trenčín, Študentská 3, 911 50 Trenčín, Slovak Republic. Email ssc2021@tnuni.sk Web www.funglass.eu/ssc2021*

**15–18 June *Cancelled***
Glastec 2020, Düsseldorf, Germany.
*Anke Seltmann, glastec, D-40474 Düsseldorf, Am Staad (Stockumer Höfe), Germany. Email seltmanna@messe-duesseldorf.de Web www.glasstec-online.com/*

**15–16 June**
Hydrogen in Glass Manufacturing conference, online.

**21–25 June**
2021 International Conference on Modern Materials & Technologies, Florence, Italy.
**26–30 June**
2021 Forum on New Materials
*CIMTEC 2021, Ms Stefania Bianchedi, Corso Mazzini 52, 48018 Faenza, Italy. Email congress@technagroup.it Web www.cimtec-congress.org*

**18–23 July**
Materials Challenges in Alternative & Renewable Energy 2020 (MCARE 2020) combined with the 4th Annual Energy Harvesting Society Meeting (EHS 2020) – RESCHEDULED, Bellevue, Washington, USA.
*The American Ceramic Society, 550 Polaris Parkway, Suite 510, Westerville, OH 43082, USA. Email customerservice@ceramics.org Web www.ceramics.org*

**1–6 August**
23rd International Conference on Composite Materials (ICCM23), Belfast, Northern Ireland.
*Prof Brian G. Falzon, School of Mechanical and Aerospace Engineering, Queen's University Belfast, Ashby Building, Stranmillis Road, Belfast BT9 5AH, Northern Ireland. Email b.falzon@qub.ac.uk Web iccm23.org/*

**9–13 August** (online only)
Nonlinear Optics, USA.
*The Optical Society, 2010 Massachusetts Ave, NW, Washington, DC 20036 USA. Email custserv@osa.org Web www.osa.org*

***16–20 August* (postponed to 2023)**
Tenth International Conference on Borate Glasses, Crystals and Melts and Third International Conference on Phosphate Materials, Corning, UK.
*Dr Randall Youngman, Corning Incorporated Email youngmanre@corning.com Web boratephosphate2020.org/*

**22–27 August**
Sol-Gel 2021, International Sol-Gel Society Conference, Lyon, France.
*Université Claude Bernard Lyon 1, Cellule Congrès -*

*SOLGEL 2021, 43, bd du 11 Novembre 1918, 69622 Villeurbanne, France. Email solgel2021@univ-lyon1.fr Web www.solgel2021.fr*

**27–30 August**
International Festival of Glass, Stourbridge, UK.
*Ruskin Glass Centre, Wollaston Road, Amblecote, Stourbridge DY8 4HF, UK. Email ifg@rmlt.org.uk Web www.ifg.org.uk*

**29 August–2 September**
XVII ECerS Conference, Dresden, Germany.
*Fraunhofer Institute for Ceramic Technologies and Systems IKTS, Winterbergstrasse 28, 01277 Dresden, Germany. Web www.ecers2021.org/*

**September**
Society of Glass Technology Annual Meeting, onlline option.
*Christine Brown, Society of Glass Technology, 9 Churchill Way, Chapeltown, Sheffield S35 2PY, UK. Email christine@sgt.org Web www.sgt.org*

**5–8 September**
International Commission on Glass (ISG) Annual Meeting, Incheon, South Korea.
*Secretariat, Room 403, Meolijae Building, 76 Bangbae-ro, Seochu-gu, Seoul, Republic of Korea. Email secretariat@icg2021.org Web www.icg2021.org/*

**14–17 September**
XXth Biennial Worldwide Congress Unified International Technical Conference on Refractories (UNITECR2021), Chicago, USA.
*The American Ceramic Society, 550 Polaris Parkway, Suite 510, Westerville, OH 43082, USA. Email customerservice@ceramics.org Web www.ceramics.org*

**13–17 September**
22nd Congress of the Association Internationale pour l'Histoire du Verre, Lisbon, Portugal.
*Web aihv.org/congress/*

**20–23 September**
Resilient Materials 4 Life (RM4L2021), Cambridge, UK.
*Email info@rm4l.com Web rm4l.com/rm4l2020*

**20–23 September**
16th Conference on Furnace Design: Operation and Process Simulation, Velke Karlovice, Czech Republic.
*Glass Service, Rokytnice 60, 755 01 Vsetin, Czech Republic. Email info@gsl.cz Web www.gsl.cz*

**26–28 September (new dates)**
FIT Show, Birmingham, UK.
*Freya Humphries, Account Manager, FIT Events Ltd, 9 Manchester Square, London, W1U 3PL, UK. Web www.fitshow.co.uk*

**4–6 October**
Sixth International Conference on Competitive Materials and Technology Processes (ic-cmtp6), Miskolc-Lillafured, Hungary.
*Email cmtp.conf@gmail.com Web www.ic-rmm1.eu/ic-cmtp/*

**5–8 October**
Vitrum 2021, Milan, Italy.
*Fiera Milano, Via Carlo Ilarione Petitti, 16, 20149 Milano MI, Italy. Email vitrum@vitrum-milano.it Web www.vitrum-milano.com*

**16–17 October**
Icon Bath 2021 Conference: Ceramics, Glass and Stained Glass, Bath UK.
*The International Institute for Conservation of Historic and Artistic Works, 3 Birdcage Walk, Westminster, London, W1H 9JJ, UK. Email IconBath2021@gmail.com Web www.iiconservation.org*

**17–21 October**
ACerS 123rd Annual Meeting with Materials Science & Technology 2021, Pittsburgh, USA.
*The American Ceramic Society, 550 Polaris Parkway, Suite 510, Westerville, OH 43082, USA. Email customerservice@ceramics.org Web www.ceramics.org*

**20–22 October**
Glass Performance Days, Tampere, Finland.
*Brown Onduso, GPD, Vehmaistenkatu 5, 33730 Tampere, Finland Email brown.onduso@gpd.fi Web gpd.fi/*

**27–30 October**
Seventh International Conference on Multifunctinal, Hybrid and Nanomaterials, Genoa, Italy.
*Web www.elsevier.com/events/conferences/international-conference-on-multifunctional-hybrid-and-nanomaterials*

**30 October–5 November**
Sixth Wuhan ICG Winter School, Wuhan, China.
*Prof. Kai Xu: SMART, Wuhan University of Technology, 1040 Heping Ave, Wuchang Qu, Wuhan Shi, Hubei Sheng, China. Email kaixu@whut.edu.cn Web www.icglass.org*

**30 October–5 November**
82nd Conference on Glass Problems, Columbus, USA.
*Donna Banks, GMIC, 550 Polaris Parkway, Suite 510, Westerville, OH 43082, USA. Email dbanks@gmic.org Web glassproblemsconference.org*

**17–18 November**
Glass and Modern Technologies - XXI, Moscow, Russia.
*Organizing Committee Email forum@steklosouz.ru Web www.steklosouz.ru*

**12–17 December (new dates)**
14th Pacific Rim Conference on Ceramic and Glass Technology (PACRIM 14) including Glass & Optical Materials Division 2021 Annual Meeting (GOMD 2021), Vancouver, BC, Canada.
*Andrea Ross, The American Ceramic Society, 550 Polaris Parkway, Suite 510, Westerville, OH 43082, USA. Email aross@ceramics.org Web www.ceramics.org*

**2022**

**Posponed to 2022 no date announced**
13th International Symposium on Crystallization in Glasses and Liquid, Paris, France.
*Laurent Cormier, CNRS, Paris, France. Email crystallization@sciencesconf.org Web crystallization.sciencesconf.org/*
ACerS Pan American Ceramics Conference, PACC-FMAs, Panama City, Panama.
*Andrea Ross, The American Ceramic Society, 550 Polaris Parkway, Suite 510, Westerville, OH 43082, USA. Email aross@ceramics.org Web www.ceramics.org*

**6–10 March**
Optical Fiber Communications Conference and Exhibition (OFC), San Diego, California USA.
*The Optical Society, USA. Email custserv@osa.org Web www.ofcconference.org*

**2–4 April**
Deco '22, The Society of Glass and Ceramic Decorating Products annual conference, Columbus, OH, USA.
*Myra Warne, Society of Glass and Ceramic Decorating Products, PO Box 2489, Zanesville, OH 43702, USA. Email info@sgcd.org Web www.sgcd.org*

**3–6 April**
TMS World Congress on Artificial Intelligence in

Materials and Manufacturing (AIM 2022), Pittsburgh, Pennsylvania, USA.
*TMS Meeting Services, 5700 Corporate Drive Suite 750, Pittsburgh, PA 15237, USA. Web www.tms.org/AIM2022*

**18–20 April**
Int. Conf. on Glass Science and Technology from the Golden Era of the Pharaohs to a Sustainable Future, Egypt.

**18–20 May**
REFRA PRAGUE 21, 21st Conference on modern refractory materials and key achievements in high temperature technologies, Prague, Czech Republic.
*Ing. Jan K. Luxa, Silikátová společnost České republiky, Novotného lávka 5, 110 01 Praha 1, Czech Republic. Email hluxa@seznam.cz Web www.silikaty.cz*

**18–21 May**
GAS 2022, Glass Art Society conference, Tacoma, WA, USA.
*Glass Art Society, 2208 NW Market Street, Suite 200, Seattle, Washington 98107, USA. Email info@glassart.org Web www.glassart.org*

**22–26 May**
2022 Glass and Optical Materials Division Annual Meeting (GOMD2022), Baltimore, USA.
*Andrea Ross, The American Ceramic Society, 550 Polaris Parkway, Suite 510, Westerville, OH 43082, USA. Email aross@ceramics.org Web www.ceramics.org*

**21–24 June**
Techtextil 2022, Frankfurt, Germany.
*Email techtextil@messefrankfurt.com Web techtextil.messe-frankfurt.com*

**3–8 July**
26th International Congress on Glass, Berlin, Germany. Hosted by the DGG in their centenary year.
*Annette Doms, Deutsche Glastechnische Gesellschaft, Siemenstrasse 45, D-63071 Offenbach, Germany. Email dgg@hvg-dgg.de Web www.hvg-dgg.de*

**4–8 July**
Ninth International Conference on Optical, Optoelectronic and Photonic Materials and Applications (ICOOPMA), Ghent, Belgium.
*Center of Materials and Nanotechnology and Department of General and Inorganic Chemistry, Faculty of Chemical Technology, University of Pardubice, Cs. Legii sq. 565, 532 10 Pardubice, Czech Republic. Email info@icoopma.com Web www.icoopma.com*

**10–15 July**
Physics of Non-Crystalline Solids, now also including Society of Glass Technology Annual Meeting Canterbury, UK.
*Christine Brown, Society of Glass Technology, 9 Churchill Way, Chapeltown, Sheffield S35 2PY, UK. Email christine@sgt.org Web www.sgt.org*

**10–14 July**
Ninth Int. Congress on Ceramics (ICC9), Krakow, Poland.
*European Ceramic Society AISBL, Av. Gouverneur Cornez 4, B-7000 Mons, Belgium. Email ecers@bcrc.be Web ecers.org/*

**20–23 September**
Glastec 2022, Düsseldorf, Germany.
*Anke Seltmann, glastec, D-40474 Düsseldorf, Am Staad (Stockumer Höfe), Germany. Email seltmanna@messe-duesseldorf.de Web www.glasstec-online.com/*

**9–13 October**
ACerS 124th Annual Meeting with Materials Science & Technology 2022, Pittsburgh, USA.
*The American Ceramic Society, 550 Polaris Parkway, Suite 510, Westerville, OH 43082, USA. Email customerservice@ceramics.org Web www.ceramics.org*

**2023**
**5–9 March**
Optical Fiber Communications Conference and Exhibition (OFC), San Diego, California USA.
*The Optical Society, USA. Email custserv@osa.org Web www.ofcconference.org*

**10–14 July**
Nonlinear Optics, Honolulu, Hawaii, USA.
*The Optical Society, 2010 Massachusetts Ave, NW, Washington, DC 20036 USA. Email custserv@osa.org Web www.osa.org*

**17–21 July**
Tenth International Conference on Borate Glasses, Crystals and Melts and Third International Conference on Phosphate Materials, Corning, USA.
*Dr Randall Youngman, Corning Incorporated Email youngmanre@corning.com Web boratephosphate2020.org/*

**August**
NCM15, Structure of Non-Crystalline Materials 15, combined with the SGT Annual Meeting, UK.
*Christine Brown, Society of Glass Technology, 9 Churchill Way, Chapeltown, Sheffield S35 2PY, UK. Email christine@sgt.org Web www.sgt.org*

**2024**
**June**
Tenth International Conference on Optical, Optoelectronic and Photonic Materials and Applications (ICOOPMA), Pardubice, Czech Republic.
*Center of Materials and Nanotechnology and Department of General and Inorganic Chemistry, Faculty of Chemical Technology, University of Pardubice, Cs. Legii sq. 565, 532 10 Pardubice, Czech Republic. Email info@icoopma.com Web www.icoopma.com*

**2–6 July**
XVIII ECerS Conference, Lyon, France.
*European Ceramic Society AISBL, Av. Gouverneur Cornez 4, B-7000 Mons, Belgium. Email ecers@bcrc.be Web ecers.org/*

**14–19 July**
Tenth Int. Congress on Ceramics (ICC10), Montreal, Canada.
*The American Ceramic Society, 550 Polaris Parkway, Suite 510, Westerville, OH 43082, USA. Email customerservice@ceramics.org Web www.ceramics.org*

**20–23 August**
Materials Challenges in Alternative & Renewable Energy 2023 combined with the Sixth Annual Energy Harvesting Society Meeting (EHS 2023), Bellevue, Washington, USA.
*The American Ceramic Society, 550 Polaris Parkway, Suite 510, Westerville, OH 43082, USA. Email customerservice@ceramics.org Web www.ceramics.org*

Glass Technol.: Eur. J. Glass Sci. Technol. A, 2021, 62 (1), 9–23

# Dissolution of sodium silicate glasses for the production of water glass – part IV: characterisation of reaction layer and glass surface

*M. Dathe,*[a] *W. Lebek,*[a] *E. Koslowski,*[b] *M. Bron,*[b] *T. Bräuniger,*[c] *I. Stoessel*[a] *& H. Roggendorf*[a,1]

[a] *Martin-Luther-University Halle-Wittenberg, Institute of Physics, Von-Danckelmann-Platz 3, Halle, 06120 Germany*
[b] *Martin-Luther-University Halle-Wittenberg, Institute of Chemistry, Von-Danckelmann-Platz 3, Halle, 06120 Germany*
[c] *Ludwig-Maximilians-University München, Department of Chemistry, Geschwister-Scholl-Platz 1 D-80539, München, Germany*

Manuscript received 23 July 2020
Revision received 26 October 2020
Manuscript accepted 11 December 2020

*Corrosion and dissolution reactions of $Na_2O.xSiO_2$ glasses (x=2·0, 2·5, and 3·3) were investigated by static and dynamic corrosion tests at pH values between 7 and 14 and temperatures of 30 and 50°C. The investigated glass compositions are close to those used for water glass production. Several types of leachants with varied pH and $SiO_2$ concentrations were applied to induce corrosion effects. One characteristic feature of the corrosion process is the development of a reaction layer on top of the corroding glasses, a process depending on leachant pH. At intermediate pH values reaction layers were observed, but not at high pH values (>12). To generate altered glass surfaces with and without reaction layers sodium silicate glasses were corroded in deionized $H_2O$, 0·1 M NaOH, and diluted liquid sodium water glasses. The resulting samples or layer materials were investigated by scanning electron microscopy, Raman microscopy, infrared spectroscopy, or $^{29}$Si-MAS-NMR spectroscopy. Especially Raman microscopy was applied to characterize depth profiles of structural changes of silicate ($Q^n$) groups and of concentration profiles of molecular $H_2O$ within reaction layers. The results are discussed with respect to influences of pH and leachant composition in order to evaluate possible corrosion mechanisms.*

## 1. Introduction

Studies on corrosion kinetics and on the influence of leachate composition on the corrosion of sodium silicate glasses were already reviewed.[(1–3)] In the following, the literature on glass surfaces with and without reaction layers and their characterisation will be evaluated.

Testing methods as well as characterisation techniques applicable to corroded glasses were reviewed by Clark *et al.*[(4)] Reports on surfaces changes of glasses with high chemical durability are abundant (e.g. Kaspar *et al*[(5)]), but reports on the alterations of quite soluble glasses are rare. Reactions affecting composition and structure of corroding glass surfaces are ion exchange, reaction layer formation and precipitation of reaction products on the surface.[(6)] Depending on pH, reaction layers on top of the dissolving glasses were observed.[(2)] Their formation was explained as a consequence of pH as well as of the solubility dependence of silica materials on pH.

Textbooks (e.g. Doremus[(7)]) on glass science describe reaction layer formation during corrosion in acidic aqueous solutions via ion exchange of alkalis versus $H^+$ or $H_3O^+$. Hench[(8)] classified five types of glass surfaces generated by different corrosion processes. Those of relevance here are:

Type IV: development of a non-protective silica enriched surface due to a combination of network dissolution and ion exchange.

Type V: extensive and predominant network dissolution.

Sodium or potassium silicate glasses are sometimes called 'soluble glasses'[(9)] since they can be dissolved totally for the production of alkaline water glasses. Usually, temperatures up to 150°C in a pressurised autoclave are applied.[(10)] In these production processes, it can be assumed that the surface layers eventually developed during these dissolution processes do not retard dissolution to a larger extent and are possibly of type IV or V.

Das[(11)] investigated the leaching of sodium and potassium silicate glasses with 85 mol% $SiO_2$ (rest alkali oxides) as a function of pH. He found that the alkali extraction rate decreases with increasing pH and attributed that to the occupation of surface sites with alkalis following the interpretation of earlier results by Douglas & El-Shamy.[(12)] The thickness of

[1] Corresponding author. Email hans.roggendorf@physik.uni-halle.de
DOI: 10.13036/17533546.62.1.007

the reaction layer reached about 8 μm when leached 18·3 h at 69°C in an aqueous solution with a pH of 1·4. It has to be mentioned that the sodium silicate glass investigated by Das had a composition liable to phase separation.(13) Sanders & Hench(14) investigated the corrosion of alkaline disilicate glasses. By infrared reflection spectroscopy they demonstrated the development of silica enriched gel layers on the surface of sodium and lithium disilicate glasses corroding in water. A recent publication by Portal & Sempere(15) investigated the formation of thin gel layers on alkaline silicate glasses with a composition similar to that investigated by Das(11) (84 mol% silica plus minor additions of potassium oxide and alumina). They applied acidic corrosion conditions and characterised the surfaces by ellipsometry and secondary ion mass spectrometry. They found two different layers with a total thickness in the range of 1 μm, one depleted with respect to sodium and the other – closer to the unchanged glass – not entirely depleted. Hahn & van Duin(16) investigated the corrosion of sodium silicate glasses in $H_2O$ in the first nanoseconds of contact by molecular dynamic simulations. They confirmed ion exchange as the earliest process. The water activity is strongest at nonbridging oxygen (NBO) sites. With the presence of modifying ions $H_2O$ dissociates into $H^+$ and $OH^-$. As a result, the NBOs are protonated and the $OH^-$ groups are captured by the $Na^+$. Additionally, the silanol density at the surface is slightly increased.

Some reports on surface characterisation of more durable glasses may contribute to the discussion. Layer formation on corroding soda–lime–silica glasses with (in mol%) $20Na_2O$, 6CaO, and $74SiO_2$ (McInnes–Dole glass) was investigated by Scholze *et al*.(17) They used ca. 30 μm thick glass foils and corroded them at 60°C in 0·1 M HCl or in 0·1 M DCl solutions. They reduced the thickness of the corroded glass samples to allow infrared analysis with dispersive spectrometers. After corrosion they analysed the chemical composition of the leachate. The glass foils were analysed by infrared spectrometry before and after corrosion. Additionally, a stepwise etching process in $HF/HNO_3$ solutions was applied to characterise elemental concentration profiles within the reaction layer. S-shaped interdiffusion profiles of $Na^+$ (counterion: $H^+$) with a thickness of up to 4 μm and a dependence of the amount of incorporated $H_2O$ content on square root of leaching time verified diffusion exchange of alkalis versus protons with protons as rate determining species. Additional water uptake was explained by $H_2O$ accompanying diffusing protons or by condensation of silanol groups to molecular $H_2O$. The application of DCl highlighted the importance of proton diffusion as rate determining.

Doremus *et al*(18) described the ion exchange process theoretically as a concentration dependent interdiffusion process and experimentally by applying the $^{15}N$ nuclear resonance reaction for measuring hydrogen concentration profiles (Lanford *et al*(19)).

Eventual effects of reaction layers formed during glass corrosion were discussed in the frame of studies on disposal of radioactive waste glass in geological repositories.(20,21) These glasses were designed for high chemical durability and layers developed under these conditions are usually not regarded as corrosion inhibitive.

Surface analysis of corroded glasses improves the understanding of glass corrosion. Reviews on surface analysis of corroded glasses were published, e.g. by Bach(22) and Pantano.(23) Generally, two different types of characterisation techniques can be applied for compositional depth profiling. The first type combines surface sputtering with analytical methods (photoelectron spectroscopy, Auger spectroscopy, or secondary ion mass spectrometry(24)). These techniques have a depth resolutions down to the nm range. The second type uses methods to scan two-dimensional surfaces, e.g. fracture surfaces perpendicular to the corroded sample surface, with microprobe analysis or scanning electron microscopy (SEM) combined with energy dispersive x-ray analysis (EDX). The second type of techniques has a lateral – in this case depth – resolution in the μm range.

$^{29}Si$ NMR spectroscopy is used to characterize the structural surrounding of Si atoms.(25) Of special interest is the $Q^n$ group distribution (e.g. Stebbins(26)) since it delivers information on the polymerisation degree of silicate networks. $Q^n$ denotes a Si atom with $n$ bridging oxygens (BO) connecting the Si atom directly to another Si atom. Thus, $Q^4$ denotes a silicon in a fully polymerized network, whereas $Q^0$ denotes a Si atom not bonded to the network via BOs. Glass corrosion may induce ion exchange $Na^+ \leftrightarrow H^+$(18) which alters the chemical surroundings of Si atoms and thus the chemical NMR signal shifts. Li *et al*(27) investigated silica gels derived from water glasses and found chemical shifts close to those of other silicate materials. These shifts changed a bit due to ageing and pH value during gelation. Longer ageing and higher pH (up to 9·9) led to lower values of the chemical shifts. Boehm *et al* showed the condensation of silanol groups in reaction layers on potassium lime silicate glasses corroded in acidic solutions by $^{29}Si$ NMR spectroscopy.(28) Gin *et al* used $^{29}Si$ NMR spectroscopy to compare pristine and corroded silicate glasses with high chemical durability.(29) They confirmed a polymerisation of the silicate network of the surface layer during corrosion of the glass. Mauro described recently connectivity in glass networks via topological constraints.(30) The method was originally developed by Phillips & Thorpe.(31) By topological constraints the translational degrees of freedom of atoms are described, which are reduced by rigid bondings to neighboring atoms. The number of constraints per network forming atom $n_c$ is calculated from the average coordination numbers of atoms in the structure considering constraints arising from rigid bondings between two neighboring atoms and rigid bonding angles between three atoms. In principle, the higher the number of rigid bondings per atom the higher $n_c$. Pignatelli *et al* correlated the topological constraint theory with silicate glass dissolution kinetics including sodium silicate glasses which had the lowest $n_c$ value and the highest corrosion rates in their comparison.[32]

Scholze *et al*[33] applied infrared spectroscopy to characterise hydroxyl groups in the 3700 to 2400 $cm^{-1}$ range. Additionally, they distinguished silanol groups from molecular water in the near infrared. Further progress in infrared reflectance spectroscopy was reported by Lynch *et al*[34] who also concluded that the condensation of silanol groups is one of the structure forming processes in reaction layer formation during leaching of lithium silicate glasses in deionised $H_2O$. Amma *et al* used the ATR method (attenuated total reflection) to investigate the water and hydroxyl group contents of leached glass surfaces.[35] Especially the structure of silicate units in glasses, gel materials and crystals was investigated in the infrared range between 700 and 1300 $cm^{-1}$.[36] It has to be noted, that not all vibrations are equally detectable in infrared or Raman spectra. Vibrations are infrared active if they lead to a change of the dipole moment of the respective unit whereas vibrations are Raman active if they induce a change in polarisability. Brawer *et al*[37] used Raman spectroscopy to investigate the structure of alkaline silicate glasses. They interpreted the spectra as superposition of the Raman active vibrations of silicate units. The spectra resemble to a certain degree those of crystalline materials of identical composition.

The application of more than one method improves understanding as Silva *et al*[38] demonstrated by combination of $^{29}Si$ NMR, infrared and Raman spectroscopy applied to characterise sodium magnesium silicate glasses. NMR and Raman results characterised the distribution of $Q^2$ and $Q^3$ groups very well, whereas the authors did not assign the bands of the infrared spectroscopy to $Q^n$ groups.

In this report, the results of the characterisation of sodium silicate glass surfaces altered by corrosion processes under conditions relevant for liquid sodium water glass synthesis are reported. Surface analysis and spectroscopic methods were applied. The structural development of the gels and the process of water incorporation will be discussed.

## 2. Experimental

Details of glass synthesis and corrosion testing were reported in Dathe & Roggendorf.[1–3] $Na_2O.xSiO_2$ glasses ($x$=2·0: NaSi2·0, $x$=2·5: NaSi2·5 (only for infrared investigations), and $x$=3·3: NaSi3·3) were prepared from batches containing $Na_2CO_3$ (p. a. quality, Merck, Germany), $SiO_2$ (quartz flour W6, Dörentrup Quarz, Germany; >99·5 wt% $SiO_2$) and $NaNO_3$ (p.a. quality, Merck, Germany). The $NaNO_3$ content of the batch (0·5 g per 100 g glass) provided oxidising melt conditions which help to avoid colouration of the glass due to dissolution of metals from the crucible. The batches were calculated to achieve 100 g of glass with one melting run and are reported in Dathe & Roggendorf[1] and contained additional $Na_2CO_3$ to compensate evaporation losses. They were melted, quenched by casting on metal plates, crushed and remolten. The glasses were melted in covered Pt80/Rh20 crucibles. The following conditions were applied:

- First melting 1 h at 1450°C.
- Second melting:
    - NaSi2·0: 1 h at 1450°C, then casting (melt temperature of 1450°C) into brass forms preheated to 460°C with a size of approximately 1·5×2·0×10 $cm^3$. Then annealing for ten minutes at 480°C and cooling down inside a muffle furnace. Resulting glass composition (analysis details reported in Dathe & Roggendorf:[1] 34·2±0·2 wt% $Na_2O$.
    - NaSi2·5: 2 h at 1450°C, casting temperature of 1500°C, other parameters identical to NaSi2·0. Resulting glass composition: 29·0±0·2 wt% $Na_2O$.
    - NaSi3·3: same as NaSi2·5. Resulting glass composition: 23·0±0·3 wt% $Na_2O$.

The glass blocks were cut to platelets with a size of 1·5×1·2×0·15 $cm^3$ with a low speed saw (Isomet, Buehler, Düsseldorf, Germany) equipped with a diamond wafering blade. A water free sawing liquid (Buehler Isocut) was applied. The exact sample size was determined with a calliper. After sawing, the platelets were first cleaned mechanically by wiping off the sawing liquid and then by an ethanol bath. After that, the samples were dried with ambient air. Before corosion tests the samples were etched 10 min in an aqueous solution with 0·5 wt% HF (prepared from concentrated HF (p.a. quality, Merck, Germany)) and 1 wt% $HNO_3$ (prepared with 65% $HNO_3$, p.a. quality, Carl Roth, Germany). Layers with a thickness of about 10 μm were etched off.

Static (no exchange of leachant) as well as dynamic corrosion tests (continuous flow of leachant through container) were applied. These test conditions were evaluated and described in Dathe & Roggendorf.[1] The main advatage of the dynamic corrosion test applied here is that they allow a better control and constancy of pH conditions without using buffer solutions. The constant flow is achieved by a peristalic pump (Ismatec Reglo-Digital MS-4/12-100; Cole-Parmer GmbH, Wertheim, Germany). Optimised corrosion conditions are listed in Table 1.

The corrosion results were characterised by the following methods:

*Table 1. Standard parameters of corrosion tests applied to sodium silicate glasses*

| | *Static corrosion* | *Dynamic corrosion* |
|---|---|---|
| T in °C | 30 (NaSi2·0, NaSi2·5),<br>50 (NaSi3·3, NaSi2·5) | 30 (all glasses),<br>50 (NaSi2·5, NaSi3·3) |
| corrosion time in d | 14 | 4 |
| volume of container in $cm^3$ | 50 | 60 |
| surface area in $cm^2$ | 4·75 | 4·75 |
| SA/V (surface area/volume) in $m^{-1}$ | 9·5 | |
| flow rate in $l.d^{-1}$ | | 2 |

- The surfaces of the glasses were characterised by scanning electron microscopy (SEM; equipment: JSM 6300, JEOL, Japan) combined with energy dispersive x-ray analysis (EDX; NORAN Instruments, Middleton, Wi, USA, Voyager 1100 X-ray). Samples prepared previously for optical microscopy or freshly broken samples were coated with carbon and then analysed. A line on the fracture surface across the gel layer into the glass core was scanned with up to 1200 measuring points by the electron beam of the SEM.
- Magic-angle spinning (MAS) spectra of $^{23}Na$ and $^{27}Si$ were recorded with a Bruker Avance-III 400 spectrometer (Bruker Corporation, USA). Bruker MAS double-resonance probes with 2·5 mm rotors were used with a rotation frequency of 10 kHz. The uncorroded glasses as well as separated reaction layers were investigated. The spectra were normalised to maximum intensity.
- For confocal Raman microscopy a Renishaw in-Via Raman microscope (Renishaw plc, UK) with a CCD camera coupled with a Leica DMI 3000 inverted microscope (Leica Microsystems GmbH, Germany) was applied. A Cobolt CW DPSS Laser with a wavelength of 532 nm and a maximum output of 45 W was applied. The samples were analysed perpendicular to the corroded surface and the focal plane of the microscope was varied. Depth profiling was achieved by mechanically shifting the focal plane. The method is described by Everell.[39,40]
- Infrared spectrometry: a Fourier-transformation infrared spectrometer Bruker Tenso 35 (Bruker Corporation, USA) was applied. Reaction layer materials generated by corrosion were scratched off the glasses with a scalpel. Pristine glasses were first crushed in an agate mortar and then milled in a ball mill. The samples were dispersed in KBr. In the case of reaction layer materials 20 mg of them were dispersed in 100 mg KBr and pressed to tablets with a diameter of 1·3 cm and a thickness of 0·1 mm. 50 mg of the pristine glass powders were dispersed in 0·45 g (NaSi2·0 and NaSi2·5) or 1 g (NaSi3·3) KBr and pressed to tablets with the same diameter. Adsorption spectra of these tablets were recorded between 7000 and 400 $cm^{-1}$. Additionally, adsorption spectra and infrared reflection spectrograms of whole corroded samples (glass and layer) were recorded.
- Interesting sections of NMR, infared, and Raman spectrograms with information on $Q^n$ group distribution were evaluated by deconvolution with Gauss distribution functions. The peak area fractions were calculated by relating the individual peak areas $A(Q^n)$ to the sum of peak areas of the Q groups:

$$\text{Peak area fraction of } Q^n = \frac{A(Q^n)}{A(Q^0)+A(Q^1)+A(Q^2)+A(Q^3)+A(Q^4)} \quad (1)$$

- To discuss the polymerisation degree of the silicate networks in glasses and reaction layers mean $\bar{n}$ values ($\bar{n}$=average number of bridging oxygens per silicaon) of the $Q^n$ groups can be calculated:

$$\bar{n} = \frac{4A(Q^4)+3A(Q^3)+2A(Q^2)+A(Q^1)}{100} \quad (2)$$

- To distinguish the different spectroscopic methods applied the indices NMR, infrared and Raman were augmented to $A$ or $\bar{n}$, e.g. $A_{NMR}$, $\bar{n}_{NMR}$...
- These different n values are compared with the $Y$ value introduced by Stevels[41] and $n_c$ values according to the topological constraint theory.[30,32] Stevels used the $Y$ number (number of bridging oxygen atoms per polyhedron) as a first approximation to discuss glass structure. Vitreous silica has $Y$=4, each Na atom creates one nonbridging oxygens according to

$$\equiv Si{-}O{-}Si\equiv + Na_2O \rightarrow 2\equiv Si{-}O{-}Na. \quad (3)$$

$Y$ and $n_c$ are related since bondings via NBOs are weak and bondings via BOs rigid.

## 3. Results

An evaluation of corrosion kinetics as well as of microstructure development revealed two pH regimes:

- Neutral to pH of about 12: here, a layer formation on top of the corroding glass and corrosion rates increasing with pH was observed.
- pH 12 and above: In this pH range, no layer formation was observed and the corrosion rates decreased with increasing pH.

To investigate the microstructure of the glass surface further, NaSi2·0 and NaSi3·3 samples corroded in both pH regimes were selected for structural and chemical characterisation. Additionally, NaSi2·5 samples were corroded in both pH regimes and analysed by infrared spectroscopy.

### 3.1. SEM micrography and EDX profiling

SEM micrographs of fracture surfaces of corroded sodium silicate samples are displayed in Figure 1. Elemental concentration profiles of Na and Si measured with energy-dispersive x-ray analysis (EDX) of the same samples are shown in Figure 2. In most cases, the samples were corroded dynamically. Exceptions are Figure 2(a) and (b): here samples obtained with static corrosion tests are shown since the reaction layers of equivalent samples of dynamic corrosion test were highly fractured. The drying under vacuum conditions contributed to the fracturing of thicker reaction layers.

The following observations were made:

- NaSi2·0, corroded in deionised $H_2O$ at 30°C: thick reaction layers were formed on top of the glass

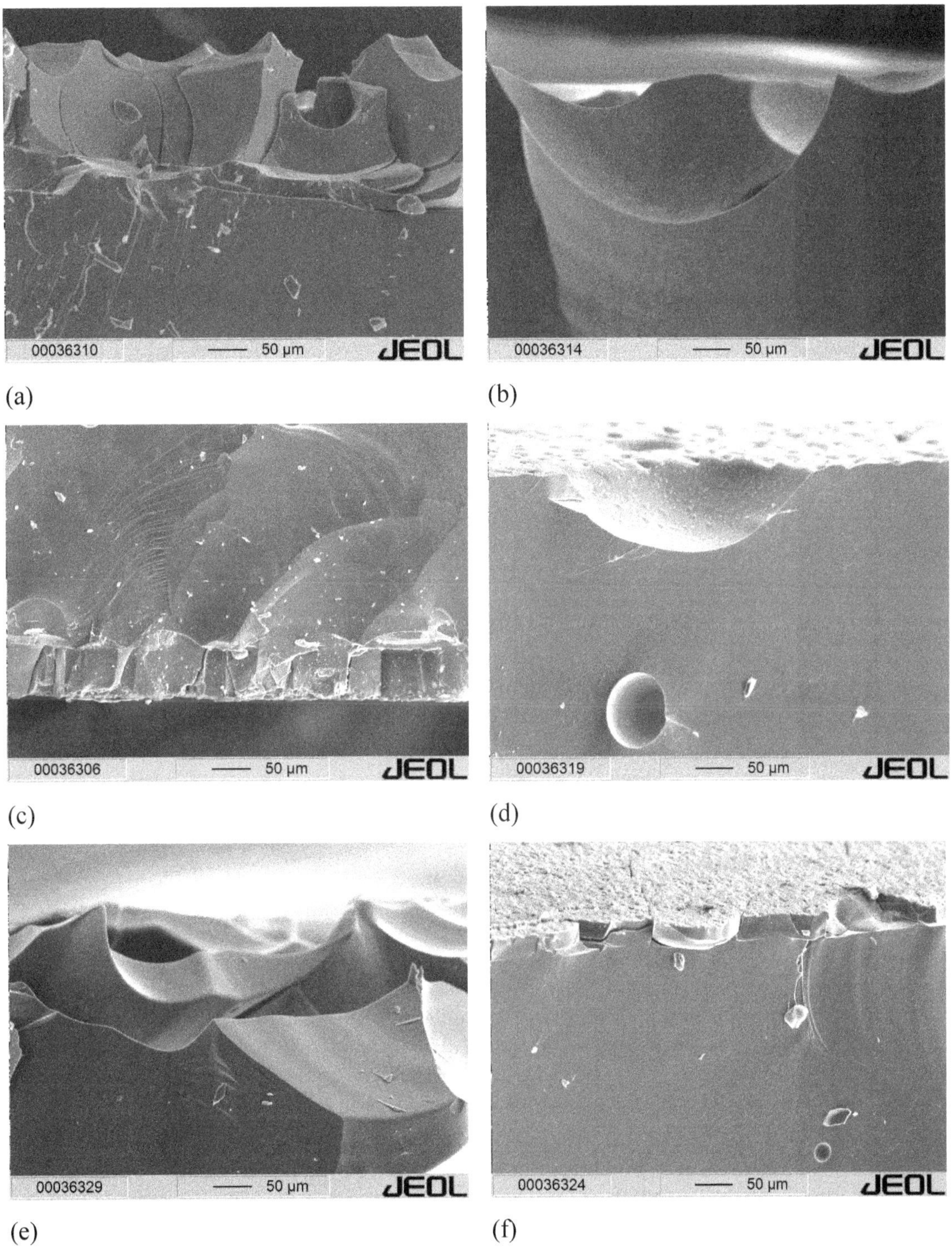


*Figure 1. SEM micrographs of fracture surfaces of corroded sodium silicate glasses; scale bar in picture legend; after 4 d of dynamic dynamic corrosion, other conditions: (a) glass NaSi2·0, deionised $H_2O$ at 30°C; (b) glass NaSi2·0, 0·1 M NaOH at 30°C; (c) glass NaSi3·3, deionised $H_2O$ at 50°C; (d) glass NaSi3·3, 0·1 M NaOH at 50°C; (e) glass NaSi2·0, sodium water glass (2 wt% $SiO_2$) at 30°C; (f) glass NaSi3·3, in sodium water glass (2 wt% $SiO_2$) at 50°C*

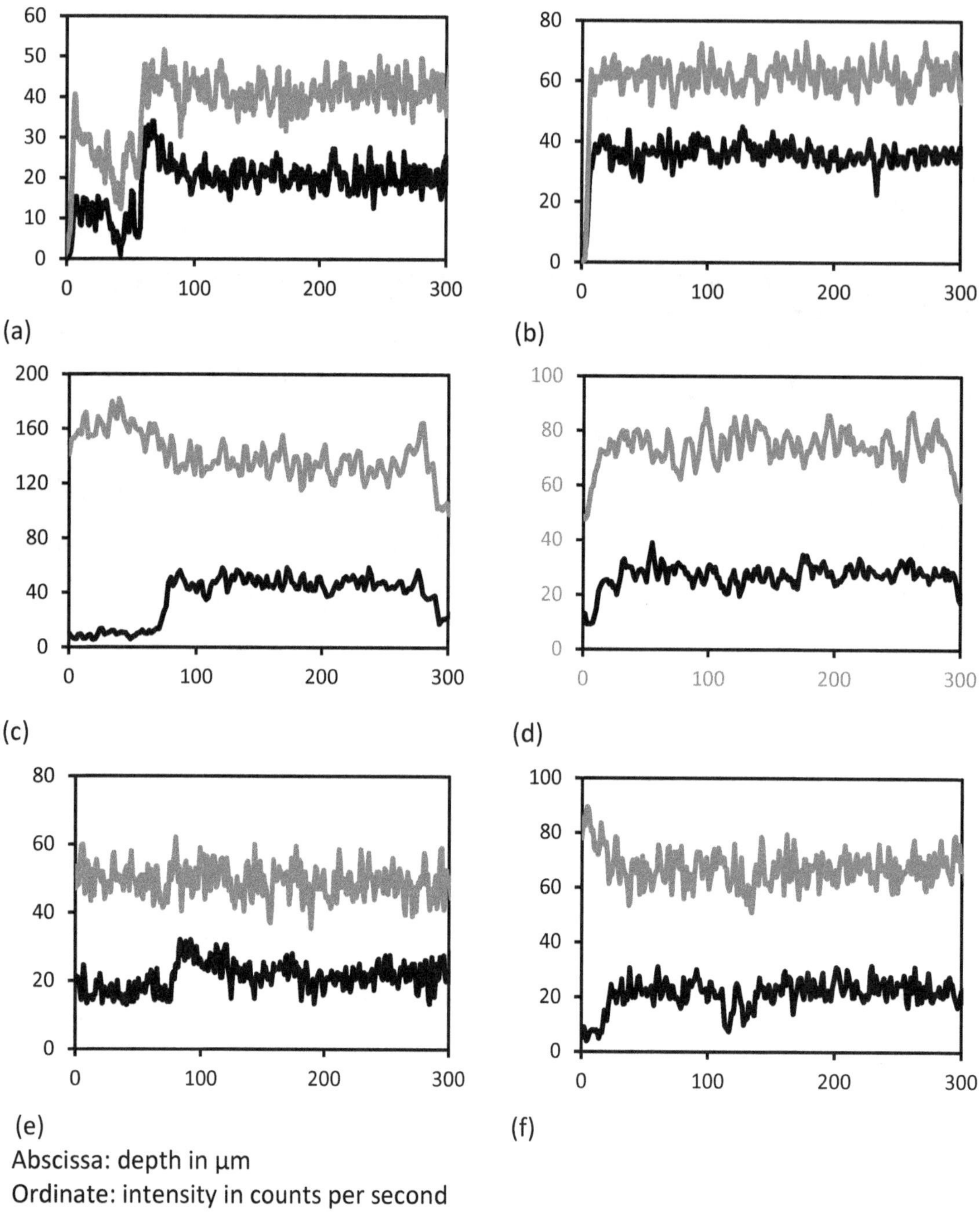


*Figure 2. EDX elemental profiles measured at fracture surfaces of corroded sodium silicate glasses; The profiles start at the surface of the reaction layer or the glass and run perpendicularly to the exposed surface; scale bar in picture legend; after 4 d of dynamic dynamic corrosion, other conditions: (a) glass NaSi2·0, static corrosion in deionised $H_2O$ at 30°C; (b) glass NaSi2·0, static corrosion in 0·1 M NaOH at 30°C; (c) glass NaSi3·3, dynamic corrosion in deionised $H_2O$ at 50°C; (d) glass NaSi3·3, dynamic corrosion in 0·1 M NaOH at 50°C; (e) glass NaSi2·0, dynamic corrosion in sodium water glass (2 wt% $SiO_2$) at 30°C; (f) glass NaSi3·3, dynamic corrosion in sodium water glass (2 wt% $SiO_2$) at 50°C; black line: Na; grey (in colour red) line: Si (always upper curve) [Colour available online]*

samples. Corrosion tests in $H_2O$ generated reaction layers with a thickness up to 150 µm (Figure 1(a)). The intensities of the Na as well as of the Si signals (Figure 2(a)) in the reaction layer were lower than the signals in the bulk glass, but not zero. Additionally, the Na:Si signal ratio was a bit lower than in the bulk glass. That means that the alkalis are leached only partially. The same was found by Scholze *et al*[42] for soda–lime–silica glasses corroded in HCl. An increase of the signal towards the outer surface can be noticed. Concentration minima of Si as well as Na hint at a subdivision of the reaction layer. Cracks developing parallel to the surface (Figure 1(a) might back that finding. A further finding: the intensity of both signals, Na and Si, is reduced within the reaction layer. This indicates a smaller density, possibly due to the incorporation of $H_2O$.

- NaSi2·0, corroded in 0·1 M NaOH at 30°C: no reaction layer was detected on top of the residual glass (Figure 1(b)) and the EDX profile (Figure 2(b), again static corrosion sample) is compatible with congruent network dissolution. Neither enrichment nor depletion of Na at the surface within the lateral resolution of EDX was observed.
- NaSi3·3, corroded in deionised $H_2O$ at 50°C: a reaction layer with a thickness of about 60 μm (Figure 1(c)) was found on top of the corroding glass. The EDX profile (Figure 2(c)) shows an enrichment of Si in this layer (type IV according to Hench[8]) and a depletion of Na. As in the case of NaSi2·0 corroded in $H_2O$, the depletion is again not a total one and the reaction layer contains significant amounts of Na.
- NaSi3.3, corroded in 0·1 M NaOH at 50°C; no reaction layer was detected by SEM on top of the residual glass (Figure 1(d)). On the other hand, the EDX profiles (Figure 2(d)) reveal a 15 μm thick zone, where both concentrations, that of Na as well as that of Si, decrease towards the surface. These results are typical of corrosion of glasses with a higher chemical durability.
- NaSi2·0, corroded in sodium water glass (2 wt% $SiO_2$) at 30°C: The EDX profile reveals that within a 80 μm deep surface zone the Na counting rate was reduced to about 70% of the bulk level (Figure 2(e)). Since no reduction of the Si counting rate was observed, this change is possibly due to an altered surface zone. On the other hand, neither an altered surface zone nor topological features explaining the concentration reduction of Na were observed in the backscattering image of the scanning electron microscope (Figure 1(e)). Therefore, the nature of this observation is not clear. Again, both signal intensities are reduced within the reaction layer indicating a smaller density.
- NaSi3·3, corroded in sodium water glass (2 wt% $SiO_2$) at 50°C: a reaction layer with a thickness of about 25 μm (Figure 1(f)) was found on top of the corroding glass. The EDX profile (Figure 2(c)) shows an enrichment of Si in this layer towards the surface (type IV according to Hench[8]) and a partial depletion of Na. The $Na_2O$ content of the reaction layer is about 6 wt%. Again, a similarity to Scholzes finding has to be mentioned.[42]

Reaction layers were generated during corrosion in $H_2O$, but not during corrosion in 0·1 M NaOH, which corresponds to the findings published in Ref. 2. The reaction layers did contain appreciable amounts of sodium, especially in the case of NaSi2·0 and NaSi3·3 corroded in a sodium silicate solution with 2 wt% $SiO_2$. The absence of a distinct reaction layer on NaSi2·0 corroded in a sodium silicate solution with 2 wt% $SiO_2$ is probably due to the higher $Na_2O$ content of that solution (molar ratio of $SiO_2$:$R_2O$=2·20) leading to a higher pH value (see Böschel[43]). Generally, the findings for NaSi3·3 are more compatible to the common perception of glass corrosion, probably because their chemical durability as well as their structure is closer to conventional silicate glasses. In the case of NaSi2·0 corroded in $H_2O$ or sodium silicate solutions, appreciable Na concentrations were found near the surface.

### 3.2. NMR spectroscopy

In Figure 3 the chemical shifts of $^{29}$Si NMR signals of corroded and pristine NaSi2·0 (Figure 3(a)) and NaSi3·3 (Figure 3(b)) are compared. The respective chemical shifts and the peak area fractions of the respective $Q^n$ peaks are listed in Table 2. Literature data were evaluated for peak assignment. The respective data are listed in Table 3. The peak areas were calculated via deconvolution with Gauss distribution functions and the peak area fractions $A(Q^n)_{NMR}$ according to Equation (1).

The assignment of the $Q^n$ groups offered some problems. These problems are solved by assuming that corrosion and layer formation as well as ion exchange $Na^+ \leftrightarrow H^+$ shift the bands towards lower (re-

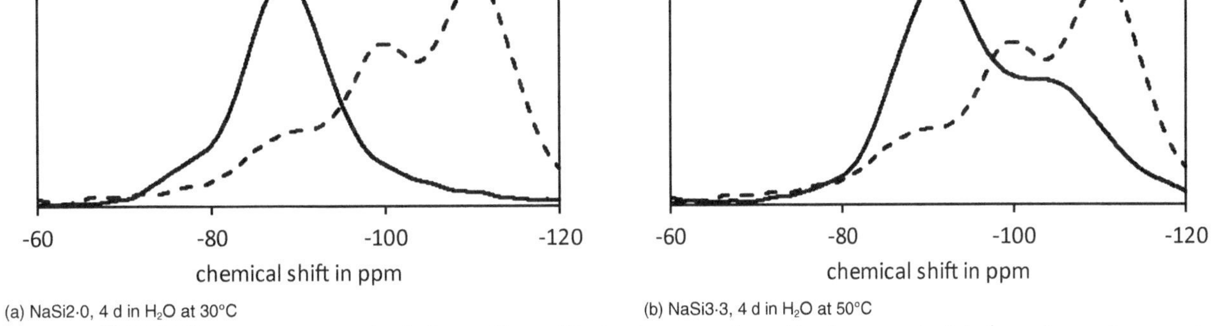


(a) NaSi2·0, 4 d in $H_2O$ at 30°C (b) NaSi3·3, 4 d in $H_2O$ at 50°C

*Figure 3. $^{29}$Si NMR spectrograms of pristine sodium silicate glasses and reaction layer materials from dynamic corrosion tests;* —— *: pristine glass;* – – – *: layer materials*

*Table 2. Chemical shifts and the peak area fraction of the respective $Q^n$ peaks calculated from spectra in Figure 3 by deconvolution; additionally $\bar{n}_{NMR}$ (according to Equation (2)) and $Y^{(41)}$ are listed*

| Property | NaSi2·0 Pristine shift in ppm | fraction in % | Corroded shift in ppm | fraction in % | NaSi3·3 Pristine shift in ppm | fraction in % | Corroded shift in ppm | fraction in % |
|---|---|---|---|---|---|---|---|---|
| $Q^1$ | | | | | | | −80 | 8 |
| $Q^2$ | −77 | 2 | 88 | 14 | | | −92 | 37 |
| $Q^3$ | −88 | 90 | −100 | 48 | −91 | 59 | −100 | 33 |
| $Q^4$ | −100 | 8 | −111 | 38 | −105 | 41 | −111 | 41 |
| $\bar{n}_{NMR}$ | | 3·06 | | 3·24 | | 3·41 | | 3·07 |
| $Y^{(41)}$ | | 2·97 | | | | 3·40 | | |

gard the negative sign) chemical shifts.[44] According to Olivier *et al*[45] such a shift due to corrosion can be caused by an enlargement of polymerised structure units, possibly by the formation of larger polymerised silicate units connected by less polymerised regions.

Mean $\bar{n}_{NMR}$ values ($\bar{n}_{NMR}$=average number of bridging oxygen per silicon according to $^{29}$Si NMR) were calculated with Equation (2) and compared with Stevels[41] *Y* values (Table 2). The chemical shifts detected by $^{23}$Na NMR spectroscopy for the same samples are shown in Figure 4(a) (NaSi2·0) and Figure 4(b) (NaSi3·3). The signal measured for the reaction layer formed on NaSi2·0 has a narrower intensity distribution than the signal of the pristine glasses. This finding indicates a change of the structural surrounding of the $Na^+$ ions, possibly due to a higher mobility of the ions or a lesser bond strength. This effect is not that pronounced in the case of the reaction layer on NaSi3·3.

### *3.3. Infrared and Raman spectroscopy*

Figure 5 shows absorption spectra of reaction layer materials generated by dynamic corrosion in $H_2O$ on NaSi2·0, NaSi2·5, and NaSi3·3 at wavelengths between 6700 and 400 cm$^{-1}$. The following bands can be recognised:

- 5230 cm$^{-1}$: combination band of $H_2O$ bending and stretching vibrations.[47] According to Scholze[33] this absorption band proofs the existence of $H_2O$ molecules in glass. The intensity of this band and of its background is not very pronounced. To enhance the visibility of this band the absorbance values of reaction layer materials on NaSi2·5 were multiplied by a factor 5 (Figure 5). In that infrared band region it was possible to analyse complete corroded samples (bulk glass and two layers) in transmission (Figure 6).
- 3800 to 2600 cm$^{-1}$: OH stretching vibrations. Vibrations with smaller wavenumbers in this infrared region are due to OH groups connected to nonbridging oxygens.[47]
- Peaks between 2800 and 3000 cm$^{-1}$: attributed to $CO_2$[48] which is probably absorbed from the atmosphere.
- 1640 cm$^{-1}$: attributed to bending vibrations of molecular water ($H_2O$).[47] The band demonstrates the existence of molecular water in the reaction layer.
- 1450 cm$^{-1}$: attributed to $CO_3^{2-}$[48] in connection with $Na^+$. The band is detected in layers recovered from NaSi2·0 and NaSi2·5, but not from NaSi3·3. The stronger $CO_2$ absorption of the more alkaline glasses corresponds to the higher residual Na content of the reaction layers on NaSi2·0 found by EDX profiling. Recently, Barrow *et al*[49] investigated the residual $CO_3^{2-}$ content of rapidly cooled sodium silicate glasses with up to 70 mol% $Na_2O$. The $CO_3^{2-}$ content rises with $Na_2O$ content of the glasses, but significant amounts are reached if the $Na_2O$ content exceeds 40 mol% which is not the case here. Additionally, no specific carbonate bands were discovered in the infrared spectra of the pristine glasses. Therefore, it is concluded that the 1450 cm$^{-1}$ band of the layer materials is due to $CO_2$ adsorbed from the atmosphere.
- 1300 to 800 cm$^{-1}$: silicate bands due to asymmetric Si–O stretching vibrations or vibrations within the tetrahedral cage ($Q^0$ at 800 cm$^{-1}$). The infrared bands between 1300 and 800 cm$^{-1}$ can be assigned to $Q^n$ groups following King *et al.*[50] The assignments of the $Q^n$ groups to the infrared bands are listed in Table 4.
- <600 cm$^{-1}$: O–Si– or Si–O–Si angle bending vibrations.[50]

In the near infrared the transparency of silicate glasses is higher. In some cases, it was possible to analyse the complete corroded glass sample in that range. The laser penetrates the uncorroded bulk of the glass sample and two reaction layers on both sides of the

*Table 3. Peak assignments of $^{29}$Si NMR data published in literature*

| | Emerson et al,[44] $Na_2O$–$2SiO_2$, glass | Emerson et al,[44] $Na_2O$–$4SiO_2$, glass | Li et al,[27] silica gel | Gin et al,[29] reaction layer | Engelhardt et al,[46] silicates |
|---|---|---|---|---|---|
| $Q^1$ | | | | | −70 to −85 |
| $Q^2$ | −85 | −85 | −85 to −101 | −75 to −95 | −75 to −95 |
| $Q^3$ | −92 | −100 | −94 to −101 | −90 to −105 | −90 to −100 |
| $Q^4$ | −105 | −109 | −105 to −110 | −105 to −120 | −105 to −120 |

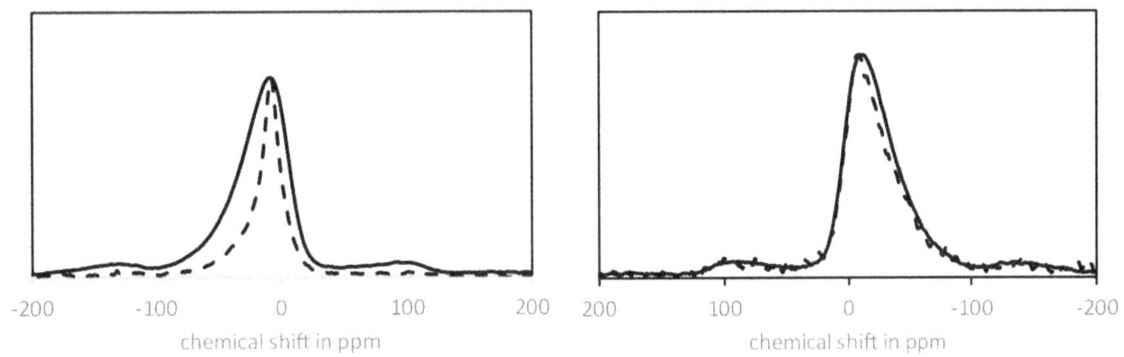


(a) NaSi2·0, 4 d in $H_2O$ at 30°C (b) NaSi3·3, 4 d in $H_2O$ at 50°C

*Figure 4. $^{23}$Na NMR spectrograms of pristine sodium silicate glasses and reaction layer materials from dynamic corrosion tests; —— : bulk glass; - - - : layer materials*

sample. In Figure 6 the absorption results of NaSi3·3 samples corroded under dynamic conditions in $H_2O$ at 50°C were compared for two different corrosion times. Samples corroded individually for 1 and 4 d were compared. After only 1 d corrosion time infrared bands at about 4500 and 5250 $cm^{-1}$ were observed. The band at 4500 $cm^{-1}$ is attributed to a combination of bending and stretching vibrations of OH groups[47] and typical of silanol groups in silicate glasses. The intensity of this band is reduced after longer corrosion times as is shown in Figure 6 for the 4 d corrosion sample. In Figure 7, the silicate bands between 1300 and 800 $cm^{-1}$ observed in layers on NaSi3·3 during corrosion in $H_2O$ are deconvoluted with Gauss distribution functions. This concentration decrease might be explained by condensation of silanol groups in later stages of the reaction layer formation. The band at 5250 $cm^{-1}$ is attributed to molecular $H_2O$.[33] Infrared reflection spectroscopy (not shown here) showed that already after 5 min of corrosion the surface of sodium silicate glasses corroded in $H_2O$ showed typical corrosion signs which did not change significantly after longer corrosion times.

The bands between 1200 and 1100 $cm^{-1}$ are due to asymmetric stretching vibrations of the Si–O–Si bond. The band at 800 $cm^{-1}$ is, according to King *et al*,[50] due to the movement of $Si^{4+}$ in the tetrahedral cage of isolated tetrahedra which are $Q^0$ units. The presence of $Q^0$ units is questionable since these units should be easily soluble in $H_2O$. In Table 4 the peak area fractions (calculated in the same way as for NMR results, Equation (1)) of the deconvoluted peaks were compared. The peak areas of infrared and Raman spectrograms cannot be evaluated quantitatively without knowing extinction coefficients. The values representing the average number of bridging oxygens per silicon were calculated with Equations (1) and (2). They are used to discuss variations of the polymerisation degree of glass and reaction layer with glass composition. A slight increase of with the

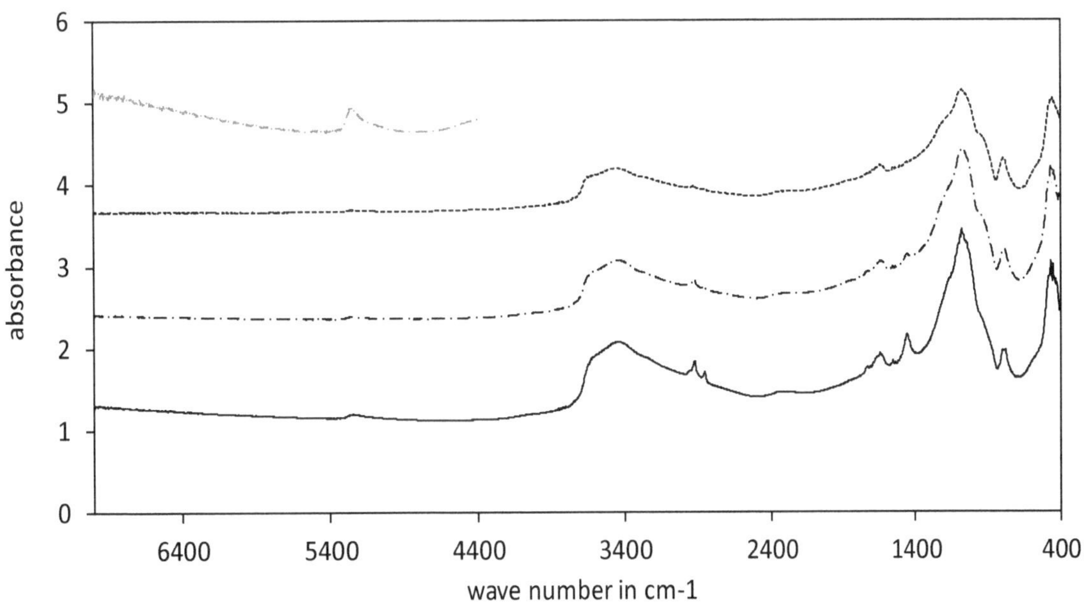


*Figure 5. Infared spectrograms of layer materials generated by 4 d of dynamic corrosion in deionised $H_2O$; the absorbances were separated by adding a constant value; from bottom: —: NaSi2·0 (30 °C); —·—: NaSi2·5 (50°C); —·—: NaSi2·5 signal multiplied by factor 5 (upper curve); ----: NaSi3·3 (50°C) [Colour available online]*

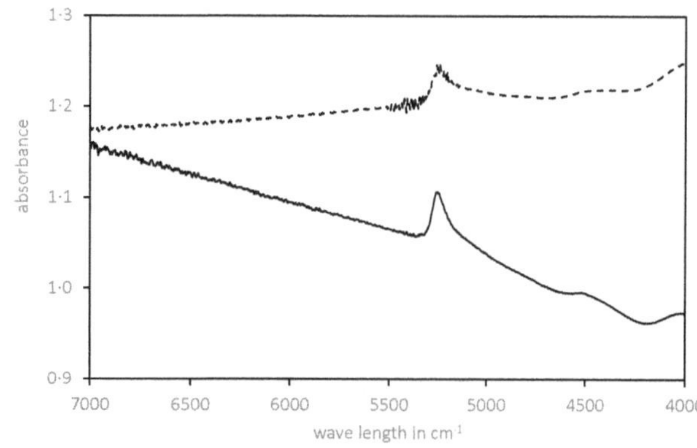


*Figure 6. Near infrared spectrogram of whole samples of NaSi3·3 after static corrosion in $H_2O$ at 50°C; after 1 d: —, after 4 d: --------*

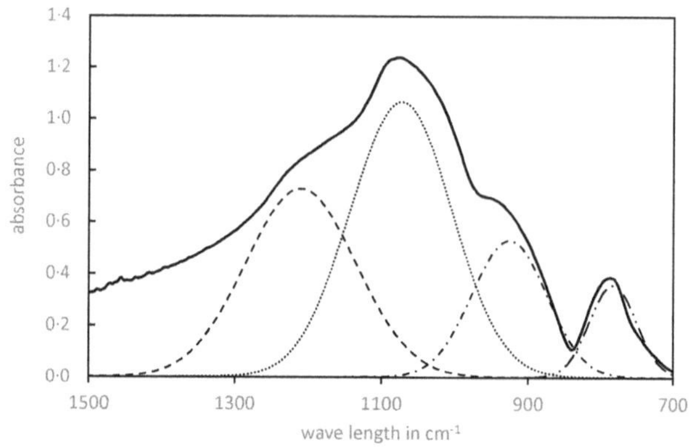


*Figure 7. Deconvolution of infrared spectrogram (—) of reaction layer on NaSi3·3 after 4 d of dynamic corrosion in $H_2O$ at 50°C; Q4: ------, Q3: ·········, Q2: -·-·-·, Q1: -··-··*

molar ratio $SiO_2$:$Na_2O$ of the glasses was observed, less than should be expected from stoichiometry. The values of the reaction layer materials are higher and almost equal for each glass, indicating a higher polymerisation degree of its silicate network compared to the glass.

Raman microscopy allows depth profiling of $Q^n$ group distribution as well as of $H_2O$ or OH group contents. The assignment of the Raman bands to $Q^n$ groups again follows King *et al*[(50)] and is listed in Table 5. Figure 8 shows Raman spectra of the silicate band region between 1200 and 700 $cm^{-1}$ measured at various depth of NaSi2·0 corroded dynamically in $H_2O$ at 30°C. In Figure 9, a similar diagram is shown for NaSi3·3. With increasing sample depth, the intensity of a band at 1030 $cm^{-1}$, which is attributed to $Q^2$, decreases, whereas the band at 1090 $cm^{-1}$ ($Q^3$) gains in relative intensity. Again, the spectra were deconvoluted with Gaussian type intensity distribution functions.

The following Raman bands were detected (see also Table 5):

- 1182 $cm^{-1}$: $Q^4$,
- 1099 $cm^{-1}$: $Q^3$,
- 1033 $cm^{-1}$: $Q^2$ (here a band range from 1000 to 950 $cm^{-1}$ was reported in King *et al*[(50)]), and
- 950 $cm^{-1}$: $Q^1$ (900 $cm^{-1}$ according to Barrow *et al*,[(49)] a band at 950 $cm^{-1}$ is assigned by Furukawa *et al*[(51)] to symmetric stretching vibrations of $Q^2$ groups, whereas the 1030 $cm^{-1}$ band was not mentioned there),
- 800 $cm^{-1}$: $Q^0$ (although it should be easily dissolved).

The intensities of the Raman spectra are influenced by absorption of laser light and scattered Raman radiation which means that the intensity of the Raman bands decrease with depth even if the concentration of the Raman active groups stays constant.[(39)] Therefore, the relative peak areas were calculated by relating the individual peak areas to the sum of peak areas of the Q groups ($Q^0$+$Q^1$+$Q^2$+$Q^3$+$Q^4$). The information depth depends also on the refractive index of the materials. The geometrical depth derived from the mechanics of the microscope has to be multiplied by the refractive index. For the estimation of the refractive index of the layer materials the wavelength of the applied laser radiation has to be taken into account ($\lambda$=532 nm). Since the main ingredients of the reaction layer are silica,[(52)] $H_2O$[(53)] and air, a mean refractive index of 1·4 was assumed for calculating the information depth. The depth profiles of the relative $Q^n$ group peak areas are shown in Figures 10 (NaSi2·0) and 11 (NaSi3·3). Additionally, the relative peak areas were taken to get relative information on $\bar{n}_{Raman}$, the average number of bridging oxygens per silicon according to Raman results, evaluated with Equation (1). For both glasses $\bar{n}_{Raman}$ increases with distance from the unaffected bulks glass, reaches a maximum in the layer and decreases again towards the surface of the reaction layer, a behaviour more pronounced for NaSi3·3.

Raman spectroscopy can also detect molecular water at 1640 $cm^{-1}$. That absorption band measured in NaSi2·0 corroded dynamically in $H_2O$ at 30°C is

*Table 4. Fraction of $Q^n$ groups and absorption band maxima in pristine glasses as well as in reaction layer materials after dynamic corrosion of sodium silicate glasses in $H_2O$ at 30 (NaSi2·0) and 50°C (NaSi2·6 and NaSi3·3); additionally $\bar{n}_{IR}$ is calculated via Equation (2)*

| *Material* | $\lambda$ ($Q^4$) | $\bar{n}_{IR}$ ($Q^4$) | $\lambda$ ($Q^3$) | $\bar{n}_{IR}$ ($Q^3$) | $\lambda$($Q^2$) | $\bar{n}_{IR}$ ($Q^2$) | $\lambda$($Q^1$) | $\bar{n}_{IR}$ ($Q^1$) | $\bar{n}_{IR}$ |
|---|---|---|---|---|---|---|---|---|---|
| NaSi20, g | 1234 | 17 | 1075 | 47 | 928 | 31 | 753 | 5 | 2·76 |
| NaSi2.0, l | 1229 | 29 | 1069 | 52 | 929 | 14 | 785 | 5 | 3·05 |
| NaSi2.5, g | 1261 | 19 | 1088 | 46 | 943 | 29 | 759 | 6 | 2·78 |
| NaSi2.5, l | 1237 | 31 | 1071 | 49 | 936 | 15 | 785 | 5 | 3·06 |
| NaSi3.3, g | 1217 | 19 | 1086 | 50 | 960 | 26 | 770 | 5 | 2·83 |
| NaSi3.3, l | 1247 | 33 | 1073 | 44 | 926 | 16 | 783 | 7 | 3·03 |

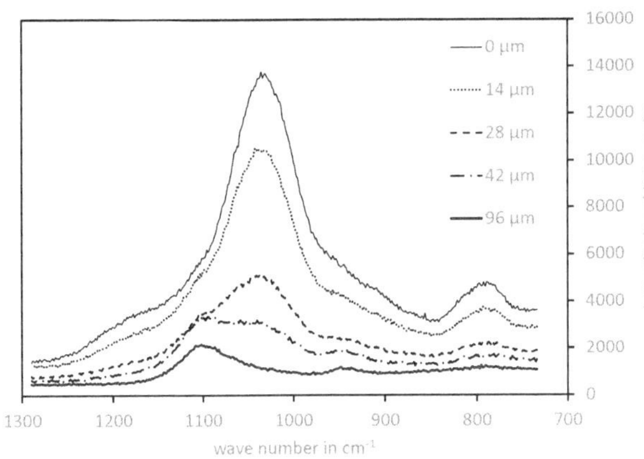


*Figure 8. Raman spectrogram of NaSi2·0 corroded 4 d at 30°C dynamically in deionised $H_2O$ at increasing depth in the wavelength region of silicate groups; sample depths at array of curves*

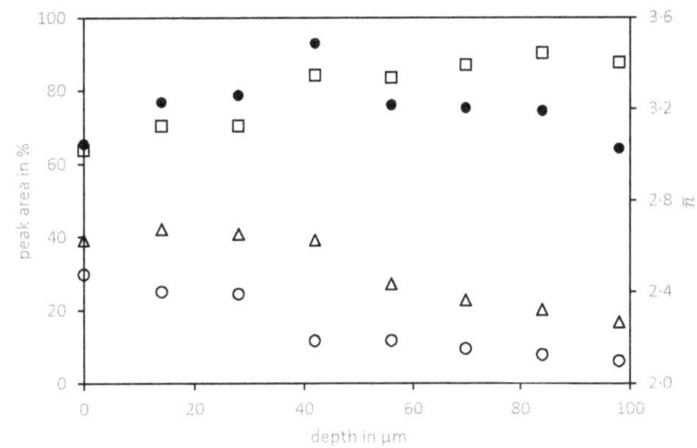


*Figure 10. Relative peak areas of Raman bands assigned to different Q groups (according to Table 5) of NaSi2·0 corroded dynamically 4 d at 30°C in deionized $H_2O$ as a function of sample depth;* ○: $Q^1$, ▵: $Q^2$, □: $Q^3$, ●: $\bar{n}_{Raman}$

shown for different sample depths in Figure 12. The signal is fitted with a Gauss distribution function after base line correction. Similar Raman spectra for NaSi3·3 are not shown here but included in Figure 13 which shows depth profiles of the $H_2O$ absorption band obtained for NaSi2·0 and NaSi3·3 corroded in deionised $H_2O$.

The following information can be gained from the depth profiles (Figures 10, 11, and 13):

- $Q^n$ groups in reaction layer on NaSi2·0 (Figure 10): a surface zone with a thickness of about 90–100 µm was found. The relative peak areas $A(Q^3)_{Raman}$ decrease when the focal plane of the Raman microscope approaches the outer surface from the bulk glass whereas $A(Q^2)_{Raman}$ and $A(Q^1)_{Raman}$ increase. A step in the depth profile of the $Q^2$ and $Q^1$ peaks at about 50 µm indicates a subdivision of the reaction layer. The peak areas of $Q^4$ are quite low compared to the other peaks.
- $Q^n$ groups in reaction layer on NaSi3·3 (Figure 11): Evident structural differences were seen within a surface layer with a thickness of about 50 µm. Within that layer similar developments as for NaSi2·0 are detected: A decrease of $A(Q^3)_{Raman}$ and an increase of $A(Q^2)_{Raman}$ and $A(Q^1)_{Raman}$ towards the outer surface of the reaction layer. Near the bulk glass the $Q^3$ signals increase and dominate. Beyond a depth of 25 µm only slight changes are observed, e.g. an increase of the relative $Q^0$ signal. The $Q^4$ signal is hardly detectable within the outer 10 µm.
- For both glasses $\bar{n}_{Raman}$ increases with distance from the unaffected bulks glass, reaches a maximum in the layer and decreases again towards the surface of the reaction layer, a behaviour more pronounced for NaSi3·3.
- $H_2O$ (Figure 13): In both samples, the $H_2O$ signal decreases to zero within the measured depth range. The reaction layers have thicknesses of 80 µm (NaSi2·0) and 40 µm (NaSi3·3). In the case of NaSi3·3 the strongly altered silica structure has a depth of about 50 µm. The $H_2O$ concentration profile has a depth of about 40 µm, which is at least comparable. This leads to the conclusion that $H_2O$ incorporation and reorganisation of silicate structure happens simultaneously.

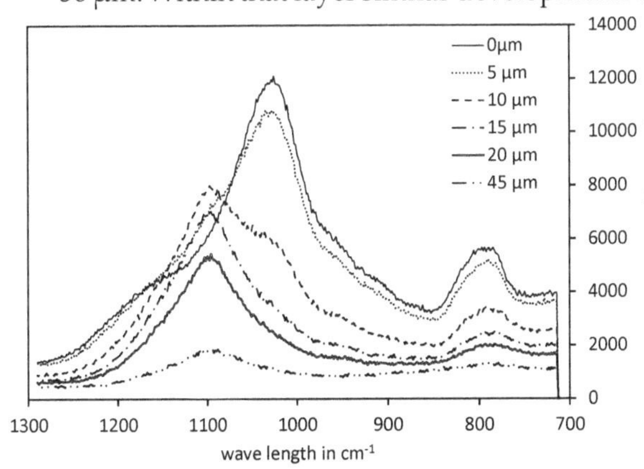


*Figure 9. Raman spectrogram of NaSi3·3 corroded 4 d at 30°C dynamically in deionised $H_2O$ at increasing depth in the wavelength region of silicate groups; sample depths at array of curves*

## 4. Discussion

The corrosion behaviour and the surface alterations of the investigated sodium silicate glasses during corrosion have similarities on one side but display differences on the other side. A look at the chemical surface characterisation by SEM combined with EDX

*Table 5. $Q^n$ group assignment in Raman and infrared spectra of silicate materials according to King et al[50] and bands found in this investigation*

| $Q^n$ | *Infrared band in cm⁻¹* | | *Raman band in cm⁻¹* | |
|---|---|---|---|---|
| *source* | *King et al[50]* | *here* | *King et al[50]* | *here* |
| $Q^4$ | 1200–1250 | 1210 | 1200 | 1190 |
| $Q^3$ | 1000–1100 | 1070 | 1050–1100 | 1100 |
| $Q^2$ | 900–1000 | 925 | 950–1000 | 1030 |
| $Q^1$ | 700–900 | 780 | 900 | 950 |
| $Q^0$ | | | 850 | 800 |

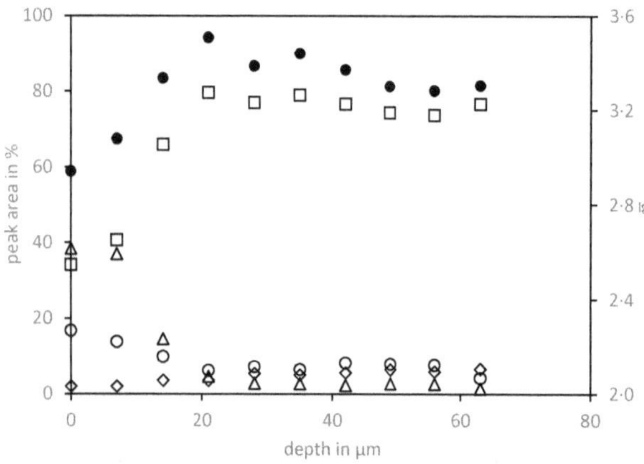


*Figure 11. Relative peak areas of Raman bands assigned to different Q groups (according to Table 5) of NaSi3·3 corroded dynamically 4 d at 30°C in deionised $H_2O$ as a function of sample depth; o: $Q^1$, ▲: $Q^2$, □: $Q^3$, ◇: $Q^4$, • : $\bar{n}_{Raman}$*

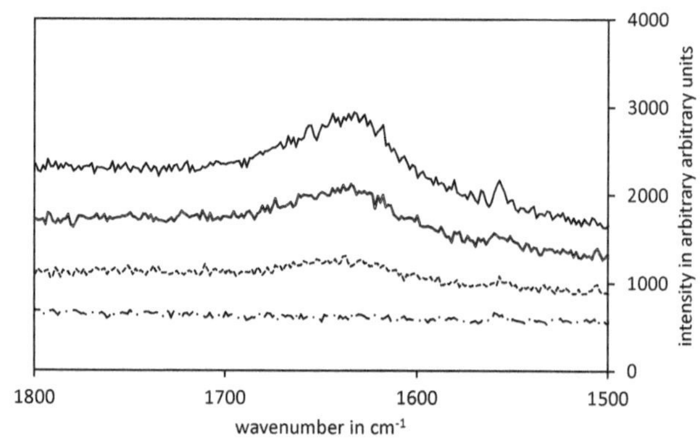


*Figure 12. Raman spectrogram of NaSi2·0 corroded 4 d at 30°C dynamically in deionised $H_2O$ at increasing depth in the wavelength region of molecular $H_2O$; depth of focal plane: —: 0 µm; ═══: 14 µm; - - - -: 28 µm; —·—·—: 96µm.*

helps to distinguish four different cases:

NaSi2·0 corroded in $H_2O$ develops extensive reaction layers possibly divided into two sublayers with surprisingly high Na concentrations. The high level of Na concentration resembles those reported by Scholze *et al*(17) but without developing the typical S-shaped concentration profile. The cause of the Na content of the layer is probably the adsorption of $Na^+$ at surface sites in a structurally altered reaction layer.

NaSi3·3 corroded in $H_2O$ develops concentration depth profiles similar to those found for typical silicate glasses.(8,17) The corroded surface is comparable to type IV according to Hench.(8)

During corrosion in 0·1 M NaOH no reaction layers were developed. That means probably, that the dissolution of the altered surface is faster than the surface alteration itself. Solubility of $SiO_2$ significantly increases with pH at these conditions.(54) On the other hand, own investigations showed that corrosion rates of sodium silicate glasses are significantly decreased(3) with increasing pH at these high pH values. Therefore, it is concluded that the high pH values on one side prevent the formation of reaction layers by fast dissolution of these materials but on the other side inhibit the surface alteration via adsorption of $Na^+$.

For the corrosion of NaSi2·0 and NaSi3·3 in diluted liquid sodium water glasses having the same molar $SiO_2$:$Na_2O$ ratio as the glass the concentration dependence reported in Dathe & Roggendorf(3) it should be remembered: The corrosion rate of NaSi2·0 declined with increasing concentration whereas the corrosion rate of NaSi3·3 increased with concentration of liquid sodium water glass. Surface analysis by SEM combined with EDX of NaSi3·3 corroded in a liquid sodium water glass with a molar $SiO_2$:$Na_2O$ ratio=3·3 with 2 wt% $SiO_2$ content revealed a relatively thin reaction layer with features typical of the corrosion of ordinary silicate glasses: depletion of Na down to an $Na_2O$ content of about 3–5 wt% and enrichment of Si towards the surface. Of course, reaction layers on more durable glasses are much thinner. The results of surface analysis of NaSi2·0 corroded in a liquid sodium water glass with a molar $SiO_2$:$Na_2O$ ratio=2·2 with 2 wt% $SiO_2$ content are more ambiguous. In a surface layer with a thickness of 80 µm the Na content was reduced to about 70% of the level of the bulk glass. Since the Si signal stayed nearly constant the high level of $Na^+$ in the surface layer seems to be real. At the moment, no explanation for that high Na content can be offered.

For two samples, NaSi2·0 corroded in deionized $H_2O$ and NaSi2·0 corroded in diluted sodium silicate solution, a smaller density of the reaction layer was found which is possibly caused by porosity. Type IV surface(8) with silica enrichment were only found for NaSi3·3 samples corroded in $H_2O$ or liquid sodium water glass.

The spectroscopic investigations yielded mainly results for reaction layer materials and pristine glasses. A stoichiometric sodium disilicate glass has Y=3,(41) a value with a structure which might theoretically consist only of $Q^3$ groups. Such a $Q^n$ group distribution can be realised in crystalline silicates but not in glasses, as Gurman pointed out.(55) He concluded that a broader $Q^n$ group distribution is more favourable due to higher entropy. More accurate $^{29}Si$ NMR measurements (e.g. Machacek *et al*(56)) showed

*Table 6. $\bar{n}_{NMR}$, $\bar{n}_{IR}$, $\bar{n}_{Raman}$, and Y of pristine glasses and layer materials determined according to Equation (2) after deconvolution of $^{29}Si$ NMR, infrared, and Raman spectra*

| | *Pristine glass* | | | | | *Reaction layer* | | |
|---|---|---|---|---|---|---|---|---|
| *Glass* | $\bar{n}_{NMR}$ | $\bar{n}_{IR}$ | $\bar{n}_{Raman}$ | *Y* | $n_c$ | $\bar{n}_{NMR}$ | $\bar{n}_{IR}$ | $\bar{n}_{Raman}$, *maximum* |
| NaSi2·0 | 3·06 | 2·76 | 3·03 | 2·97 | 2·55 | 3·24 | 3·05 | 3·41 |
| NaSi2·5 | | 2·78 | | 3·21 | 2·73 | | 3·06 | |
| NaSi3·3 | 3·41 | 2·83 | 3·30 | 3·40 | 2·89 | 3·07 | 3·03 | 3·51 |

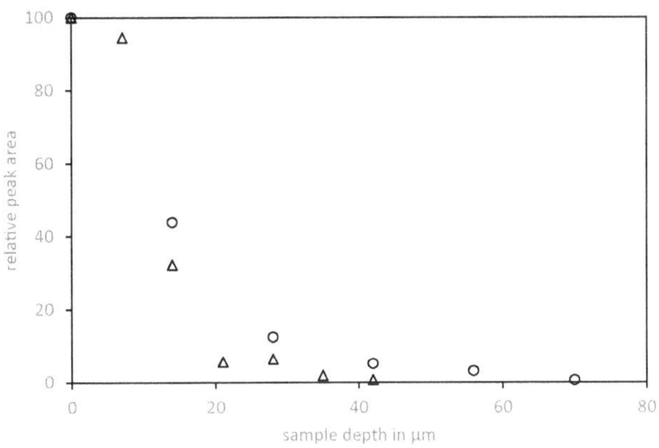


*Figure 13. Relative peak areas of Raman bands assigned to molecular $H_2O$ at 1640 $cm^{-1}$ of sodium silicate glasses corroded 4 d dynamically in deionised $H_2O$ as a function of sample depth;* **o**: *NaSi2·0 at 30°C,* **▲**: *NaSi3·3 at 50°C.*

that sodium disilicate glass contains about 80% $Q^3$ groups, the rest being $Q^2$ or $Q^4$. A $Na_2O.2{\cdot}5SiO_2$ has $Y$=3·20 and a $Na_2O.3{\cdot}3SiO_2$ glass would yield $Y$=3·40. For the latter glass this is confirmed by $^{29}$Si NMR, infrared and Raman results.

The $\bar{n}$ values obtained (Equation (2)) by $^{29}$Si NMR, infrared, and Raman spectroscopy of the pristine glasses and of the reaction layer materials are compared in Table 6 with $Y$ values calculated according to Stevels[41] and $n_c$ according to the topological constraint theory.[30,32] A comparison of $\bar{n}_{NMR}$ with Stevels $Y$ value shows that the peak area distribution can be evaluated quantitatively: They allow to establish relative concentration relations between the $Q^n$ units. To yield absolute concentrations the silicon densities of the reactions layer materials have to be known.

The finding for $^{29}$Si NMR spectroscopy on reaction layer materials is different: at first sight, the change of $\bar{n}_{NMR}$ indicates a polymerisation in the case of NaSi2·0 and a depolymerisation in the case of NaSi3·3. A closer look at the $^{29}$Si–NMR spectrum of reaction layer material generated by NaSi3·3 shows that it suffers from low intensities, since the layer mass available for analysis was quite low which is concealed by the normalisation of the spectra. To overcome the differences between $^{29}$Si NMR and the other spectroscopic methods the following should be considered: The layer materials investigated by $^{29}$Si NMR were scratched off the surface and their origin is not exactly known with respect to the layer depth. Possibly, the reaction layer materials from NaS3·3 originated mainly from the already depolymerised surface zone (see Raman results).

Infrared spectroscopy yields valuable information on structural groups in glasses and reaction layer materials. Silicate groups and $OH^-$ or $H_2O$ are in the focus. The silicate bands allow to identify $Q^n$ group distributions on a semiquantitative level. The calculated $\bar{n}_{Infrared}$ values are about 10% lower than those found by $^{29}$Si NMR. The shifts of $\bar{n}_{Infrared}$ between pristine glass and reaction layer indicates for all three glasses a polymerisation of the silicate network.

The bands typical of $OH^-$ groups and molecular water proof the existence of these species, which was expected according to the findings of Scholze.[33] Also, the sensibility of the reaction layer for atmospheric $CO_2$ was shown by the existence of the peaks between 3000 and 2800 $cm^{-1}$. Since $Na_2O$ is a prerequisite for these peaks, the existence of that compound in the reaction layer is demonstrated again.

The average number of bridging oxygen per silicon $\bar{n}_{Raman}$ determined by Raman spectroscopy are also quite close to the $Y$ numbers[41] calculated from composition. This indicates that the Raman bands of the silicate groups can be evaluated quantitatively. An exact evaluation yields minor deviations for NaSi2·0, which are still within the measurement errors, and a good accordance for NaSi3·3. The main advantage of Raman microscopy is that it allows depth profiling of structural units and molecular $H_2O$ in reaction layers if their thickness exceeds 10 µm. This is the case for corrosion of sodium silicate glasses in $H_2O$. That technique has only limited potential to investigate corrosion layers on glasses with higher chemical durability submitted to short time corrosion tests. The thickness of layers on such corrosion products rarely exceeds 1 µm.[20]

The shifts of $\bar{n}_{Raman}$ with depth from surface to pristine glass allows a more detailed discussion. The Raman profiles shown in Figures 10 and 11 indicate a polymerisation in the center of the reaction layer and a depolymerisation near the surface. A polymerisation reaction during layer formation was already described by Melnikova *et al.*[57] The depolymerisation is probably due to the hydrolysis of the silicate network during dissolution.

If the mass loss by corrosion is mainly caused by the dissolution of the reaction layer, the thermodynamics of the glass cannot be related directly to the corrosion rate. Therefore, the linear dependence of the corrosion rate on the enthalpy of glass hydrolysis described in previous results[2] must be motivated indirectly. A possible reason is that the microstructure of the reaction layer depends on glass composition. Sodium silicate with higher $Na_2O$ contents have a higher degree of $Na^+$ diffusion and probably a smaller density. This may accelerate transport reactions.

At very high pH values, e.g. for corrosion in 0·1 M NaOH, the dissolution of the reaction layer is supposed to be faster than its development by ion exchange and silicate polymerisation. In Dathe & Roggendorf[3] the argument was put forward that diffusion through a boundary layer is an essential step in the corrosion process. Then the leachate in direct contact with the glass surface has about the pH of a concentrated water glass, which is between 12 and 13.[10] Then the pH of the leachant is higher outside the diffusion layer. Higher pH means higher solubility of

$SiO_2$ and thus, precipitation is additionally avoided.

Reactions layers provide abundant adsorption sites for $Na^+$. The absence of the reaction layer shifts the adsorption equilibrium

$$\equiv Si–O–H^+ + Na^+ \leftrightarrow \equiv Si–O–Na^+ + H^+ \quad (4)$$

towards more complete coverage (right side of Equation (4)) and the adsorption of $Na^+$ thus provides a diffusion barrier or shields the surface against the attack of $OH^-$.

Under these conditions, the corrosion rate is reduced with increasing pH values as reported in Dathe & Roggendorf.[3] The role of $Na^+$ might be ambiguous. Its concentration increases with NaOH concentration, but pH measurements show that the $Na^+$ activity is smaller due to the buffer action of silicate solutions. When a reaction layer with abundant silanol groups exists, the $Na^+$ activity may not be sufficient to shield the silicate structure. When the reaction layer is formed as fast as she is dissolved, the maximum of the mass loss rate is reached. Therefore, the dissolution of sodium silicate glasses in $H_2O$ is a faster than dissolving in more concentrated NaOH solutions.

In the case of reaction layer formation, the corrosion process is more complex. Then the $Na^+$ has to cross the reaction layer, too, and the pH at the dissolution zone is governed by the interchange of $Na^+$ and $H^+$. In Dathe & Roggendorf[3] it was reported that the thickness of the reaction layer increases with time. Thus, diffusion through the layer will slow down, possibly resulting in a lower pH of the leachate. Overall, the pH values in the dissolution zone during corrosion in $H_2O$ should be lower than during corrosion in 0·1 M NaOH, which stabilises the reaction layer. Nevertheless, pH in the leachate will still be lower and dissolved $SiO_2$ will become less stable.

To summarise, the following reactions during corrosion of sodium silicate glasses at neutral to intermediate pH conditions are postulated:

1. Ion exchange $Na^+ \leftrightarrow H^+$.
2. Diffusion of $Na^+$ into solution. In the case of reaction layer formation the diffusion continues through the layer.
3. pH > 12:
   a. Dissolution of $SiO_2$ is faster than surface alteration by ion exchange.
   b. $Na^+$ is adsorbed at silanol groups and shields the surface. With increasing $Na^+$ concentrations the dissolution rate of the glass is reduced.
4. pH < 12:
   a. Formation of a reaction layer by polymerisation of silicate network in surface zone partially depleted from $Na^+$. The bonding of sodium to the layer is weaker than the bonding in the glass network.
   b. Dissolution of silica due to a hydrolysis reaction at the outside of reaction layer accompanied[2] by depolymerisation of the silicate network.

From these mechanisms, two topics of future research can be derived. The first is the adsorption of $Na^+$ at the surface. In Roggendorf[58] such an adsorption was shown for borosilicate glasses intended for the disposal of high level radioactive waste. The second is the determination of pH values near the dissolution zone of the glasses. The measurement of pH gradients on a μm scale is quite ambitious. The application of simulation techniques is possibly more promising. A simulation process for glass corrosion was presented e.g. by Nunes de Carvalho *et al.*[59]

## 5. Summary

Corrosion and dissolution reactions of $Na_2O.xSiO_2$ glasses ($x$=2·0, 2·5, and 3·3) were investigated by static and dynamic corrosion tests at pH values between 7 and 14 and temperatures of 30 and 50°C. The investigated glass compositions are close to those used for water glass production. Several types of leachants with varied pH and $SiO_2$ concentrations have been applied to induce corrosion effects. One characteristic feature of the corrosion process is the development of a reaction layer on top of the corroding glasses, a process depending on leachant pH. At intermediate pH values reaction layers were observed, but not at high pH values (>12). To generate altered glass surfaces and reaction layers sodium silicate glasses were corroded in deionised $H_2O$, 0·1 M NaOH, and diluted liquid sodium water glasses. The resulting samples or layer materials have been investigated by scanning electron microscopy, Raman microscopy, infrared spectroscopy or $^{29}Si$ MAS-NMR spectroscopy. Especially Raman microscopy was applied to characterise depth profiles of structural changes and concentration profiles of molecular $H_2O$ within reaction layers. The results are discussed with respect to influences of pH and leachant composition.

## Acknowledgements.

The investigations reported here were funded by the Deutsche Forschungsgemeinschaft (contract Ro 880/1 and Ro 880/2). We also thank the grants DFG INST 271/285-1, within the Forschergruppe AK Prof. Bron, for providing the Raman microscope. The experimental help of S. Brinke and Th. Pfeiffer is gratefully acknowledged.

## References

1. Dathe, M. & Roggendorf, H. Dissolution of sodium silicate glasses for the production of water glass. Part 1. Study of experimental parameters. *Phys. Chem. Glasses: Eur. J. Glass Sci. Technol. B*, 2018, **59** (5), 241–50.
2. Dathe, M. & Roggendorf, H. Dissolution of sodium silicate glasses for the production of water glass. Part II. Dependence of corrosion process on pH in basic aqueous media. *Phys. Chem. Glasses: Eur. J. Glass Sci. Technol. B*, 2018, **59** (6), 301–10.

3. Dathe, M. & Roggendorf, H. Dissolution of sodium silicate glasses for the production of water glass. Part III Dependence of corrosion process on silica concentration of liquid. *Phys. Chem. Glasses: Eur. J. Glass Sci. Technol. B*, 2019, **60** (1), 21–31.
4. Clark, D. E. & Zoitos, B. K. Corrosion testing and characterisation. In *Corrosion of glass, ceramics and ceramic superconductors - Principles, testing, characterization and applications*. Edited by D. E. Clark & B. K. Zoitos, Noyes Publications, Park Ridge, NJ, USA 1992, 51–102.
5. Kaspar, T. C., Reiser, J. T., Ryan, J. V. & Wall, N. A. Non-destructive characterisation of corroded glass surfaces by spectrometric ellipsometry. *J. Non-Cryst. Solids*, 2018, **481**, 260–6.
6. Conradt, R. Chemical durability of oxide glasses in aqueous solutions - a review. *J. Am. Ceram. Soc.*, 2008, **91**, 728–35.
7. Doremus, R. H. *Glass Science*. John Wiley & Sons, New York, 1994.
8. Hench, L. L. & Clark, D. E. Physical chemistry of glass surfaces. *J. Non-Cryst. Solids*, 1978, **28**, 83–105.
9. Weldes, H. H. & Lange K. R. Properties of soluble silicates. *Ind. Eng. Chem.*, 1969, **61**, 29–44.
10. Roggendorf, H., Grond, W., & Hurbanic, M. Structural characterization of concentrated alkaline silicate solutions by $^{29}$Si-NMR spectroscopy, FT-IR spectroscopy, light scattering, and electron microscopy - molecules, colloids and artefacts. *Glastech. Ber.: Glass Sci. Technol.*, 1996, **69**, 216–31.
11. Das, C. R. Diffusion-controlled attack of glass surfaces by aqueous solutions. *J. Am. Ceram. Soc.*, 1980, **63**, 160–5.
12. Douglas, R. W. & El-Shamy, T. M. M. Reactions of glasses with aqueous solutions. *J. Am. Ceram. Soc., 1967*, **50**, 1–8.
13. Thach, L. T. Phase separation and devitrification products of the binary system sodium oxide - silica. *Glass Technol.*, 1965, **6**, 161–5.
14. Sanders, D. M. & Hench, L. L. Mechanisms of glass corrosion. *J. Amer. Ceram. Soc.*, 1973, **56**, 373–7.
15. Portal, S. & Sempere, R. Study of alkali silicate glass corrosion using spectroscopic ellipsometry and secondary ion mass spectrometry. *Phys. Chem. Glass.*, 2003, **44**, 303–7.
16. Hahn, S. H. & van Duin, A. C. T. Surface reactivity and leaching of a sodium silicate glass under an aqueous environment: A ReaxFF molecular dynamics study. *J. Phys. Chem. C.*, 2019, **123**, 15606–17.
17. Scholze, H., Helmreich, D. & Bakardjiev, I. Investigations of the behaviour of soda-lime-silica glasses in dilute acids. *Glastech. Ber.*, 1975, **48**, 237–47.
18. Doremus, R. H. Interdiffusion of hydrogen and alkali ions in glass surface. *J. Non-Cryst. Solids*, 1975, **19**, 137–44.
19. Lanford, W. A., Burman, C, Doremus, R. H., Mehrotra, Y. & Wassick T. Nuclear reaction analysis of glass surface: The study of the reaction between water and glass. In *Advances in materials characterization*. Edited by D. R. Rossington, R. A. Condrate & R. L. Snyder, Springer, New York, 1983, 549–70.
20. Conradt, R., Roggendorf, H. & Scholze, H. Investigations on the role of surface layers in HLW glass corrosion. *Mater. Res. Soc. Symp. Proc.*, 1986, **50**, 203–10.
21. Maeda, T., Hotta, K., Usui, H. & Banba, T. Characteristics of alteration layers formed on simulated HLW glass under silica-saturated solutions. *J. Austral. Ceram. Soc.*, 2012, **48**, 90–5.
22. Bach, H. Advanced surface analysis of silicate glasses, oxides and other insulating materials: a review. *J. Non-Cryst. Solids*, 1997, **209**, 1–18.
23. Pantano, C. G. What do we know about glass surfaces? *Ceram. Eng. Sci. Proc.*, 2001, **22**, 137–48.
24. Richter, T., Frischat, G. H., Borchardt, G. & Scherrer, S. Initial stages of glass corrosion in water. *Glastech. Ber.*, 1990, **63**, 300–8.
25. Stallworth, P. E. & Bray, P. J. Nuclear magnetic resonance. In *Glass Science and Technology: Advances in Structural Analysis*. Eds D. R. Uhlmann & N. J. Kreidl, Academic Press, Boston, 1990, **Vol. 4B**, 77–149.
26. Stebbins, J. F. Identification of multiple structural species in silicate glasses by $^{29}$Si NMR. *Nature*, 1987, **330**, 465–7.
27. Li, J., Hayakawa, S., Shirosaki, Y. & Osaka, A. Revisiting structure of silica gels from water glass: an $^{1}$H and $^{29}$Si MAS and CP-MAS NMR study. *J. Sol-Gel Sci. Technol.*, 2013, **65**, 135–42.
28. Boehm, T., Leissler, J. & Chudek, J. A. Investigations of glass corrosion in acid solutions with $^{29}$Si-MAS NMR. *Glass Sci. Technol.*, 1995, **68**, 400–3.
29. Gin, S., Jollivet, P., Fournier, M., Angeli, F., Frugier, P. & Charpentier, T. Origin and consequences of silicate glass passivation by surface layers. *Nature Commun.*, 2015, **6** (6360), 1–8.
30. Mauro, J. C. Topological constraint theory of glass. *Am. Ceram Soc. Bull.*, 2011, **90** (4), 31–7.
31. Phillips, J. C. & Thorpe, M. F. Constraint theory, vector percolation and glass formation. *Solid State Commun.*, 1985, **83**, 699–702.
32. Pignatelli, I., Kumar, A., Bauchy, M. & Sant, G. Topological control on silicates dissolution kinetics. *Langmuir*, 2016, **32**, 4434–9.
33. Scholze, H. Zur Frage der Unterscheidung zwischen $H_2O$-Molekeln und OH-Gruppen in Gläsern und Mineralen. *Die Naturwisenschaft*, 1960, **47**, 226–7.
34. Lynch, M. E., Folz, D. C. & Clark, D. E. Use of FTIR reflectance spectroscopy to monitor corrosion mechanisms on glass surfaces. *J. Non-Cryst. Solids*, 2007, **353**, 2667–74.
35. Amma, S., Kim, S. H. & Pantano, C. G. Analysis of water and hydroxyl species in soda lime glass surface using attenuated total reflection (ATR)-IR spectroscopy. *J. Am. Ceram. Soc.*, 2016, **99**, 128–34.
36. Almeida, R. M. & Marques, A. C. Characterization of sol-gel materials by infrared spectroscopy. In *Handbook of sol-gel science and technology*. Edited by L. Klein, M. Aparicio & A. Jitianu, Springer, Berlin, 2018, 1121–51.
37. Brawer, S. A. & White, W. B. Raman spectroscopic investigation of the structure of silicate glasses. I. Binary alkali silicates. *J. Chem. Phys.*, 1975, **63**, 2421–32.
38. Silva, A. M. B., Queiroz, C. M. Agathopoulos, S., Correia, R. N., Fernandez, M. H. V. & Oliveira, J. M. Structure of $SiO_2$–MgO–$Na_2O$ glasses by FTIR, Raman and $^{29}$Si MAS NMR. *J. Mol. Struct.*, 2011, **986**, 16–21.
39. Everell, N. Depth profiling with confocal Raman microscopy, Part I. *Spectroscopy*, 2004, **19** (10), 22–7.
40. Everell, N. Depth profiling with confocal Raman microscopy, Part II. *Spectroscopy*, 2004, **19** (11), 16–27.
41. Stevels, J. M. The structure and the physical properties of glass. *Handbuch der Physik, Thermodynamik der Flüssigkeiten und Festkörper*. Edited by S. Flügge, Springer, Berlin, 1962, **Vol. 13**, 510–645.
42. Scholze, H. Significance of the leached layer for chemical durability: investigations with a soda-lime-silica glass (title translated from german). *Glastech. Ber.*, 1985, **58**, 116–24.
43. Böschel, D., Janich, M. & Roggendorf, H. Size distribution of colloidal silica in sodium silicate solutions investigated by dynamic light scattering and viscosity measurements. *J. Colloid Interf. Sci.*, 2003, **267**, 360–8.
44. Emerson, J. F., Stallworth, P. E. & Bray, P. J. High-field silicon-29 NMR studies of alkali silicate glasses. *J. Non-Cryst. Solids*, 1989, **113**, 253–9.
45. Olivier, L., Yuan, X., Cormack, A. N. & Jäger, C. Combined $^{29}$Si double quantum NMR and MD simulation studies of network connectivities of binary $Na_2O.SiO_2$ glasses. *J. Non-Cryst. Solids*, 2001, **293–295**, 53–66.
46. Engelhardt, G. & Koller, H. $^{29}$Si NMR of inorganic solids. In *NMR Basic Principles and Progress book series*. Edited by B. Blümich, Springer, Berlin, 1994, **Vol. 31**, 1–29.
47. Davis, K. M. & Tomozawa, M. An infrared spectroscopic study of water-related species in silica glasses. *J. Non-Cryst. Solids*, 1996, **201**, 177–98.
48. Brooker, R. A., Kohn, S. C., Holloway, J. R. & McMillan, P. F. Structural controls on the solubility of $CO_2$ in silicate melts: Part II: IR characteristics of carbonate groups in silicate glasses. *Chem. Geol.*, 2001, **174**, 241–54.
49. Barrow, N., Packard, M. Vaishnaw, S., Wilding, M.C., Bingham, P. A., Hannon, A. C., Appler, M. & Feller, S. MAS-NMR studies of carbonate retention in a very wide range of $Na_2O$–$SiO_2$ glasses. *J. Non-Cryst. Solids*, 2020, **534**, 119958
50. King, P. L., McMillan, P. F. & Moore, G. M. Infrared spectroscopy of silicate glasses with application to natural systems. In: King, P. L. (ed.). Molecules to planets: infrared spectroscopy in geochemistry, exploration geochemistry and remote sensing. Short course: Mineralogical Association of Canada, Quebec, 2004, 93–134.
51. Furukawa, T. R., Fox, K. E. & White, W. B. Raman spectroscopic investigation of the structure of silicate glasses. III. Raman intensities and structural units in sodium silicate glasses. *J. Chem. Phys.*, 1981, **75**, 3226–37.
52. Malitson, I. H. Interspecimen comparison of the refractive index of fused silica. *J. Opt. Soc. Am.*, 1965, **55**, 1205–9.
53. Hale, G. M. & Querry, M. R. Optical constants of water in the 200-nm to 200 μm wavelength region. *Appl. Opt.*, 1973, **12**, 555–63.
54. Iler, R. K. *Chemistry of silica*. Wiley-Interscience, New York, 1979.
55. Gurman, S. J. Bond ordering in silicate glasses: a critique and a resolution. *J. Non-Cryst. Solids*, 1990, **125**, 151–60.
56. Machacek, J, Gedeon, O. & Liska, M. Group connectivity in binary silicate glasses. *J. Non-Cryst. Solids*, 2006, **352**, 2173–9.
57. Mel'nikova, I. M. & Zhdanov, S. P. Vacuum thermal dehydration of leached binary alkali silicate glasses and products of their crystallization. *Glass Phys. Chem.* (*Fiz. Khim. Stekla*), 1998, **24**, 463–70.
58. Roggendorf, H. Corrosion of the borosilicate glass R7T7 in a concentrated NaCl brine - experiments and surface characterization. *Ceram. Trans.*, 1991, **23**, 675–84.
59. Nunes de Carvalho, J., Kirkby, N., Cleaver, J. A. S. & Holmes, P. A. A cellular automaton technique for the solution of a model of glass corrosion. *Glass Technol: Eur. J. Glass Sci Technol.*, 2011, **52** (3), 88–96.

Glass Technol.: Eur. J. Glass Sci. Technol. A, 2021, 62 (1), 24–33

# Some secrets of Renaissance Venetian opaque red glass revealed by analyses and glassmaking treatises

*Mario Bandiera,*[a,b,1] *Marco Verità,*[c] *Sandro Zecchin*[d] *& Marcia Vilarigues*[a,b]

[a] *VICARTE, Research Unit Vidro e Ceramica para as Artes, FCT/UNL, Campus Caparica, 2829-516 Caparica, Portugal*
[b] *Dep. de Conservacao e Restauro, FCT/UNL, Campus Caparica, 2829-516 Caparica, Portugal*
[c] *LAMA laboratory, Università Iuav, Venice, Italy*
[d] *Independent scholar, Mirano, Venice, Italy*

Manuscript received 18 February 2020
Revision received 1 November 2020
Manuscript accepted 11 December 2020

*Opaque copper red glass has been produced since the birth of the glassmaking technology. Since the first/second c. AD the so-called red brown glass was produced and was constantly made up to nowadays. Despite several analytical studies, many technological aspects of this particular glass are still unclear. Opaque red brown glass was widely made also in Venice. Samples of Renaissance Venetian glass were analysed and abundant historical sources of this period including recipes of Venetian glassmakers are available. This combination of analyses and texts represents a unique opportunity to shed light on the production technology of red brown glass. The aim of this work is to better understand some aspects of this glass, such as the raw materials used, the glassmaking procedure, the control of the final colour, through a comparison between chemical analyses of Renaissance Venetian samples and the indications of historical sources. The information obtained for Venetian glass could be useful for the interpretation of the technology of red opaque glass made in other places and periods for which specific written documents are lacking.*

## 1. Introduction

The production of opaque red copper glass dates back to the beginning of the glassmaking technology. An extensive literature is devoted to the analysis of this colour and it is possible to identify three phases in its history. In the second millennium BC opaque red glass was produced using high contents of copper (CuO 5–10%), negligible concentrations of lead and iron and in some cases a significant antimony content ($Sb_2O_3$ 0·8–2·8%).[1,2]

From the middle of the first millennium BC a new opaque red glass with high concentrations of copper (CuO 5–10%), lead (PbO 15–50%) and antimony ($Sb_2O_3$ 1–4·8%) began to be manufactured. This glass called *sealing wax* red, was the main opaque red glass made from the fourth century BC to the first century AD in the Mediterranean basin and used also in central and northern Europe. Chemical and mineralogical analyses established that micrometric cuprite crystals ($Cu_2O$) in dendritic shape are the colouring/opacifying agent of this glass.[1,3–5]

After the middle of the first century BC, the glassmaking technology changed dramatically with the invention of the blowing pipe and the production of huge amounts of glass in primary centres. In this period, the place of sealing wax red was taken by a new type of glass called red brown. In general, the term red brown glass defines a group of glasses duller than sealing wax, made in a range of hues from bright red to dark brown. It was extensively used to make mosaic tesserae, sectilia in the opus sectile technique, beads, ornaments as well as to make blown vessels.[4,6–8] Opaque red has been considered one of the most difficult colours to obtain. Describing a recipe for red brown glass in the first printed book on glassmaking technology[9] Antonio Neri wrote in 1612: 'It is an intriguing colour and a great care is necessary to melt and to work it properly'. Nevertheless, this glass has been widely and continuously produced since the Roman time until nowadays in the Mediterranean area and in Europe.

The abundant literature devoted to ancient opaque red brown glass does not help to resolve this enigma. It concerns mainly chemical analyses highlighting the recurrent presence of elements such as tin, antimony, zinc, lead in different concentrations, associated to copper and iron.[2,6,10–12] It is still unclear whether

[1] Corresponding author. Email m.bandiera@campus.fct.unl.pt
DOI: 10.13036/17533546.62.1.001

these elements were added deliberately through specific raw materials or entered the melt as secondary components of the raw materials.

The nature of the colouring agent in ancient red brown glass has been debated for decades, and although metallic copper nanoparticles (Cu°) have been ascertained to be the main colouring agents by authoritative scholars,[13,14] some analyses highlighted the presence of both cuprite crystals and metallic copper in Roman red brown glass.[15]

The most difficult point in the manufacture of red brown glass is to obtain the appropriate redox conditions to allow the formation of elemental copper particles in the melt. Metallic copper has a low solubility in molten glass; for this reason, it is dissolved in the melt in oxidised form ($Cu^{2+}$, $Cu^{+}$).[13] In order to shift the system toward metallic copper, the partial pressure of oxygen in the melt has to be cut down by adding reducing agents or creating a reducing atmosphere inside the furnace.[16] Iron, tin and antimony in the proper oxidation state are the most commonly used reducing agents.[13,17] Moreover, tin, phosphorous and other elements promote the nucleation rate and accelerate the formation of the copper particles and the appearance of the red colour.[17]

In general, the lack of information on the methods and procedures associated with the manufacture of ancient red brown glass do not allow us to know in which way it was obtained. In this frame, Renaissance Venetian red brown glass represents a unique source of information, since both analyses and glassmaking information are available. During the Renaissance the Venetian glass reached an extraordinary quality that allowed Venice to become the most advanced glass manufacturing centre in Europe. The abundant documents concerning glass production in Venice and in the island of Murano were extensively investigated by Luigi Zecchin.[18] Among them, some manuscripts belonging to Muranese glassmakers of the 15th and 17th century report recipes to prepare the glass melt and the technology to obtain several colours, including red brown glass. Furthermore, the chemical analyses of a number of 15th–17th century opaque red Venetian glass samples are available.

The aim of this work is to better understand some aspects of the production technology of this puzzling glass, such as the raw materials used, the glassmaking process and the control of the final colour through a comparison of chemical analyses and the indications of the historical sources of Renaissance Venetian red brown glass.

## 2. Experimental procedure

Twenty Venetian Renaissance red brown glass samples were analysed for their quantitative chemical composition in different laboratories with different methods.

Samples labelled: MG, MH, RBM and GAR, are archaeological sherds of blown objects, monochrome and polychrome rods and pressed glass. They were excavated in the Venetian lagoon in a location at the west edge called Fusina. In this area, the embankment of the river Brenta built with natural clays around 1325 to divert the river and impede the silting of the Lagoon, was reinforced since the 15th century also with wastes from ceramic and glass furnaces of Venice and Murano. The embankment was abandoned in 1610 when the course of the Brenta was changed again.[19]

Three samples labelled MG and two samples MH were analysed by laser ablation inductively coupled plasma mass spectrometry (LA-ICP-MS) at the CNRS laboratories in Orleans (France) and published in Refs 20 and 21, respectively.

Seven samples labelled RBM were analysed by electron probe microanalysis (EPMA) and three samples labelled GAR were analysed by x-ray fluorescence (XRF).

Two mosaic tesserae (SPC-R7 bright red and SPC-R9 brown analysed by EPMA) belonged to the altar piece of the Cappella Lando in the Church of San Pietro di Castello (Venice). The mosaic is signed 1575, but a document dated August 1566 attests that *Nicolò verier* (Nicolò, glassmaker) was paid for supplying the glass slabs for the mosaic work.[22]

Finally, three samples PG-SCV 171, PG-V67 and PG-V74 are 17th century archaeological sherds from the excavation of the Monastery of Sta. Clara-a-Velha (Portugal) and were analysed by EPMA. Comparison of their analyses with the compositional database of Venetian Renaissance glass attested their genuine Venetian provenance.[23]

EPMA and XRF analyses were performed at the Stazione Sperimentale del Vetro, Murano-Venice (Italy). EPMA analyses were carried out with a Cameca SX-50 equipped with three wavelength-dispersive spectrometers (PET, LiF and TAP crystals). The operating conditions: accelerating potential 15 kV, beam current 20 nA for major and minor components or 100 nA for trace elements, respectively. A 40×50 mm scanning electron beam and limited counting time (10 s for major and minor elements, 20–30 s for trace elements) were employed to ensure that no significant alkali drift (ion migration) occurred during the irradiation.

XRF analyses were performed by energy dispersive X-ray micro-fluorescence spectrometry (EDXRF) using a Bruker M4 Tornado spectrometer equipped with a rhodium x-ray tube (maximum power 30 W) with polycapillary lens optics (spot size ~30μm) and a Peltier cooled energy dispersive silicon drift detector (SDD) with a 30 $mm^2$ sensitive area and an energy resolution ~140 eV for Mn $K_\alpha$.

For LA-ICP-MS analyses, a VG Laserlab-A was used, comprising a pulsed Nd:YAG laser operating at 1064 nm (spot size 50–200 μm), the repetition rate of the laser may vary from 1 to 10 Hz and the maximum energy of the beam is 2 mJ. The laser beam is focused onto the sample surface through the window of a quartz sample cell. The cell is flushed by approximately 1·2 l/min of argon carrier gas. ICP-MS instrument was a Plasma Quad2+ EDR.

Further details concerning the analytical methods used for quantitative chemical analysis (device setting, limits of detection and accuracy of the analyses) are given in respectively: LA-ICP-MS in Gratuze *et al*,[24] EPMA in Verità *et al*[25] and XRF in Tesser *et al*.[26]

Three GAR samples were available for further investigations. They were analysed at the Lisbon University, Conservation and Restauration department, for optical microscopy observation (Zeiss Axioplan 2 Imaging system -HAL 100- coupled to a Nikon DXM1200F digital camera and ACT-1 software under different illumination modes -bright field and dark field-), FORS and μ-Raman spectroscopy and at the CENIMAT laboratory for x-ray diffraction XRD. For the FORS analyses a MAYA 200 PRO from Ocean Optics spectrophotometer with a single beam dispersive optical fiber was used together with a 2048 CCD Si detector that allows operating in the 200–1050 nm range. The light source is an HL-200-HP 20 W halogen from Ocean Optics, with a single optical path between 360 and 2500 nm. The spectra were taken on the glass surface, in reflectance (R) mode, with a 45°/45° configuration (illumination angle/acquisition) and a ca. 2 mm diameter of the area analysed. Spectra were obtained between 380 and 1050 nm, with an integration time of 8 ms per scan and 15 scans. A Spectralon surface was used as a reference for calibration. The identification of the crystalline phases was conducted by μ-Raman spectroscopy with a Labram 300 Jobin Yvon spectrometer equipped with a solid-state 50 mW laser, operating at 532 nm. The spectra were recorded as an extended scan, and the system was calibrated using a silicon standard. The laser beam was focused either with a 50× or a 100× Olympus objective lens and its power at the surface of the samples was controlled with neutral density filters (optical densities 0·3 and 0·6). Raman data were treated using the LabSpec 5 software, and all spectra are presented as acquired without any baseline correction or other treatment. RUFF database was useful for the identification of some crystalline phases. x-ray diffraction was used in samples where Raman spectroscopy did not allow the identification of the colouring agent, The measurements were performed by X'Pert PRO MPD, equipped with an x-ray tube with Cu anode (1·54059 A, $K_\alpha$, 45 kV, 40 mA) and an X'Celerator detector (ultra-set x-ray 1D detector based on real time multi steps (RTMS) technology). The scans were collected in the angular range of 3°– 89° 2θ with a 0·03° virtual step size and 30 s/step counting time. The XRD data were processed by the X'Pert HighScore (PANalytical©). In order to avoid any contamination of resin peaks, the analyses were performed on bulk samples.

Finally, in Paris at the Centre de Recherche et de Restauration des Musées de France (C2RMF) samples were analysed by field emission scanning electron microscopy (FEG-SEM JEOL 7800F with a EDS BRUKER Quantax 400 system for x-ray microanaly-

*Table 1. Quantitative chemical composition (wt% of oxides) of the analysed samples. n.a.: not analysed. Searched for and not detected: Sb, As, Ba, Zn, Co, Ni. Glass type: VB, vitrum blanchum; CR, cristallo glass*

| *Analytical technique* | *Sample* | *Object* | *Dating* | *Glass type* | $SiO_2$ | $Al_2O_3$ | $Na_2O$ | $K_2O$ | CaO | MgO | $SO_2$ | $P_2O_5$ | Cl | $TiO_2$ | MnO | $Fe_2O_3$ | CuO | PbO | $SnO_2$ |
|---|---|---|---|---|---|---|---|---|---|---|---|---|---|---|---|---|---|---|---|
| EPMA | SPC-R9 | mosaic brown | 1566 | VB | 54·5 | 1·05 | 13·1 | 2·48 | 7·47 | 2·85 | 0·19 | 0·23 | 0·53 | 0·05 | 0·52 | 6·45 | 2·25 | 5·50 | 2·85 |
| EPMA | SPC-R7 | mosaic red | 1566 | VB | 63·0 | 1·56 | 12·3 | 2·58 | 8·85 | 3·43 | 0·21 | 0·34 | 0·58 | 0·08 | 0·62 | 4·00 | 1·70 | 0·35 | 0·42 |
| XRF | GAR-R3 | bead | 15th-17th | CR | 62·7 | 0·70 | 14·4 | 2·25 | 3·45 | 1·20 | 0·07 | 0·10 | 0·84 | 0·03 | 0·04 | 3·60 | 3·45 | 3·60 | 3·55 |
| EPMA | RBM-2H | sherd | 15th-17th | CR | 67·5 | 0·75 | 14·6 | 2·65 | 5·60 | 1·80 | 0·21 | 0·12 | 0·90 | 0·04 | 0·27 | 2·40 | 2·65 | 0·25 | 0·27 |
| LA-ICP-MS | MG-VeR1 | sherd | 16th-17th | CR | 68·0 | 0·80 | 14·4 | 2·90 | 6·10 | 1·50 | n·a· | 0·12 | 0·60 | 0·05 | 0·23 | 2·40 | 2·50 | 0·22 | 0·21 |
| LA-ICP-MS | MG-VeR3 | sherd | 16th-17th | VB | 63·7 | 1·05 | 13·8 | 1·75 | 9·61 | 3·25 | n·a· | 0·26 | 0·73 | 0·14 | 0·10 | 3·80 | 1·15 | 0·36 | 0·34 |
| EPMA | RBM-4E | sherd | 15th-17th | VB | 60·2 | 1·05 | 12·5 | 3·70 | 8·90 | 2·90 | 0·22 | 0·30 | 0·83 | 0·08 | 0·47 | 3·30 | 1·65 | 2·00 | 1·90 |
| XRF | GAR-R2 | sherd | 15th-17th | VB | 56·0 | 1·05 | 13·8 | 2·22 | 9·00 | 3·20 | 0·20 | 0·27 | 0·70 | 0·10 | 0·48 | 4·05 | 1·10 | 3·85 | 4·00 |
| LA-ICP-MS | MH-R1 | sherd | 16th | VB | 63·4 | 1·10 | 13·0 | 2·80 | 8·40 | 2·50 | n·a· | 0·20 | 0·70 | 0·10 | 0·40 | 3·80 | 1·60 | 1·00 | 1·00 |
| EPMA | RBM-4D | sherd | 15th-17th | VB | 60·7 | 1·10 | 12·9 | 2·40 | 10·4 | 3·20 | 0·17 | 0·30 | 0·75 | 0·07 | 0·55 | 5·50 | 0·70 | 0·60 | 0·70 |
| EPMA | RBM-4M | sherd | 15th-17th | VB | 61·4 | 1·15 | 13·2 | 3·80 | 8·90 | 1·00 | 0·21 | 0·30 | 0·85 | 0·07 | 0·42 | 3·20 | 1·65 | 1·90 | 1·90 |
| EPMA | PG-V67 | applied dots | 17th | VB | 56·5 | 1·23 | 13·4 | 2·22 | 8·90 | 3·10 | 0·22 | 0·28 | 0·63 | 0·10 | 0·70 | 3·17 | 1·15 | 5·22 | 3·16 |
| XRF | GAR-R1 | sherd | 15th-17th | VB | 57·7 | 1·25 | 13·3 | 2·12 | 8·30 | 3·00 | 0·10 | 0·25 | 0·60 | 0·07 | 0·40 | 3·30 | 1·00 | 4·20 | 4·45 |
| EPMA | PG-SCV171 | applied dots | 17th | VB | 58·0 | 1·26 | 13·2 | 2·28 | 8·15 | 3·15 | 0·19 | 0·28 | 0·73 | 0·04 | 0·52 | 3·20 | 0·92 | 5·10 | 3·00 |
| EPMA | PG-V74 | applied dots | 17th | VB | 57·3 | 1·30 | 12·1 | 1·77 | 9·20 | 3·45 | 0·20 | 0·32 | 0·60 | 0·09 | 0·82 | 3·90 | 0·70 | 4·20 | 4·00 |
| EPMA | RBM-5L | sherd | 15th-17th | VB | 62·0 | 1·30 | 13·4 | 2·35 | 9·40 | 3·30 | 0·18 | 0·28 | 0·50 | 0·08 | 0·65 | 4·80 | 0·90 | 0·40 | 0·50 |
| EPMA | RBM-5B | sherd | 15th-17th | VB | 62·0 | 1·30 | 12·0 | 2·45 | 9·80 | 3·40 | 0·32 | 0·35 | 0·80 | 0·07 | 0·15 | 5·60 | 1·45 | 0·20 | 0·10 |
| EPMA | RBM-2G | sherd | 15th-17th | VB | 58·3 | 1·45 | 11·2 | 3·70 | 8·80 | 3·20 | 0·18 | 0·38 | 0·70 | 0·07 | 0·90 | 4·70 | 0·85 | 2·50 | 3·10 |
| LA-ICP-MS | MG-VeR2 | sherd | 15th-17th | VB | 58·0 | 1·60 | 11·6 | 3·85 | 9·27 | 2·80 | n·a· | 0·27 | 0·65 | 0·17 | 0·80 | 5·00 | 1·05 | 2·65 | 2·30 |
| LA-ICP-MS | MH-R2 | Chevron rod | 15th-16th | VB | 56·7 | 1·80 | 10·2 | 3·10 | 9·55 | 2·65 | n·a· | 0·20 | 0·50 | 0·10 | 0·40 | 4·60 | 0·70 | 4·80 | 4·70 |

sis),[8] to identify the colouring/opacifying particles after platinum coating of samples. The 2 μm electron beam (current 1 nA) was scanned during analysis to prevent alkali drift.

## 3. Results

### 3.1. Chemical analysis

The quantitative chemical composition of the Venetian red brown glass samples is reported in wt% of the oxides in Table 1.

The base glass composition (reduced composition) was calculated by subtracting to the composition of Table 1 the content of elements added to colour the glass (copper, iron, manganese, tin and lead) and then normalising the composition to 100 wt%.

The base glass of the red brown samples is of the soda–lime–silica type. Their potassium ($K_2O$ 1·9–4·4%), magnesium (MgO 1·1–4%) and phosphorous ($P_2O_5$ 0·06–0·43%) contents indicate that they were melted using soda-plant ash as a flux. No mixed alkali or potash glass was found, which confirms the historical sources attesting that the potash-bearing plant ash used in northern Europe was forbidden in Murano.[27] This prohibition is reported in an additional chapter of the *Capitulare de Fiolariis* (the gild of glassmakers) in 1271 and issued by the *Maggior Consiglio* of Venice first in the 1306.[28] The reasons for this restriction are not explained, but are reasonably related to the poor quality of the potash ashes as glass flux because of their high content of colouring contaminants (iron and manganese), of their variability in composition (unstable ratio: K-carbonate/Ca-carbonate) and of their low content of alkali.[29]

Three recipes for transparent glass were in use in Venice during the Renaissance period. Common glass was slightly coloured in green to yellow natural hues. *Vitrum blanchum* was a well decolourised glass made at least since the 14th century. Around the middle of the 15th century the *cristallo* glass was invented in Venice, as an important improvement of *vitrum blanchum*. The main secret of *cristallo* glass was the removal of the insoluble colouring contaminants from the plant ash by a process of dissolution, decantation and evaporation.[27]

The base glass of the red brown samples shows a strong similarity with Venetian glass composition. By reporting the calcium versus potassium contents in a diagram (Figure 1) in which their concentrations (black diamonds) are compared with traditional Venetian glass, one observes that most of the red samples are in the area of *vitrum blanchum* and common glass, except three samples (GAR-R3, RBM-2H and MG-VeR1) which show a good correspondence with *cristallo* glass. Similar considerations are found for magnesium, silica and sodium concentrations.

The potassium content slightly exceeding 4% in samples RBM-4M, RBM-2G and MG-VeR2, could indicate an addition of the potash bearing tartar as indicated in some Venetian recipes which will be discussed below.

By reporting alumina versus manganese oxide contents (Figure 2) the correspondence of analysed

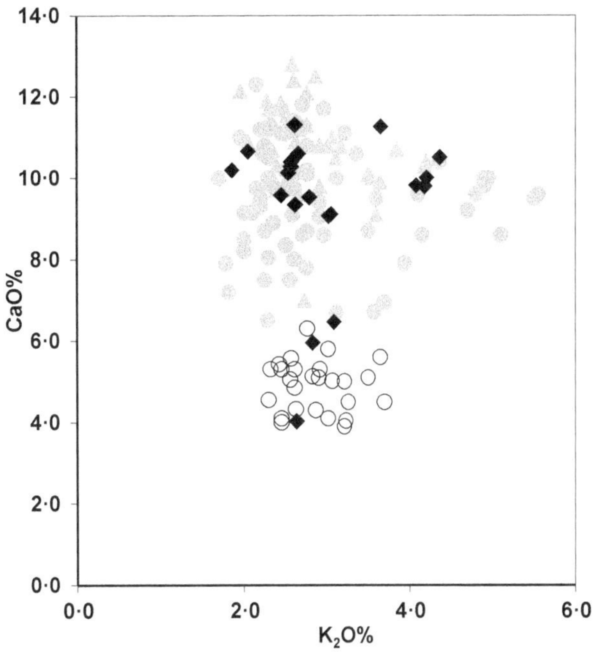


*Figure 1. Calcium versus potassium contents (wt% of oxides) of the base glass of analysed red samples (black diamonds) compared with Venetian transparent glass (cristallo: open circles; vitrum blanchum: grey circles; common glass: grey triangles)*

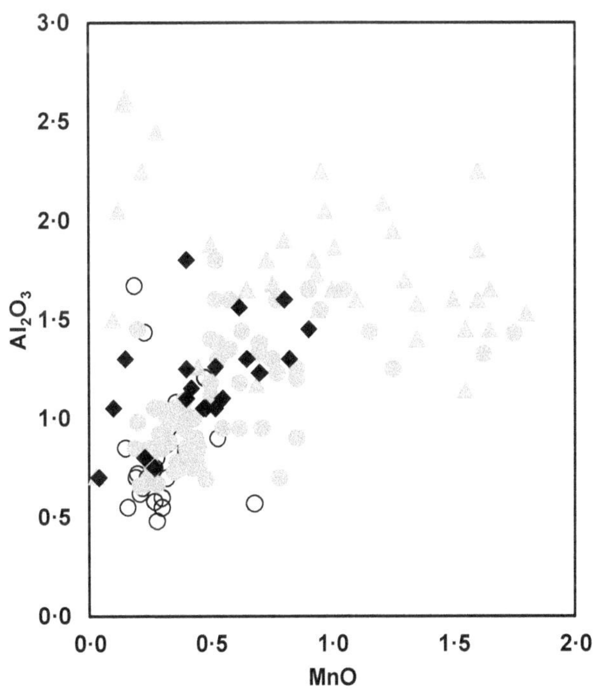


*Figure 2. Alumina versus manganese content (wt% of oxides) of the base glass of analysed red samples (black diamonds) compared with Venetian transparent glass (symbols as in Figure 1)*

red brown samples with Venetian Renaissance glass is confirmed. In particular, the alumina content for most of the red samples is in the range 1–2% typical of common and *vitrum blanchum* glass. The lower alumina content of the *cristallo* type samples ($Al_2O_3$ 0·7–0·8%) indicates the use of a better, less contaminated silica source. The manganese content (MnO 0·04–0·90%) in the red samples is in a range of concentration similar to that of *cristallo* (0·15-0·7%), *vitrum blanchum* (0·2-1·7%) and common glass (0·03–2·7%) suggesting that this element was already present in the transparent glass melt (decolourant) and not voluntary added to colour red brown glass.

No evident relationships exist between the two crucial colouring elements iron and copper (Figure 3) and their contents vary in a wide range ($Fe_2O_3$ 2·5–6·5%; CuO 0·7–3·5%).

The Fe/Cu ratio highlights the existence of three groups:

- high Fe/Cu ratio (5 to 8 times more iron than copper) in a group of six samples showing high iron ($Fe_2O_3$ 4–5·5%) and low copper contents (CuO 0·7–1·0%);
- intermediate Fe/Cu ratio (2 to 4 times more iron than copper) in a group of eleven samples showing an intermediate copper content (CuO 0·9–2·2%);
- low Fe/Cu ratio (around 1, identical iron and copper concentrations) in a group of three samples, showing a high copper content (CuO 2·5–3·5%).

Lead and tin in a wide range of concentrations (PbO 0·2–5·2%; $SnO_2$ 0·2–6·3%) were detected in all the samples. A strict correlation was found between these two elements (Figure 4) with Pb/Sn ratio in a very tight range, around 1 for most of the samples

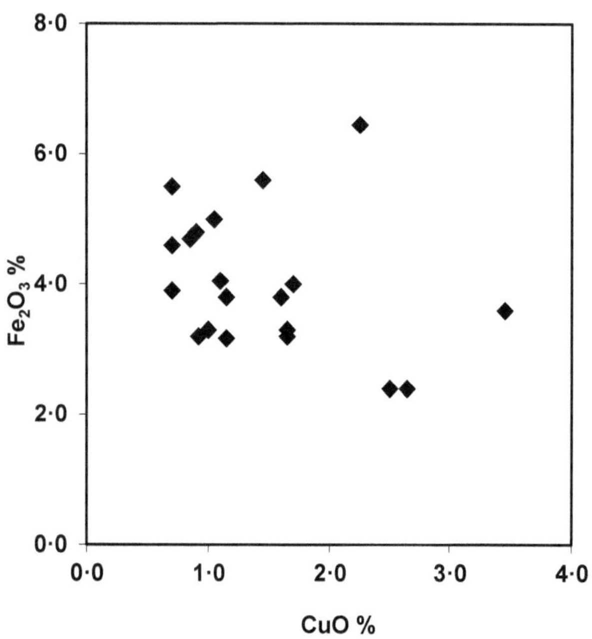


*Figure 3. Iron versus copper content (wt% of oxides) of the analysed red samples*

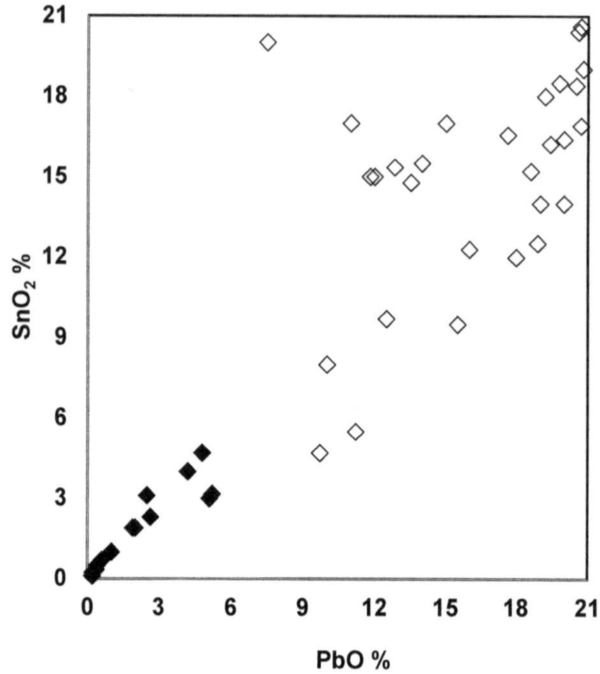


*Figure 4. Tin versus lead content (wt% of oxides) of Venetian white opaque glass (open diamonds) and red brown samples (black diamonds)*

and around 2 in four samples (SPC-R9, RBM-5B, PG-V67 and PG-SCV171). In any case, the lead and tin contents in red brown glass is much lower than in white opaque Venetian glass (PbO and $SnO_2$ 10 to 30%) and in general rare tin oxide crystals are detected in red brown samples of this group.

The lead and tin content are low in a group of seven samples (0·2–0·65%) and in sample MH-R1 (1%), and is higher in a second group of twelve samples (2–5%). The tin and lead contents in the first group do not necessary correspond to voluntary additions of these elements. In fact, the analyses of common and *vitrum blanchum* transparent Venetian glass frequently reveal the presence of lead and tin in concentrations up to 0·4 and 0·6% respectively, attributed to the recycling of not well selected glass cullet (fragments of opaque white glass contaminating the colourless transparent glass).[27]

The addition of bone ash is excluded since the low phosphorous content ($P_2O_5$<0·38%), in the range of Venetian colourless glass[27] and no other elements influencing the colour of the red brown glass (antimony, zinc, cobalt) were detected by the analyses.

Further investigations were performed in three GAR samples in order to identify the colouring particles. Using FORS analyses an inflection point around 580 nm was identified in all the investigated samples, which suggests the presence of metallic copper.[30] By means of FEG-SEM nanoparticles extremely rich in copper, showing a spherical and cubic shape were observed. Their size range between 40 nm and 130 nm in sample GAR-1, 40 and 180 nm in GAR-2 sample (in this sample particle sizes range between 150 and 250 nm) and between 40 and 300 nm in sample GAR-3. These particles were identified as metallic copper by XRD analysis exhibiting three weak peaks at $d$=2·09, 1·81 and 1·28.

Using Raman spectroscopy, the intense peak at 634 nm and a weak peak at 775 nm indicate the presence of cassiterite crystals ($SnO_2$).[31]

### 3.2. The recipe books

Some Venetian glassmaking treatises were considered to examine the recipes for the manufacture of opaque red brown glass.

The Trattatelli are three booklets attributed to the Tuscan glassmaking environment kept in the State Archives of Florence, highlighting in some recipes the influence of the Venetian technology. Their dating is ascribed to between late 14th early 15th centuries. Published in the 19th century,[32] the recipes were commented by Luigi Zecchin.[33] Recipes from these treatises are indicated with TR in the text.

The Montpellier manuscript dated 1536 entered the Bibliotheque de l'Ecole de Medicine de Montpellier (France) at the end of the 18th century. Before, it was in the Libreria of the Albani family in Rome. It was partially commented by Luigi Zecchin;[34] recipes are indicated with M.

The Anonimo recipe book written by an unknown Venetian glassmaker around 1560 is probably a copy of two Muranese treatises. The first dates to the second half of the 15th–early 16th century. It was published and commented in Ref. 35; these recipes are indicated with A.

The *Darduin* manuscript dated 1644 is a collection of recipes on glassmaking and glass colouring drawn up by several family members since the late 15th century. The final part includes some recipes dated to the late 17th–early 18th century;[36] recipes are indicated with D.

The so-called Brunoro manuscript is the recipe book of the Venetian glassmaker Gaspare Brunoro found in Gadnsk (Poland). Dated 1645, it includes also Northern-European recipes demonstrating that the Venetian glassmakers working abroad experimented also with other techniques;[37] recipes indicated with B.

The first published book on glassmaking, *L'Arte Vetraria* by Antonio Neri, 1612,[9] includes a number of recipes of Venetian origin commented by Luigi Zecchin;[38] Recipes indicated with N.

From these manuscripts 35 recipes for the production of red brown glass used to make mosaic tesserae, enamels for metals and glass rods (beads) were identified. Recipes devoted to translucent red glass were not considered.

### 3.3. The raw materials in the recipe books

The amounts of molten glass vary widely, from one pound in recipe B310 for enamels to more than 100 pounds in B155 for mosaic glass.

#### 3.3.1. Glass

Most of the recipes prescribe the preparation of a crucible with molten transparent glass and the addition of the colouring components to the melt. The melt is made with glass cullet or cotisso (molten glass poured in water and dried), and stored ready to be re-melted. Only recipes M92 and D202 prescribe to prepare the red brown glass starting by melting a batch of raw materials.

The recipes attest the use of common and vitrum blanchum glass, since the use of low-iron, well decolourised cristallo glass does not improve the quality of opaque red glass (B126: 'the low quality common glass is the best', and M132: 'with cristallo the red brown glass will not be neither beautiful nor good'). Nevertheless, the use of cristallo moils (cullet contaminated by iron flakes from the blow pipe) is indicated in recipes D67 and D155, and recipes A42 and D199 indicate cristallo glass mixed with lead silicate glass.

Some recipes (TRB2, M92 and A34) use a potash glass and others (TRB11 and TRB20) a mixed alkali glass where tartar, i.e. the deposit of wine barrels, is the K-bearing compound. Different types of tartar are mentioned. It was used as a potash source in a completely calcined form as in M92 or TRB2, also preliminary purified by leaching to obtain K-carbonate (N11). Tartar in an unprocessed form as in A42 or in a semi-calcined form (N11 black tartar) was used as a reducing agent (carbonaceous particles) in red brown glass.[39]

Lead-tin calx is an intermediate product prepared in the glass furnaces to make white opaque glass or to be used as an opacifier (clarifier) of transparent coloured glass. It is the main white opacifier encountered in the Renaissance Venetian recipes. To prepare the calx, the treatises prescribe to heat a mixture of the two metals at a temperature above the melting point until a powder is obtained.[40] In general, these recipes recommend to prepare a calx with a Pb/Sn ratio of 1 (for instance recipes B13, D96 and N93) or 0·5 (D97). Very important is recipe D97 which prescribes to fire the mixture at a high temperature to obtain white calx (probably, fully oxidised lead and tin) and at a lower temperature to obtain grey calx (partially oxidised lead and tin).

Bone ash (the horns or bones of animals burnt and ground to powder; Ca-phosphate) was also used as a glass opacifier. Nevertheless, it has also a nucleating effect, which can help the separation of the colouring particles from the melt since phosphorous helps the formation of nuclei and the growth of crystalline particles.[17] The use of high amounts of bone ash are reported in three recipes for red brown glass (TR2, TR20, A38).

### 3.3.2. Copper and iron

Thirty out of the thirty-five opaque red recipes investigated prescribe the use of iron together with copper to obtain the colour. In some recipes other reducing compounds (lead-tin calx, tin calx and tartar) are added together with iron. Finally, in five recipes (A38, B314, D202, M30 and M128) the reducing agent is tin alone.

The copper source is red copper ($Cu_2O$, cuprite) generally prepared by firing metallic copper sheets in a small furnace or in the annealing area of the furnace. In some cases, also sulphur is added. The firing can take from three days using a high temperature process (TR-C7) up to 20 days (D296). At the end of the process, the sheets are covered by a red product which is separated by beating and ground to a red powder.[39]

Two sources of iron are indicated: the flakes produced from the beating on the anvil of incandescent iron and unprocessed or calcined steel. These products are mixtures of metallic iron and different iron oxides. Their undefined reducing power affects their efficiency in the formation of metallic copper in the melt making it almost impossible to define the exact amounts to be added to obtain the red brown glass. This is why most of the recipes are empirical and suggest to improve the final colour by small additions of iron compounds.

### *3.4. The process*

In general, the process prescribes to prepare first a pot of molten glass to which the colouring components are added. The amounts of glass indicated in the recipes vary, depending on the application. For enamels production amounts of the order of one to few pounds (one pound could correspond to 300 or 477 g) of glass are indicated, while for mosaic production amounts of more than 100 pounds of glass are indicated. Only recipes D155 and B61 recommend to mix glass grit (cullet and cotizzo respectively) with copper, iron, lead-tin calx and melt the mixture in a crucible.

In fourteen recipes comparable amounts of copper and iron (or slightly higher iron) are used, while in three recipes the iron addition is up to twice and in four recipes copper is added in larger amounts, up to four times. In nine recipes the amounts of the two colourants are not indicated.

In thirteen recipes iron is added first. The addition of copper occurs after a certain time (30 min, 1, 8 or 16 h) and should be made bit by bit but not too much. After adding iron, recipe N58 prescribes to check the colour of the transparent glass, which should result yellow and not opaque black. Only then the addition of red copper occurs. After few hours from the last copper addition the melt is ready (homogeneously red brown coloured and free of bubbles) to be worked.

Only two recipes prescribe the addition of copper first and then iron, without any other indication; in six recipes iron and copper are preliminary mixed and added together, while in nine recipes the sequence is not specified.

Some recipes report the addition also of other compounds, as lead silicate glass 'for better results' (recipes A34 and M96) or of unprocessed tartar (M120 and B317).

Many recommendations are reported in the recipes: a continuous monitoring of the melt, or, once the colour is obtained, the melt should be worked shortly to avoid the loss of the opaque red colour (M92 and B126). The melt must be mixed frequently and carefully to avoid dark cords (tortioni neri). Once copper and iron have entered the melt, the colour might not appear, in this case further additions of them are required until colour is obtained.

Recipe N58 is the most complete one: "A batch made of cristallo frit (20 parts), colourless glass cullet (1 part) and lead-tin calx (2 parts) is first melt. Then, a mixture 1 to 1 of calcined steel and flakes obtained by beating red hot iron is added and the melt will boil terribly; wait five, six hours. Do not add too much iron, the glass must not become black but transparent dark yellow. Then add calcined red copper in three-four times (the quantity is not specified). On mixing the glass becomes immediately red. Frequent gatherings are recommended to control the colour. Let the melt cool to the working temperature leaving the mouth of the furnace open. The melt must be worked within about ten hours, otherwise it will loss colour and become black; in this case, add more iron."

### *3.5. Colour control*

Recipe M110 for mosaic tesserae gives instructions about how to obtain a sequence of seven hues from bright red to brown, confirming that Muranese glassmakers were capable to control this colour. The recipe prescribes to melt a pot of common glass and add comparable amounts of powder of iron flakes from the anvil and red copper. The first bright red hue is obtained by keeping the furnace hot. The second, more brown hue (morello) is obtained by adding more iron and copper and decreasing the temperature of the furnace. By successive additions of copper and iron and a progressive lowering of the furnace temperature, it is possible to obtain 6–7 shades from red to brown.

Recipe pairs B173 (red) and B174 (brown) and D129 - D131 are clearly derived from M110.

Recipes B173, M74, M110, M132 explain how to

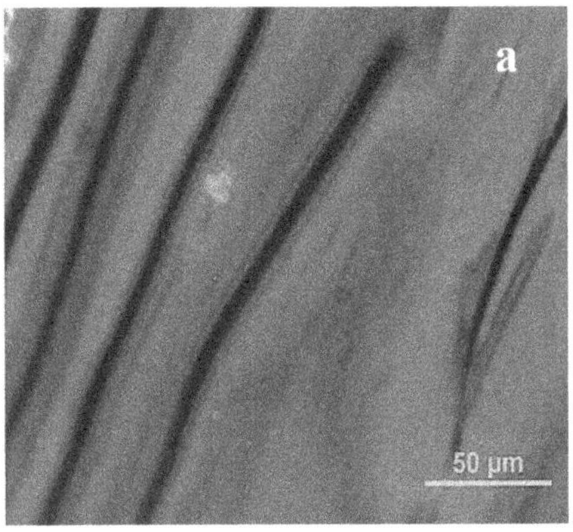


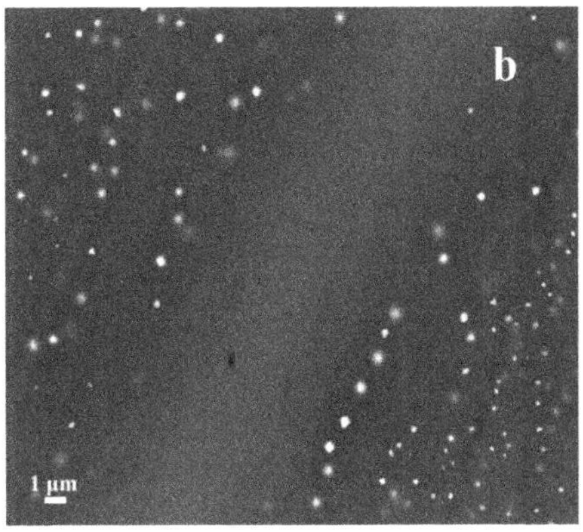


*Figure 5. Optical (a) and FEG-SEM micrographs (b) of transparent dark cords in the red opaque sample GAR-3. No particles are observed in the cord [Colour available online]*

avoid the formation of a disturbing optical defect called *tortioni neri* (black cords). The recipes claim that these cords can be removed by adding iron or copper. Instead, their formation can be avoided by slow, gradual additions of copper (M74 and M132). It is quite probable that the *tortioni* correspond to the deep green transparent cords that appear black by visual inspection and are frequently present in red brown glass objects (Figure 5). In general, the analyses of transparent dark glass do not detect any difference in chemical composition with the surrounding opaque red glass. This suggests that these layers are formed of insufficiently reduced glass where no metallic copper particles are present.[41]

## 4. Discussion and conclusions

Eighteen recipes were excluded from the following discussion because their indications are not compatible with the analyses of Venetian red brown samples. They prescribe the use of potash or mixed alkali glass, the use of lead silicate glass, the addition of calcined tin without lead, etc. Consequently, all the recipes from the *Trattatelli* are not considered. The 17 recipes compatible with the analyses are summarised in Table 2.

Two groups of recipes are observed in Table 2. In one group the addition of lead-tin calx is recommended and in the other one it is not prescribed. A similar distinction appears in the analyses in Table 1, reporting glasses containing significant amounts of lead and tin and others where these elements are present in small amounts (less than 1%).

Lead-tin calx was not added to opacify the glass (insufficient addition and rare cassiterite crystals), but it should have a different purpose. A possible explanation is to be found in a recipe (D97) to prepare lead-tin calx. It reports a white, more oxidised lead-tin calx and a grey one with less oxidised lead and tin. It is likely that the lead-tin calx added to opaque red glass was intended to help copper reduction and the formation of metallic particles.[39,42] As mentioned above, it is possible that lead and tin were added unintentionally through glass cullet contaminated by opaque white glass in samples where their content is lower than 1%. In this case, lead and tin are fully oxidised and would not affect the formation of colour.

Indeed, as mentioned in the recipes, common glass or *vitrum blanchum* are confirmed by the analyses to be the most frequently used base glass in the

*Table 2. Venetian recipes compatible with the analytical data. Amounts of lead-tin calx, iron (flakes or steel) and red copper prescribed in the recipes to be add to 100 parts of translucent glass to obtain red opaque glass*

| *Recipe* | *Lead tin calx* | *Red copper* | *Iron flakes* | *Steel* | *Raw tartar* | *Notes* |
|---|---|---|---|---|---|---|
| D213 | 5·5 | 4·1 | 2·8 | 2·8 | | amounts for 100 parts of glass cullet; the sequence is not indicated |
| M120, B310, B317 | x | x | x | | x | to be mixed together and added to the melt (iron and copper in comparable amounts); raw tartar as last |
| B61 | 8·3 | 8·3 | | 8·3 | | amounts mixed with 100 parts of glass cullet and melted together |
| D155 | 5·5 | 5·5 | 2·8 | 2·8 | | amounts mixed with 100 parts of glass cullet and melted together |
| A62, M110, B173, B174, D129, D131 | | x | x | | | iron and copper together in comparable amounts |
| M74, M132, D67 | | x | x | | | iron first, than copper (comparable amounts) |
| D156 | | x | x | | | iron first, than copper (half of the iron amount) |
| B126 | | x | | x | | iron first the amounts are decided by the glassmaker looking at the glass colour |

production of Venetian red brown glass, and only in three samples the analyses revealed a *cristallo* glass composition.

The potassium content slightly exceeds 4% (a value considered as a limit in Venetian glassmaking) in four samples ($K_2O$ 4–4·3%). It could indicate small additions of tartar to the base glass as indicated in some Venetian recipes where potassium was deliberately added in the form of tartar. Reported for the first time in the Trattatelli, tartar was added in a calcined form (K-carbonate) as a potash-bearing raw material, or unprocessed (raw tartar) or partially calcined tartar (black tartar), acting as a K-bearing compound and as a reducer compound, due to uncombusted particles.

The analyses reveal that the copper to iron ratio ranges widely. Iron contents are always higher than copper (usually from twice to five times) and only in three analyses these two elements are present in comparable concentrations. Also the recipes report that iron additions are higher (up to twice) or equal to copper additions. This is probably due to the sources of iron indicated in the recipes: the reducing power of the steel and of the iron flakes from the anvil cannot be predicted. To understand this technical aspect, systematic laboratory reproductions are ongoing. They are addressed to test what is suggested in the recipes (type of colouring compounds, their addition sequence: first iron or copper or together, thermal treatments, control of the final colour, etc.) and what it has been found in analyses reported in the literature (amounts of iron and copper, iron to copper ratio). Some preliminary results are going to be published in Bandiera *et al.*[43]

A comparison of the glass colour and compositional differences is possible for the two mosaic tesserae (SPC-R7 red and SPC-R9 brown). Their base glass composition is quite similar, suggesting that they are made in a same workshop; but the colourant additions are different. The red tesserae shows lower contents of iron and copper ($Fe_2O_3$ 4%; CuO 1·7%) as compared to the brown one ($Fe_2O_3$ 6·5%; CuO 2·25%). These results are in agreement with Venetian recipes which recommend to keep the melt at a lower temperature and add more iron and copper to obtain a brown hue. Another difference concern the lead and tin content: their concentrations are low in the red tessera (PbO 0·3%; $SnO_2$ 0·4%) and high in the brown one (PbO 5·5%; $SnO_2$ 2·8%). Nevertheless, no indications are found in the recipes concerning the addition of lead tin calx to obtain darker colours.

Through the indications of the Venetian manuscripts emerged that the manufacture of red brown opaque glass was obtained by adding to a molten transparent colourless glass compounds of iron (reducing agent) and copper (colourant). Despite these similarities, the results were based on an empirical process, in which each melting require distinct amount of colouring and reducing agent. This because several parameters changed during each melting, such as the nature and composition of the copper- and iron-bearing compounds, the glass melt temperature, the redox of furnace atmosphere (colour of the flame, with or without smoke, etc.). Indeed, although a common procedure is proposed by the manuscripts, the exact amounts of the copper and iron-bearing compounds are rarely mentioned. Hence, the technique to produce opaque red brown glass and the control of the final colour was largely dependent on the skills and the experience owned by the glassmakers.

In the Venetian glassmaking as well in the Roman period the production of red brown glass occurred by re-melting an already made transparent glass in the form of new glass (*cotisso*) or cullet. Several Venetian recipes recommend the use of recycled glass, which is preferred to the well-decolourised crystal glass. The procedure to add the colouring raw materials directly to the molten glass is highlighted by the manuscripts, a detail not mentioned before in previous research.

Furthermore, the nature of the probable copper- and iron-bearing were identified. In the recipes a pure red copper powder (cuprous oxide), obtained by calcination process of metal copper, was used. This excluded the use of bronze or brass by-product as supposed in other periods.

The indications of the Venetian recipes to use iron metallurgical by-products, confirm what was detected (metallic iron inclusions associated to different iron oxides) by analysing Late Antique Byzantine red brown glass. It could be interesting to understand if a sort of continuity to produce opaque red glass existed among these historical periods.

## 5. Conclusion

The study on Venetian Renaissance glass analyses and recipes offers a rare opportunity to improve our knowledge about the manufacture of red brown glass for a better understanding of the glassmaking technology, colour control and raw materials used. With the proper precautions, these results may help to interpret the analyses of red brown samples made in different periods and places for which no written sources are available.

## Acknowledgements

The authors acknowledge the Portuguese Foundation for Science and Technology (FCT-MCTES) for PhD grant PD/BD/135053/2017, the research unit VICARTE (UIDB/00729/2020) for supporting this project. Roberto Falcone of SSV for the XRF analyses, Patrice Léhuédé of C2RMF for FEG-SEM investigations, Joana Pinto of CENIMAT for the XRD measurements are kindly acknowledged.

## References


1. Freestone, I. C. Composition and microstructure of early opaque glass. In *Early Vitreous Materials*. Edited by M. Bimson & I. C. Freestone, British Museum, London, 1987, 173–91.
2. Brill, R. & Cahill, N. D. A red opaque glass from Sardis and some thoughts on red opaques in general. *J. Glass Stud.*, 1988, **30**, 16–27.
3. Bimson, M. Opaque red glass: a review. In *Early Vitreous Materials*. Edited by M. Bimson & I. C. Freestone, British Museum, London, 1987, 98–107.
4. Hughes, M. J. Technical study of opaque red glass of iron age. *Proc. Prehist. Soc.*, 1972, **38**, 98–107.
5. Brun, N., Mazerolles, L. & Pernot, M. Microstructure of opaque red glass containing copper. *J. Mater. Sci. Lett.*, 1991, **10** (23), 1418–20.
6. Freestone, I. C., Stapleton, C. P. & Rigby, V. The production of red glass and enamel in the Late iron Age, Roman and Byzantine periods. In: *Through a Glass Brightly - Studies in Byzantine and Medieval Art and Archaeology Presented to David Buckton*. Edited by C. Entwistle, Oxbow Books, 2003, p. 142–54.
7. Verità, M., Arena, M. S., Carruba, A. M. & Santopadre, P. Roman glass: art and technology in a 4th century A.D. opus sectile in Ostia (Rome). *J. Cult. Herit.*, 2008, **9**, 16–20.
8. Bandiera, M., Verità, M., Lehuédé, P. & Vilarigues, M. The technology of copper-based red glass sectilia from the 2nd century AD Lucius Verus villa in Rome. *Minerals*, 2020, **10**, 875 (in press).
9. Neri, A. *L'arte Vetraria 1612*. Edited by F. Abbri, Giunti, Firenze, 2001.
10. Heck, M., Hoffmann, P. Coloured opaque glass beads of the Merovingians. *Archaeometry*, 2000, **42** (2), 341–57.
11. Silvestri, A., Tonietto, S., Molin, G. & Guerriero, P. The palaeo-Christian glass mosaic of St. Prosdocimus (Padova, Italy): Archaeometric characterisation of tesserae with copper- or tin-based opacifiers. *J. Archaeol. Sci.*, 2014, **42** (1), 51–67.
12. Fiori, C. Production technology of Byzantine red mosaic glasses. *Ceram Int.*, 2015, **4** (2), 3152–7.
13. Weyl, W. A. *Coloured glasses*. Society of Glass Technology, Sheffield, 1951.
14. Bamford, C. R. *Colour Generation and Control in Glass*. Elsevier Scientific, Amsterdam, 1977.
15. Barber, D., Freestone, I. C. & Moulding, K. Ancient copper red glasses: investigation and analysis by microbeam techniques. In *From Mine to Microscope – Advances in the Study of Ancient Technology*. Edited by A. J. Shortland, I. C. Freestone & Th. Rehren, Oxbow Books, 2009, p. 898–9.
16. Tress, H. J. Ruby and related glasses from the standpoint of the chemical potential of oxygen in glass. Part 1. Chemical potential of oxygen in glass. *Glass Technol.*, 1962, **3** (1), 28–36.
17. Volf, M. B. *Chemical Approach to Glass*. Elsevier, Amsterdam, 1984.
18. Zecchin, L. *Vetro e vetrai di Murano*. Arsenale, Venezia, 1987, 1989, 1990.
19. Canal E. *Archeologia della laguna di Venezia*. Cierre, Verona, 2013.
20. Moretti, C. & Gratuze, B. Vetri rossi al rame e avventurina. Confronto di analisi e ricette. *Riv. Stn. Sper. Vetro*. 1999, **3**, 147–60.
21. Moretti, C. & Hreglich, S. Tecniche di produzione dei vetri opachi impiegate dai vetrai veneziani tra il XV ed il XX secolo. *Riv. Stn. Sper. Vetro*. 2005, **5**, 28–32.
22. Zucchetta, E. & Verità, M. La pala d'altare in mosaico della Cappella Lando in San Pietro di Castello a Venezia: inedite notizie d'archivio, tecnica di esecuzione, intervento, indagini scientifiche. *Atti XVIII Convegno Scienza e Beni Culturali*. Edited by G. Biscontin & G. Driussi, Bressanone, Italy, 2002, p. 813–21.
23. Lima, A., Medici, T., Pires de Matos, A. & Verità, M. Chemical analysis of 17th century Millefiori glasses excavated in the Monastery of Sta. Clara-a-Velha, Portugal: Comparison with Venetian and façon-de-Venise production. *J. Archaeol. Sci.*, 2012, **39** (5), 1238–48.
24. Gratuze, B., Giovagnoli, A., Barrandon, J. N., Telouk, Ph. & Imbert J. L. Apport de la methode ICP-MS couplee a l'ablation laser pour la caracterisation des archeomateriaux. *Revue d'Archéométrie*, 1993, **17**, 89–104.
25. Verità, M., Basso, R., Wypyski, M. T. & Koestler, R. J. X-ray microanalyses of ancient glassy materials: a comparative study of wavelength dispersive and energy dispersive techniques. *Archaeometry*. 1994, **36**, 241–51.
26. Tesser, E., Verità, M., Lazzarini, L., Falcone, R., Saguì, L. & Antonelli, F. Glass in imitation of exotic marbles: An analytical investigation of 2nd century AD Roman sectilia from the Gorga collection. *J. Cult. Herit.*, 2020, **42**, 202–12.
27. Verità, M. Venetian soda glass. In *Modern methods for analysing archaeological and historical glass*. Edited by K. Janssens, John Wiley & Sons, Chichester, 2013, p. 515–36.
28. Zecchin, L. L'organizzazione dell'arte dei "fioleri". In *Vetro e vetrai di Murano*. Arsenale, Venezia, 1989, **Vol. II**, 13–18.
29. Wedepohl, K. H. Chemical composition of medieval glass from excavations in West Germany. *Glastech. Ber.: Glass Sci. Technol.*, 1997, **70** (8), 246–56.
30. Aceto, M., Agostino, A., Fenoglio, G., Idone, A., Gulmini, M., Picollo, M., Ricciardi, P. & Delaney, J. K. Characterisation of colourants on illuminated manuscripts by portable fibre optic ultraviolet-visible-NIR reflectance spectrophotometry. *Anal. Methods*, 2014, **6** (5), 1488–500.
31. Colomban, P. & Truong, C. Non-destructive Raman study of the glazing technique in lustre potteries and faience (9–14th centuries): Silver ions, nanoclusters, microstructure and processing. *J. Raman Spectrosc.* 2004, **35** (3), 195–207.
32. Milanesi, G. *Dell'arte del vetro per musaico. Tre Trattatelli dei secoli XIV e XV, ora per la prima volta pubblicati*. Bologna: Romagnoli G, editor.1864. Anastatic reprint. Forni, Bologna, 1968.
33. Zecchin, L. Ricette vetrarie toscane del Quattrocento; Ricette vetrarie d'ispirazione muranese nel Quattrocento; Ancora ricette vetrarie del Quattrocento. In *Vetro e vetrai di Murano*. Arsenale, Venezia, 1990, p. 213–26.
34. Zecchin, L. Ricettario di Montpellier. In *Vetro e vetrai di Murano*. Arsenale, Venezia, 1987, p. 249–76.
35. Moretti, C. & Toninato, T. *Ricette vetrarie del Rinascimento. Trascrizione da un manoscritto anonimo veneziano*. Marsilio, Venezia, 2001.
36. Zecchin, L. Il Ricettario Darduin, un codice vetrario del Seicento, trascritto e commentato. In: *Vetro e vetrai di Murano*. Arsenale, Venezia, 1986.
37. Moretti, C., Salerno, C. S. & Tommasi-Ferroni, S. *Ricette vetrarie muranesi. Gaspare Brunoro e il manoscritto di Danzica*. Nardini, Firenze, 2004.
38. Zecchin, L. Prete Antonio Neri vetraio del Seicento; Il primo libro di Neri; Il libro di prete Neri; lettere a prete Neri; Antonio Neri e le Conterie; Terminologia muranese nel libro di Neri. In: *Vetro e vetrai di Murano*. Arsenale, Venezia, 1987, p. 155–77.
39. Moretti, C. *Glossario del vetro veneziano. Dal Trecento al Novecento*. Marsilio, Venezia, 2001.
40. Moretti, C. La pratica chimica dei vetrai del Rinascimento. *Riv. Stn. Sper. Vetro*, 2012, **6** (12), note 52.
41. Verità, M. Le tessere vitree dei mosaici medievali a Roma. Tecnologia e degrado. In *Mosaici medievali a Roma attraverso il restauro dell'ICR 1991-2004*. Edited by A. Andaloro, & C. D'Angelo, Gangemi, Roma, 2017, p. 437–77.
42. Tite, M., Pradell, T., Shortland, A. Discovery, production and use of tin-based opacifiers in glasses, enamels and glazes from the Late Iron Age onwards: a reassessment. *Archaeometry*, 2008, **50** (1), 67–84.
43. Bandiera, M., Verità, M., Zecchin, S. & Vilarigues, M. Riproduzioni in laboratorio del vetro rosso bruno opaco. Forthcoming in Atti XX Giornata di Studio sul vetro AIHV, 18th–19th may, 2019, Ravenna, Italy.

*Glass Technol.: Eur. J. Glass Sci. Technol. A*, 2021, **62** (1), 34–40

# The effect of the colemanite on the melting properties of rock glass wool

*Qingwei Wang,*[1,2,*] *Jie Gao,*[1,3] *Jin Liu*[1] *& Lida Luo*[1,2*]

[1] *State Key Laboratories for Modification of Chemical Fiber and Polymer Materials, Donghua University, Shanghai 201620, People's Republic of China*
[2] *Engineering Research Center of Advanced Glass Manufacturing Technology, Ministry of Education, Shanghai 201620, People's Republic of China*
[3] *Etimaden Asia Pacific Limited, 999077, Hong Kong*

Manuscript received 14 August 2019
Revision received 9 January 2020
Manuscript accepted 9 January 2020

*Colemanite, as a new environmentally friendly raw material for providing boron oxide, was used to promote the melting of glass during rock wool production. The sintering properties were improved dramatically when 3·0 wt% of $B_2O_3$ was added into the rock wool through the addition of colemanite. More importantly, the high-temperature viscosity of the quenched glasses exhibited significant variation among the samples doped with different concentrations of $B_2O_3$. Since thermal insulation is the most important application of rock wool, the structural properties, such as the tensile strength and chemical stability, were also investigated in this study.*

## 1. Introduction

Asbestos used as a building insulation material is harmful to health when inhaled, and therefore, some non-asbestos producers, especially in the developed countries in the West, have encouraged the replacement of asbestos with rock wool or glass wool.[(1–3)] Rock wool manufactured directly from basalt, dolomite and coke is widely used as a building insulation material[(3,4)] and supplementary cementitious material[(5)] due to its chemical resistance, thermal stability and low thermal conductivity.[(6)] In 2016, over 7 million tons of rock wool were produced worldwide, of which 2·52 million tons were produced in China.[(3)] Due to the demand for improved fire protection and reduced energy consumption, rock wool will be increasingly used in the near future,[(7)] and therefore, it is important to reduce the melting temperature and energy required for the preparation of rock wool.

Boron oxide have the advantages of a low thermal expansion coefficient, a low density, a particularly low melting point and a high boiling point that male it an excellent cosolvent[(8)] and have led to its wide use in the glass and ceramics industry to lower the melting point,[(9,10)] as well as in nuclear industry, agriculture[(12)] and other field. Borax and boric acid as the boron source are generally used to reduce the melting point of glass and ceramics[(13,14)] but are more expensive than colemanite ($2CaO.3B_2O_3.5H_2O$). Colemanite is a natural ore that contains boron oxide and calcium oxide, and Frost *et al*[(15)] reported that approximately 60% of the world supply of boron ores is located in Turkey, mainly in Anatolia.[(16)] Even though there have been few studies of boron-containing raw materials, due to its high boron oxide content, colemanite may be a candidate material for use in reducing the sintering temperature and raw material cost in the glass fibre industry.[(17–19)] There have been few reports on the use of colemanite in the rock wool industry. However, based on its successful application in the fibreglass industry, the rock wool compositions with colemanite will be beneficial by reducing the temperature and energy consumption in rock wool production.

In this study, the effects of the colemanite with different boron oxide concentrations on rock wool were studied. High temperature viscosity and differential scanning calorimetry (DSC) measurements were carried out, the sintering properties were examined, and the mechanical and chemical properties of the obtained rock wool were investigated.

## 2. Experimental

### 2.1. Raw material

The colemanite powder with a particle size of 75 μm was provided by the Eti Mine Works G.M. Company (Ankara, Turkey). The chemical composition of the colemanite powder is shown in Table 1. As shown in Figure 1, the other raw materials were basalt, dolomite, coke and slag, which were provided by Dingtai Lanxing Company (Hejian city, Hebei Province, China). The chemical composition of these materials

* Corresponding authors. Email wqwq888@dhu.edu.cn & lld@dhu.edu.cn
DOI: 10.13036/17533546.62.1.008

*Table 1. The chemical and mineral composition of four samples with different concentrations of colemanite (wt%)*

| *Oxides* | $SiO_2$ | $Al_2O_3$ | $Fe_2O_3$ | $CaO$ | $MgO$ | $Na_2O$ | $TiO_2$ | $SO_3$ | $K_2O$ | $B_2O_3$ |
|---|---|---|---|---|---|---|---|---|---|---|
| Basalt | 52·28 | 16·82 | 8·99 | 7·61 | 3·97 | 2·54 | 1·43 | 0·32 | 3·2 | 0 |
| Slag | 32·71 | 15·78 | 0·43 | 39·98 | 7·5 | 0 | 1·54 | 1·95 | 0 | 0 |
| Dolomite | / | / | / | 30·4 | 21·7 | / | / | / | / | / |
| Colemanite | 5 | / | / | 27 | 2 | 0·5 | / | / | / | 40 |
| D-0 | 39·93 | 14·72 | 5·02 | 25·27 | 9·76 | 1·37 | 1·32 | 0·87 | 1·73 | 0·00 |
| D-08 | 39·32 | 14·46 | 4·93 | 25·37 | 9·63 | 1·36 | 1·30 | 0·85 | 1·70 | 0·80 |
| D-16 | 38·71 | 14·20 | 4·84 | 25·46 | 9·49 | 1·35 | 1·28 | 0·84 | 1·67 | 1·60 |
| D-30 | 37·66 | 13·74 | 4·69 | 25·61 | 9·26 | 1·32 | 1·23 | 0·81 | 1·62 | 3·00 |

Remarks: About 1 wt% concentration of iron oxide would be reduce to metal iron. The actual content of iron oxide in rock wool range from 3·5% to 4·5%

is also described in Table 1. All of the large lumpy raw materials shown in Figure 1 were ground into fine powder for uniform mixing during the experiment. The colemanite and other raw materials were mixed at the mass ratios shown in Table 2. The compositions of the four samples used in the experiment are specified in Table 1 and are denoted as D-0, D-08, D-16 and D-30, meaning that 0, 0·77, 1·53 and 2·80 wt% $B_2O_3$ was added into the rock wool with colemanite in these samples.

## 2.2. Sample preparation

A certain amount of the rock wool batch with colemanite was melted at 1550°C for 5 h to produce a cullet in an air atmosphere followed by water quenching; then, the glass fragment was incubated in the drying oven for 24 h. It was found that the batch was very difficult to melt, and a mass of bubbles was observed during the glass melting process. Then, the cullets were remelted into the platinum crucible placed on the top of a fibre drawing machine (Shanghai, SGF-Pt). The continuous rock wool fibres were drawn at 1280–1320°C depending on their viscosity (log at a rate of 20 m/s and with a diameter of 10~13 μm) and then all of the fibres were wound on a roller.

## 2.3. Characterisation techniques

The morphology of the samples was studied by scanning electron microscopy (Hitachi, SU8010, Japan). The thermal behaviour of the cullet was characterised by a differential scanning calorimeter (Netzsch, STA409PC,Gernamy) at a heating rate of 5°C/min in air atmosphere. The temperature dependence of the viscosity was measured by a high temperature viscometer (Theta Industries, Inc., Rheotronic II, USA). The glass network was characterised by FT-IR spectrometer (Thermo Fisher Scientific Inc., Nicolet 6700, USA). The tensile strength was measured by a single-fibre electronic tensile strength tester (Xinxian Instrument Co.Ltd, XQ-1A, China), and each final result represents the average value of five measurements.

*Figure 1. The pictures of the raw material of rock wool (a) basalt; (b) dolomite; (c) coke; (d) slag [Colour available online]*

*Table 2. The formula of the rock wool*

| | *B-0* | *B-08* | *B-16* | *B-30* |
|---|---|---|---|---|
| Basalt/g | 266 | 266 | 266 | 266 |
| Dolomite/g | 112 | 112 | 112 | 112 |
| Slag/g | 175 | 175 | 175 | 175 |
| Coke/g | 147 | 147 | 147 | 147 |
| Colemanite /g | 0 | 10·05 | 20·35 | 39·10 |
| Glass weight/g | 483·56 | 491·10 | 498·83 | 512·98 |

Remarks: half batch was used to melt to prepare the cullet, about 30 g batch was used to test the DSC, others were pressed to be small cylinders to measure the melting properties.

## 3. Results and discussion

The D-series raw materials were ground into 80 mesh to prepare the cylindrical sample that was fabricated for investigating the effect of the different concentrations of colemanite on the initial sintering properties. Figures 2–4 show digital photographs of the batches at different sintering stages. It is observed that the colour of the samples changed from black to grey and then to yellow due to coke combustion until the samples collapsed and melted. There was nearly no softening or deformation before 1100°C, indicating that little glass phase was generated when the temperature is lower than 1100°C prior to the decomposition of colemanite. All of the samples collapsed at nearly 1150°C and were melted to become flat at 1200°C. Comparing the samples shown in

*Figure 2. Sintering images of samples without $B_2O_3$ in different temperatures [Colour available online]*

Figure 4, it is found the D-30 sample with 3·0 wt% $B_2O_3$ is easier to flatten on the substrate and exhibits better high temperature fluidity. This result occurs because the melting point of the $B_2O_3$ is nearly 400°C, enabling it to contribute to the material transfer by the dissolution/precipitation mechanism by exploiting the $B_2O_3$'s liquid phase.

The coke played a significant role in the sintering process of the rock wool raw materials. In addition to being used as a solid fuel to melt other raw materials, the coke is also applied to control the melting atmosphere, in particular to reduce the iron (III) oxide to iron (II) oxide and then to metal iron. The results presented in Table 1 clearly demonstrate this point because the iron oxide concentration in the rock wool should be greater than 4·5 wt% according to the chemical composition of the raw material while the actual percentage found in the sample is in the range of 3·5–4·5 wt% due to the fluctuation in the content of the raw material or to the controllable content of the coke.

## 3.2. Melting properties of quenched glasses

The quenched glasses were prepared after the sintering of the raw materials. Figure 5 presents the results of the thermo gravimetric analysis of the D series batches. It is observed that a slight mass loss occurred at about 400 and 450°C, and the mass loss rates of D-16 and D-30 are higher than those of D-0 and D-8. This difference is due to the decomposition of the crystal water of colemanite in the D-16 and D-30 samples as shown in equation (1) that is much less pronounced in the D-0 and D-8 batches with much lower colemanite mass fraction.[20] The main weight loss occurred in the 600~700°C range and can be attributed to the combustion of coke, considering that the two decomposition temperature ranges of dolomite are found at 750~792°C and 819~912°C.

$$2CaO.3B_2O_3.5H_2O \xrightarrow{\sim 400°C} 2CaO.3B_2O_3 + 5H_2O\uparrow(gas) - Q \quad (1)$$

$$C(solid) + O_2 \xrightarrow{550\text{–}650°C} CO_2\uparrow(gas) + Q \quad (2)$$

$$2CaO.3B_2O_3 \xrightarrow{\sim 600°C} 2CaO(amorphous) + 3B_2O_3 - Q \quad (3)$$

Among the four batches, the D-30 sample exhibits a significantly lower weight loss, and to verify this result, the test was replicated for six times and presented the same results. It was inferred that the CaO decomposed from colemanite must be the amorphous one which is easily directly reacted with coke and the oxygen in the air or with the carbon combustion products to form calcium carbonate, and the reaction equation is shown in Equations (2), (3) and (4). More importantly, the calcium carbonate decomposed at approximately 900°C to emit carbon dioxide. Therefore, the colemanite improved the releasing temperature of $CO_2$ and affected the heat

*Figure 3. Sintering images of samples with 0·8 wt% concentration of $B_2O_3$ in different temperatures [Colour available online]*

*Figure 4. Sintering images of samples with 3.0 wt% concentration of $B_2O_3$ in different temperatures [Colour available online]*

of the reaction, while no other effects of colemanite on the appearance and weight loss were observed.

$$CaO(amorphous) + CO_2 \xrightarrow{\sim 600°C} CaCO_3 + Q \quad (4)$$

$$nCaMg(CO_3)_2 \xrightarrow{\sim 770°C} (n-1)MgO + MgCO_3.nCaCO_3 + (n-1)CO_2(gas)\uparrow - Q \quad (5)$$

$$MgCO_3.nCaCO_3 \xrightarrow{>900°C} MgO + nCaO + (n+1)CO_2(gas)\uparrow - Q \quad (6)$$

$$CaCO_3 \xrightarrow{\sim 900°C} CaO + CO_2(gas)\uparrow - Q \quad (7)$$

An examination of Figure 6 shows that clear weight loss are present at approximately 770°C due to the decomposition of dolomite just as shown in Equation (5), (6) and (7),[21] but the endothermic and exothermic peaks of the colemanite were slight due to the limited concentration. Additionally, the exothermic peak of the coke combustion will be masked by the reaction between CaO and $CO_2$. The exothermic peak of D-0 at approximately 600°C was clearly observed for the samples without colemanite as shown in Figure 6.

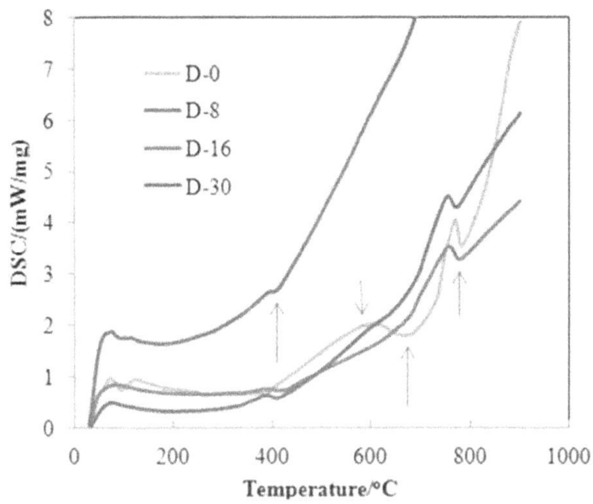


*Figure 6. DSC Analysis of series batches with different contents of colemanite [Colour available online]*

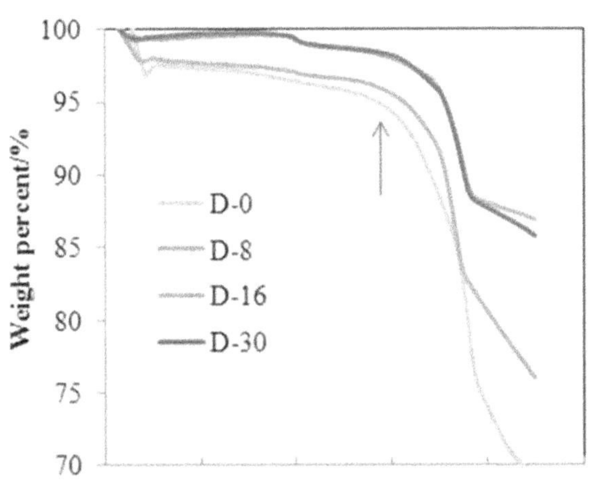


*Figure 5. Thermogravimetric analysis of series batches with different contents of colemanite [Colour available online]*

Figures 7 presents the high temperature viscosity data for the four D-series batches. It is observed that

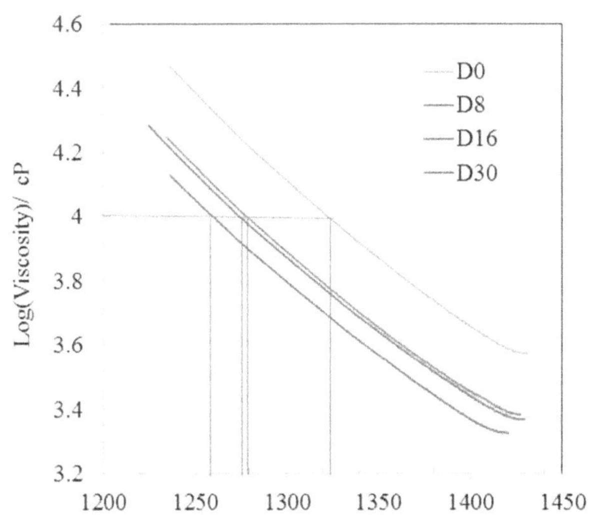


*Figure 7. High temperature viscosity of the samples with different concentrations of colemanite [Colour available online]*

*Figure. 8 The picture of the foaming samples when measured the high temperature viscosity [Colour available online]*

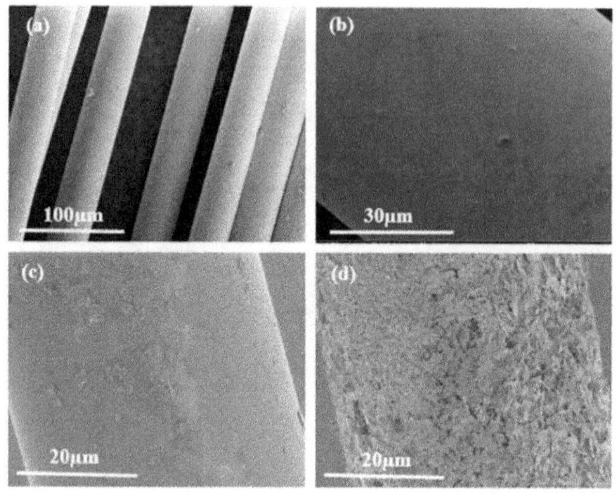


*Figure 9. The SEM micrograph of D-30 samples with 3·0 wt% $B_2O_3$: (a) pristine fibres, ×400; (b) pristine fibres, ×3000; (c) acid corrosion: 1 mol/L HCl 24 h; (d) alkali corrosion: 1 mol/L NaOH 24 h*

the melting temperature is not greater than 1450°C, and the colemanite can significantly decrease the high temperature viscosity. The difference for the viscosity curve of the samples with different concentrations of the colemanite are nearly same. Considering that the fibres were drawn at approximately 1280–1320°C, we can compare the temperature difference at the same viscosity ($\log v(cP)$=4·0), and the results show the temperature of the samples without the colemanite is higher by 46, 51 and 68°C compared with that of the samples with the addition of 0·8, 1·6 and 3·0 wt% of $B_2O_3$, respectively. The boron oxide that was introduced through colemanite is present as a network outer body, such as $[BO_3]$ at approximately 1400–1500°C, and this triangular structure reduces the viscosity to a certain extent.[17] The testing samples easily foam during the testing process, as clearly demonstrated in Figure 8, and the colemanite acting as a fluxing agent does indeed decrease the melting temperature of the rock wool.

## 3.3. Effect on the properties of rock wool

To investigate the rock wool's actual application environment, the rock wool made from D-30 was separately placed into 1 mol/L HCl and 1 mol/L NaOH solutions for 24 h at a stable temperature of 40°C; the diameters of the rock wool particles were in the range of 28–36 μm. Figure 9 shows the SEM micrograph of the pristine rock wool fibres with and without treatment with the acid or alkali solution. A few small particles were found on the surface of the rock wool after the treatment with the HCl solution. The chemical composition of these particles were analysed by EDS, and the results are presented in Table 3. It was observed that the hydrochloric acid reacted with the chemical structure of the rock wool's surface through ion exchange.[22] The alkali metal and alkaline earth metal ions in the basalt fibre easily exchange ions with the hydrogen ion in the solution just as shown in Equations (8) and (9), resulting in the removal of some ions inside the fibre.

$$-Si-O-Na^{-}+H^{+}\rightarrow -Si-O-H^{-}+Na^{+} \qquad (8)$$

$$-Si-O-K^{-}+H^{+}\rightarrow -Si-O-H^{-}+K^{+} \qquad (9)$$

Significant erosion of the rock wool surface was observed after the sodium hydroxide treatment compared with the hydrochloric acid treatment. In addition to the reaction of sodium hydroxide with the surface of the rock wool, the whole structure of the rock wool was also greatly damaged and had a layered appearance. The hydroxyl ions of the sodium hydroxide solution reacted with the Si–O and Al–O tetrahedral structure as shown in Equations (10) and (11),[23] which is the primary structure of the rock wool, resulting in the fracture of the rock wool.

$$[\equiv Si-O-Si\equiv]+2OH^{-}\rightarrow 2[\equiv Si-O]^{-}+H_2O \qquad (10)$$

$$[\equiv Al-O-Al\equiv]+2OH^{-}\rightarrow 2[\equiv Al-O]^{-}+H_2O \qquad (11)$$

It is well-known that the rock wool consists of glass fibres and basalt fibres. Due to the low content

*Table 3. EDS components analysis of rock wool treated in acid and alkali environment*

| Elements | O | Na | Mg | Al | Si | K | Ca | Ti | Fe | B |
|---|---|---|---|---|---|---|---|---|---|---|
| Figure 9 (1) | 41·83 | 3·95 | 3·34 | 9·23 | 25·76 | 2·84 | 3·15 | 2·03 | 7·46 | 0·41 |
| Figure 9 (2) | 42·33 | 4·16 | 3·85 | 8·11 | 23·44 | 5·11 | 4·13 | 2·91 | 5·96 | |
| Figure 9 (3) | 40·45 | 2·64 | 3·79 | 4·33 | 25·49 | 1·07 | 8·78 | 5·23 | 7·55 | 0·67 |
| Figure 9 (4) | 41·43 | 3·55 | 5·64 | 4·15 | 24·78 | 2·11 | 7·45 | 4·11 | 6·64 | 0·14 |

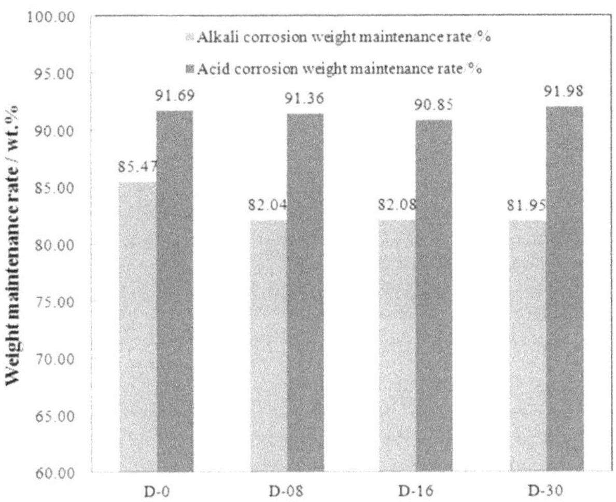


*Figure 10. The alkali and acid resistance of rock wool fibre samples treated in 1 mol/L HCl or NaOH solution for 24 h [Colour available online]*

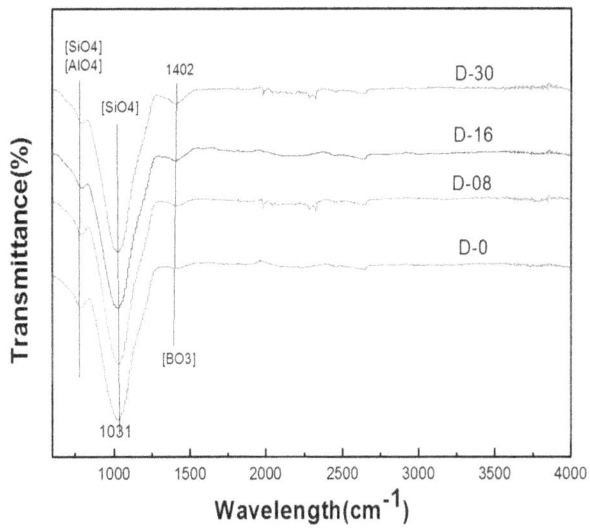


*Figure 12. FT-IR spectra of rock wool samples with different concentrations of colemanite [Colour available online]*

of alkali and high concentration of $Fe_2O_3$ in the rock wool, the glass network consisted mainly of $[AlO_4]$ tetrahedron and $[SiO_4]$ tetrahedron.(24) The ferrous and ferric oxide existed either as $[FeO_4]$ or $[FeO_6]$, depending on the external environment and atmosphere. This provides the opportunity for boron to change the chemical and mechanical properties of rock wool.

The chemical stability was investigated to identify the ion exchange and the fracture of the Si–O tetrahedral structure. During the test, rock wool samples (10 g) from the D-0, D-08, D-16 and D-30 batches were weighed by high precision balance before and after the alkali and acid resistance. Figure 10 shows that the rock wool's weight maintenance ratio ranged from 81·95–85·47% after the alkali corrosion and from 90·85–91·98% after the acid corrosion. The D-30 samples show the lowest weight maintenance in the alkali solution but exhibit the highest weight maintenance in the acid solution. It appears that colemanite is less important for acid resistance than for alkali resistance. The most important reason for this phenomenon is that the rock wool of D-30 batch has a higher content of boron oxide which appeared as a layered structure consisting of boron oxide triangles, and this structure can more easily react with the hydroxide ions.

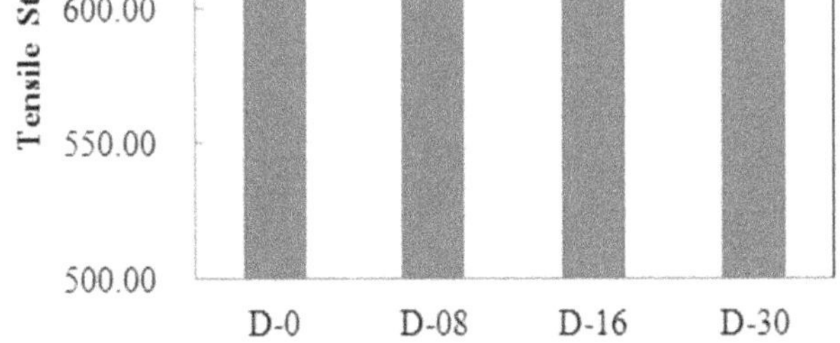


*Figure 11. Tensile strength of the fibre samples with different contents of colemanite [Colour available online]*

The results for the tensile strength of rock wool fibre are presented in Figure 11. It is observed that the tensile strength increased to some extent with increasing concentration of $B_2O_3$, albeit with large random fluctuations. Comparison of the FT-IR spectra of the four different batches of the D-series shown in Figure 12 indicates that the redshift of the peak absorption wavelength follows the increasing colemanite concentration. The absorption band in the 750–650 $cm^{-1}$ region is ascribed to the Al–O bending vibration of the $[AlO_4]$ unit, Si–O bending vibration of the $[SiO_4]$ tetrahedron unit, and the B–O bending vibration of the planar $[BO_3]$ unit.(23) The absorption peak located at 1031 $cm^{-1}$ might be due to the B–O asymmetric stretching vibration of the $[BO_4]$ tetrahedron unit and the Si–O stretching vibration of the $[SiO_4]$ tetrahedron unit. The absorption band at approximately 1402 $cm^{-1}$ is assigned to the B–O stretching vibration of the $[BO_3]$ triangle unit.

## 4. Conclusion

In summary, we have demonstrated the sintering properties of the raw material doped with different concentrations of colemanite that promotes sintering through the dissolution–precipitation mechanism while the coke plays a significant role by providing solid fuel and controlling the melting atmosphere. The quenched glasses' melting performance was significantly influenced by the colemanite, and the special structure of $[BO_3]$ reduced the high temperature viscosity to a certain extent. The chemical stability of the rock wool is affected by ion exchange and the fracture of the Si–O tetrahedral structure in the acid and alkali environment, providing guidance for the possible application of the rock wool in practical use.

## Acknowledgement

The financial support of Etimaden Asia Pacific Limited is gratefully acknowledged, as is supported by the National Science Foundation of China (NSFC, No. 51873032) and the Fundamental Research Funds for the Central Universities (223201900003).

On behalf of all authors, the corresponding author states that there is no conflict of interest.

## References

1. Cui, Y., Ma, J., Ye, W., Han, Z. X., Dong, F. Q., Deng, J. J. & Zhang, Q. B. Chrysotile and rock wool fibers induce chromosome aberrations and DNA damage in V79 lung fibroblast cells. *Environ. Sci. Pollut. Res.* 2018, **25** (23), 22328–33.
2. Luoto, K., Holopainen, M., Kangas, J., Kalliokoski, P. & Savolainen, K., Dissolution of short and long rockwool and glasswool fibres by macrophages in flowthrough cell culture. *Environ. Res.*, 1998, **78** (1), 25–37.
3. Siligardi, C., Miselli, P., Francia, E. & Gualtieri, M. L. Temperature-induced microstructural changes of fiber-reinforced silica aerogel (FRAB) and rock wool thermal insulation materials: A comparative study. *Energy Build.*, 2017, **138,** 80–7.
4. Guo, J. L., Du, Q. & Lu, L. Application of rock-wool in outer-wall external thermal insulation system. *Adv. Mater Res.*, 2013, **753–755**, 512–15.
5. Cheng, A., Lin, W. T. & Huang, R. Application of rock wool waste in cement-based composites. *Mater. Des.*, **2011,** *32* (2), 636–42.
6. Seraji, M. M., Kianersi, S., Hosseini, S. H., Davarpanah, J. & Elahi, S. Performance evaluation of glass and rock wool fibres to improve thermal stability and mechanical strength of monolithic phenol-formaldehyde based carbon aerogels. *J. Non-Cryst. Solids*, 2018, **491**, 89–97.
7. Wang, P. Q., He, L., Tan, D. J. & Wang, Y. *Research Status and Prospect of Rock Wool in China*. CRC Press-Taylor & Francis Group: Boca Raton, 2017, 167–70.
8. Krogh-Moe, J. The structure of vitreous and liquid boron oxide. *J. Non-Cryst. Solids*, 1969, **1** (4), 269–84.
9. Saitoh, A., Tricot, G., Rajbhandari, P., Anan, S. & Takebe, H. Effect of $B_2O_3/P_2O_5$ substitution on the properties and structure of tin borophosphate glasses. *Mater. Chem. Phys.*, 2015, **149**, 648–56.
10. Xu, Z. S., Liu, X. F., Jiang, C., Ma, D. W., Ren, J. J., Cheng, C. & Qiu, J. R. Effect of topological structure on photoluminescence of PbSe quantum dot-doped borosilicate glasses. *J. Am. Ceram. Soc.*, 2018, **101** (4), 1508–15.
11. Brehault, A., Patil, D., Kamat, H., Youngman, R. E., Thirion, L. M., Mauro, J. C., Corkhill, C. L., McCloy, J. S. & Goel, A. Compositional dependence of solubility/retention of molybdenum oxides in aluminoborosilicate-based model nuclear waste glasses. *J. Phys. Chem. B*, 2018, **122** (5), 1714–29.
12. Bhasker, T. V., Gowda, N. K. S., Pal, D. T., Bhat, S. K., Krishnamoorthy, P., Mondal, S., Pattanaik, A. K. & Verma, A. K. Influence of boron supplementation on performance, immunity and antioxidant status of lambs fed diets with or without adequate level of calcium. *PLoS One*, 2017, **12** (11), 14.
13. Prentice, L. H., Tyas, M. J. & Burrow, M. F. The effects of boric acid and phosphoric acid on the compressive strength of glass-ionomer cements. *Dent. Mater.*, 2006, **22** (1), 94–7.
14. Alsaadi, M. & Erklig, A. A comparative study on mode I and mode II interlaminar behaviour of borax and SiC particles toughened S-glass fabric/epoxy composite. *Arab. J. Sci. Eng.*, 2017, **42** (11), 4759–69.
15. Frost, R. L., Xi, Y. F., Scholz, R., Belotti, F. M. & Candido, M. Infrared and Raman spectroscopic characterisation of the borate mineral colemanite - $CaB_3O_4(OH)_3 \cdot H_2O$ - implications for the molecular structure. *J. Mol. Struct.*, 2013, **1037**, 23–8.
16. Derun, E. M. & Kipcak, A. S., *Neutron Absorption Capabilities of Turkish Colemanite Minerals*. Int Assoc Computer Science & Information Technology Press-Iacsit Press: Singapore, 2011, 64–8.
17. Pavlyukevich, Y. G., Levitskii, I. A. & Mazura, N. V. Use of colimanite in glass fibre production. *Glass Ceram.*, 2009, **66** (9–10), 345–9.
18. Abi, C. B. E. Effect of borogypsum on brick properties. *Constr. Build. Mater.*, 2014, **59**, 195–203.
19. Waclawska, I. Thermal behaviour of mechanically amorphised colemanite. 1. Thermal decomposition of ground colemanite. *J. Therm. Anal.*, 1997, **48** (1), 145–54.
20. Muhammed, Y. D. & Ahmet, H. S. High temperature resistance of concretes with GGBFS, waste glass powder, and colemanite ore wastes after different cooling conditions. *Constr. Build. Mater.*, 2019, **196** (30), 66–81.
21. Zhou, F., Yang, L., Zhang, L., Chao, J. & Zhang, H. Investigation of decomposition of dolomite and distribution of iodine migration during the calcination-digestion process of phosphate ore. *Hydrom.*, 2019, **188** (1), 174–81.
22. Carter, M. M., McIntyre, N. S., King, H. W. & Pratt, A. R. The aging of silicate glass surfaces in humid air. *J. Non-Cryst. Solids*, 1997, **220** (1), 127–38.
23. Li, H., Gu, P., Watson, J. & Meng, J. Acid Corrosion resistance and mechanism of E-glass fibres: boron factor. *J. Mater. Sci.*, 2014, **48** (1), 3075–87
24. Filho, J. C., Zilio, S. C., Messias, D. N., Pilla, V., Silva, A. C., Dantas, N. O. & Andrade A. A. Effects of aluminum substitution by potassium in the $P_2O_5$–$Al_2O_3$–$Na_2O$–$K_2O$ phosphate glasses. *J. Alloy. Compd.*, 2020, **815** (1),152359.

Glass Technol.: Eur. J. Glass Sci. Technol. A, 2021, 62 (1), 41–44

# Effect of temperature change on sulphuric acid–water glass slurry

Zuochun Li, Guihe Wang[1] & Feng Huang

School of Engineering and Technology, China University of Geosciences Beijing, Beijing 100083, P.R. China

Manuscript received 4 June 2019
Revision received 8 January 2020
Manuscript accepted 9 January 2020

*Sulphuric acid–water glass slurry is a type of chemical slurry. The gel time of sulfuric acid–water glass slurry is affected by many factors. However, few researchers have addressed the influence of temperature. Through a series of laboratory tests, the viscosity and gel time of sulfuric acid–water glass slurry were studied at temperatures of 20, 30, 40, 50, 60, 70, and 80°C, with sulfuric acid concentrations of 4·75, 5·0, 5·25, and 5·5%. In the tests, the viscosity of the slurry was reduced from 7 to 3 mPas with a temperature increase of 20–80°C. Gel time increased before the temperature reached 40°C. Above 40°C, the gel time began to decrease. Different concentrations of sulphuric acid slurry resulted in various reactions at sudden temperatures. Abrupt temperature changes caused the slurry gel time to increase over the previous temperature; as the temperature continued to increase, slurry gel time began to decrease. The greater the concentration of sulfuric acid, the greater the strength of the gel, however, as the temperature increased, the strength of the gel decreased continually. Temperature thus has a major influence on the strength of the gel relative to the concentration of sulphuric acid.*

## 1. Introduction

Acidic water glass can be gelled in a neutral or acidic environment. The gel is not alkaline and has no alkali dissolution.[1] The acidic water glass slurry in the fine sand formation also has a better effect than the cement slurry. In recent years, it has been applied in the underground reinforcement project in China. For example, the use of acidic water glass grouting reinforcement in the tunnel project of Fuxingmen-Xidan Station in Beijing Metro not only solves the problem of difficult grouting in fine sand formation, but also effectively controls ground subsidence.[2]

Research on acidic water glass slurry has yielded useful results. For example, Xu studied the effect of pH on the gel time of acidic water glass grouting materials.[3] Li analysed the gelation characteristics of the sulphuric acid–water glass system and found that the gel time has a strong association with the pH value of the system.[4] Xiao studied the effect of calcium carbonate on the gelation time and strength after adding acidic water glass slurry.[5] Because there is such is a wide temperature difference between the north and south of China and between summer and winter, temperature is a key factor affecting the performance of the slurry. Research was carried out on the mechanism of temperature-modified water glass solidified loess. Analysis of chemical composition, mineral composition, and microstructure of loess after solidification shed light on the influence of temperature.[6] Wu found that the effect of temperature on gel time is related to the pH value of silica sol. The gelation speed of acid silica gel solution increases with temperature.[7] Temperature has a major influence on the gelation time of the slurry, and the gelation time is short when the temperature is high, showed researchers.[8]

In this paper, the influence of temperature on the gelation time and viscosity of sulphuric acid water glass slurry was studied by controlling the change of temperature. The rule of temperature change of sulphuric acid glass slurry with temperature was obtained, which provided a reference for the application of acidic water glass slurry.

## 2. Experimental

Water glass is an alkali metal silicate with the chemical formula $R_2O.nSiO_2.mH_2O$, where $R_2O$ refers to an alkali metal oxide such as $Na_2O$, $K_2O$, and $Li_2O$. The water glass used in this experiment was sodium silicate water glass. Since the water glass solution is alkaline, the water glass solution was mixed with sulphuric acid to cause a displacement reaction. During the reaction, the free sodium ions present in the solution and the sodium ions adsorbed on the surface of the silica colloid interacted with the oppositely charged sulfate ions, thereby deactivating the colloidal particles and destabilising the sol. The sol flocculated to form a gel. The reaction of water

[1] Corresponding author. Email 1991010718@cugb.edu.cn
DOI: 10.13036/17533546.62.1.004

*Table 1. Instrumentation parameters used in experiment*

| *Instrument* | *Type* | *Parameter* <br> *Rated power (kW)* | *Range (°C)* | *Indication error (°C)* | |
|---|---|---|---|---|---|
| Thermostatic water bath | DZKW-D-1 | 0·3 | RT+10~100 | ≤±2 | |
| | | Range (mPa s) | Accuracy (mPa s) | | |
| Six-speed viscometer | ZNN-D6 | Newtonian fluid | Non-Newtonian fluid | 1–25 | >25 |
| | | 0–300 | 0–150 | ±1 | ±4% |

*RT is room temperature. For this experiment RT was 20°C

glass and sulphuric acid is shown in Equation (1):

$$Na_2O.nSiO_2+H_2SO_4+H_2O=Na_2SO_4+nSi(OH)_4 \quad (1)$$

## 3. Materials

The reagents and instruments used in this experiment consisted mainly of water glass, concentrated sulphuric acid, a thermostatic water bath (Shanghai Keheng Industrial Development Co., Ltd., Shanghai, China), and a six-speed viscometer (Qingdao HaiTongDa Special Purpose Instrument Co., Ltd., Qingdao, China). The parameters of the thermostatic water bath and six-speed viscometer are shown in Table 1.

Concentrated sulphuric acid was obtained from Beijing Chemical Works (Beijing, China), and water glass was supplied by Hongxing Guangsha Chemical Building Materials Co., Ltd. (Beijing, China). The parameters of water glass and concentrated sulphuric acid are shown in Table 2.

## 4. Slurry gel time and viscosity experiment

### *4.1. Procedure*

First, a water glass was prepared with a Baume degree of 30 and a dilute sulphuric acid solution with a mass fraction of 4·5%; these were stored separately. Then, 48 mL of dilute sulphuric acid and 40 mL of water glass were weighed to a total of seven parts. The seven solutions were placed in water at 20, 30, 40, 50, 60, 70, and 80°C, and the diluted sulphuric acid was poured slowly into the water glass solution along the cup wall after the required temperature was reached. At the same time, the two solutions were mixed thoroughly by continuously stirring with a glass rod, at which time the recording time was started. When the slurry in the beaker was poured and did not flow, the final setting time was recorded. The above steps were repeated for experiments at different temperatures. After completion, slurry volumes were reconfigured and the viscosity of the slurry at each temperature was measured using a six-speed viscometer. The above experiment was then repeated under conditions of sulphuric acid concentrations of 4·75, 5·0, 5·25, and 5·5%.

### *4.2. Data analysis*

When the slurry with sulphuric acid concentration of 4·75, 5·0, 5·25, and 5·5% was reacted at 80, 70, 70, and 60°C, a white gel was instantaneously generated in the slurry. Because the reaction mechanism is different from that at low temperatures, gelation time and viscosity will be discussed separately. Through data analysis, the curve of slurry gel time and viscosity with temperatures of different sulphuric acid concentration, gelation time of slurry at different temperatures, curve of viscosity with sulphuric acid concentration, and the temperature profile of the mutation of each concentration slurry were obtained (see Figures 1–5).

The gel time of the five different concentrations of slurry was measured at 40°C, and the time was longer with the increase in temperature (Figure 1). The change of 4·5% sulphuric acid concentration group is the most obvious, and the gel time increased from 1290 to 1890 s; the growth rate was 46·5%. Above 40°C, the gel time gradually shortened as the temperature increased. However, in the experiment, not all the concentration curves show a trend of increasing first and then decreasing at 20–80°C. As the temperature rose, the change in the gel time of different sulphuric acid concentration slurry was abrupt. At the point of mutation, the gel time increased and, as the temperature increased, the gel time began to decrease again. As the concentration of sulphuric acid in the slurry increased, the temperature of the gel in the slurry also decreased. When the concentration of sulphuric acid was 4·75%, the temperature of the mutation was 80°C. When the concentration of sulphuric acid increased to 5·5%, the temperature of the mutation was 60°C. Before the sudden temperature change in the slurry, the gel time showed a trend of increasing first and then decreasing, and reaching a maximum at 40°C.

As can be seen from Figure 2, as the concentration of sulphuric acid became greater, the slurry, having a high sulphuric acid concentration at the same temperature, had a shorter gelation time than the slurry with a lower sulphuric acid concentration. When the sulphuric acid concentration increased from 4·5 to 4·75%, the change in gel time was most pronounced. At 40°C slurry, the sulphuric acid concentration increased from 4·5 to 4·75%; the gel time changed the most, from 1890 to 870 s (the rate of change was 54·0%). When the concentration of sulphuric acid was

*Table 2. Reagents and parameters*

| *Reagent* | *Parameter* |
|---|---|
| Concentrated sulfuric acid | Quality score is 98% |
| Water glass | The modulus is 3·3 and Baume is 40°Bé |

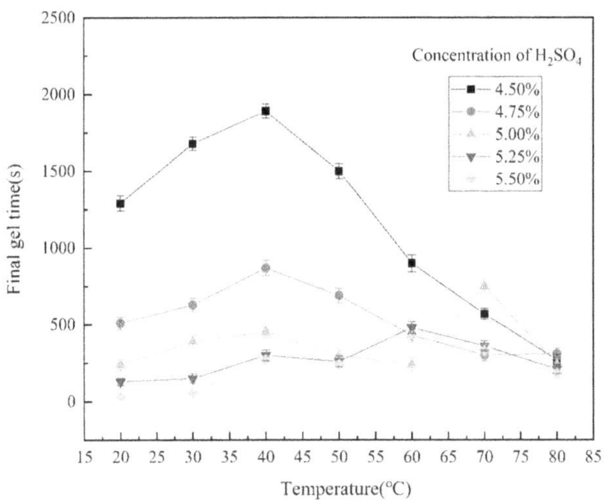


*Figure 1. Curve of slurry gel time with temperature for different sulfuric acid concentrations [Colour available online]*

4·5%, the effect of temperature change on gel time was the most obvious. As the concentration increased, the effect of temperature on gel time began to decrease.

According to Figure 3, as the temperature rises, the viscosity of the slurry declines continuously. Among the 5·25% sulphuric acid concentration group, the viscosity changed the most, from 7·5 to 3 mPa s (the change rate was 60·0%). The viscosity and temperature of the slurry are inversely related.

As shown in Figure 4, as the concentration of sulphuric acid in the slurry increased, the viscosity of the slurry at the same temperature was not significantly affected. The variation of the slurry in the 50°C group is the largest, and the value increased from 3·5 to 4·5 mPa s (the rate was 28·6%). When the slurry temperature was 30°C, the change in sulphuric acid concentration had little effect on the viscosity of the slurry and remained at 6 mPa s. Comparing Figure 4 with Figure 5, it can be seen that the effect of temperature change on the viscosity of the slurry

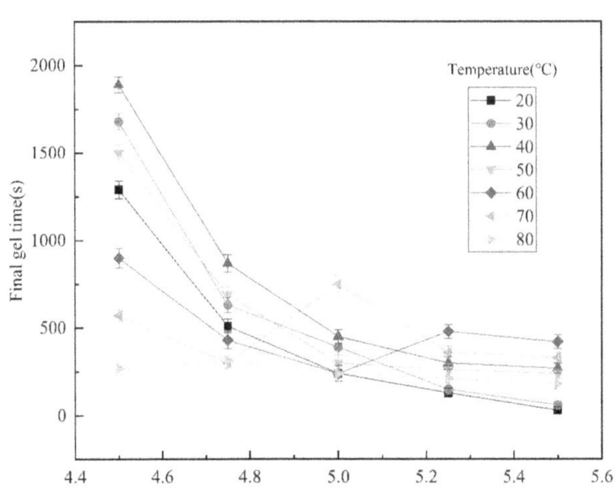


*Figure 2. Curve of gelation time of slurry at different temperatures with sulfuric acid concentrations [Colour available online]*

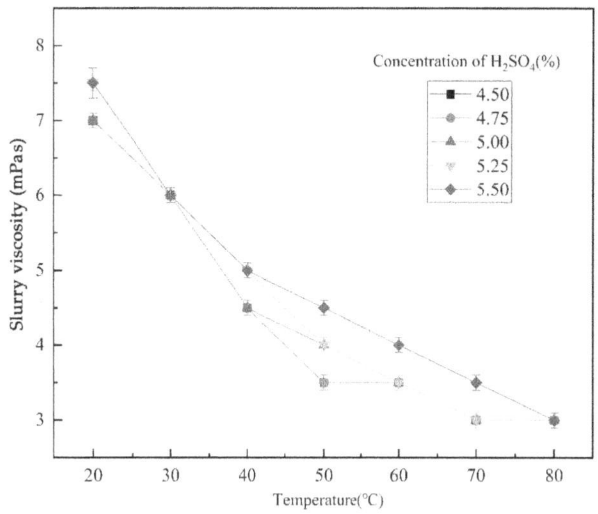


*Figure 3 – Curve of slurry viscosity with temperature for different sulfuric acid concentrations [Colour available online]*

is much greater than the effect of sulphuric acid concentration, so temperature is the main factor affecting the viscosity of the slurry.

From the above, it can be known that during the configuration of the acidic water glass slurry, the influence of temperature on the performance of the slurry must be fully considered, especially when the slurry is configured in the area where the temperature changes regularly, the temperature will affect the actual grouting effect. Through laboratory tests, it was found that the gel time of the acidic water glass slurry increased first and then decreased with increasing temperature. So, when configuring the slurry, the influence of the surrounding environment on the gel time of the slurry should be fully considered, such as formation temperature and atmospheric temperature. These laws provided a reference for the application of the acidic water glass slurry in actual engineering.

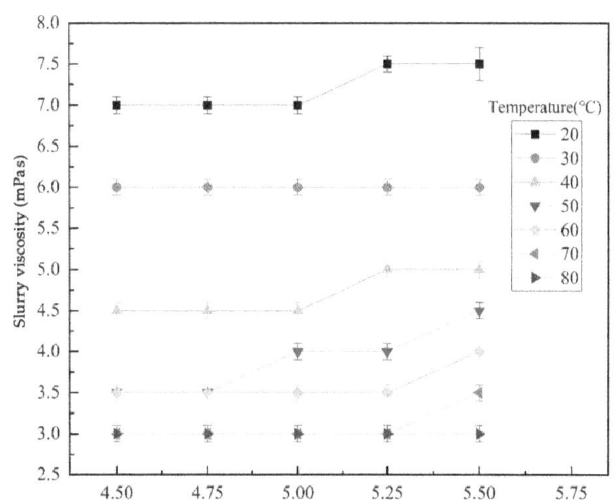


*Figure 4. Curve of slurry viscosity at different temperatures as a function of sulfuric acid concentration [Colour available online]*

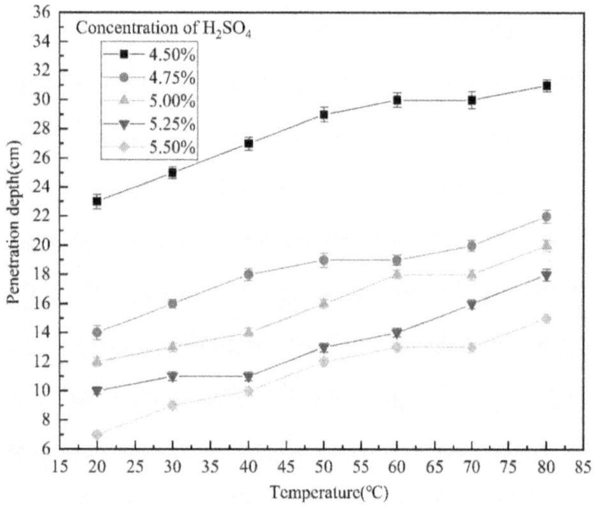


*Figure 5 – Curve of penetration depth with temperature under different sulfuric acid concentrations [Colour available online]*

## 5. Pure slurry gel strength experiment

The measurement of the compressive strength of the pure slurry was carried out using a Vicat device. First, the prepared slurry was poured into the test apparatus. After the slurry was finally set, the apparatus was subjected to a plastic strength test that directly measured the penetration depth $h$ of the iron cone. The larger the $h$ value, the lower the strength of the slurry. Figure 5 shows the penetration depth of the gel as a function of temperature and sulphuric acid concentration.

As the temperature increased, the penetration depth of the gel increased, indicating that the strength of the gel gradually decreased. When the sulphuric acid concentration was increased from 4·5 to 4·75%, the strength of the gel increased most significantly. When the concentration of sulphuric acid is low, the concentration of sulphuric acid is the main factor affecting the strength of the gel; when the concentration of sulphuric acid is more than 4·75%, the effect of temperature change on the gel is more obvious than the effect of sulphuric acid concentration on the strength of the gel.

## 6. Conclusion

By studying the influence of temperature changes on the gelation time and slurry performance of sulphuric acid water glass slurry, the following conclusions were obtained:

Below 40°C, the gel time of the slurry with different sulphuric acid concentration became longer as the temperature increased. Above 40°C, the gel time of the slurry began to decrease. The temperature at which the flocculent gel was instantaneously generated in the slurry after mixing sulphuric acid and water glass was the sudden change point of the slurry. At this temperature, the gel time of the slurry increased from the previous temperature, but as the temperature continued to increase, the slurry gel time began to shorten again.

The viscosity of the slurry with different sulphuric acid concentration decreased with the increase in temperature, which is the main factor affecting the viscosity of the slurry relative to the concentration of sulphuric acid.

At the same temperature, the higher the concentration of sulphuric acid, the shorter the gelation time and the higher the strength of the resulting gel.

When the temperature was raised, the slurry at a higher sulphuric acid concentration first appeared to instantaneously produce a flocculent gel, which had a negative effect on the strength of the gel. Therefore, when the temperature is high, it is recommended that a lower concentration of sulphuric acid is selected to prevent the occurrence of a mutation reaction.

In the subsequent study, the equation between temperature and the gelation time of acidic water glass slurry will be established to predict the behavior of the solution in different situations. And the temperature-modified acidic water glass slurry will be applied in the field grouting test to give stronger guidance to actual engineering.

## Acknowledgements

Zuochun Li, Guihe Wang, and Feng Huang are with the School of Engineering and Technology, China University of Geosciences (Beijing), Beijing 100083, P.R. China. This work was supported by the Research on Key Technologies of Groundwater Control in Deep Subway Construction in Beijing (grant no. 2016-K4-056).

## References

1. Sun, Y. M. & Hua, P. Prospect & current situation of water glass in chemical grouting material. *J. Jilin Water Res.*, 2005, **17** (9), 13–14.
2. Xie, L.Y. & Fu, Q. Grouting concretion technology of fine silty sand layer in Xidan sub-railway station. *J. Selected Records of Anchoring and Grouting Engineering in China*, 1995,158–60.
3. Xu, C. J. Effects of pH value on gelation time of acidic water glass grouting materials. *J. Chem. Mater. Construct.*, 2009, **25** (5), 22–4.
4. Li, X. Characteristics of gelation process of $H_2SO_4$-water glass system. *M. J. Chem. Ind. Eng. (China)*, 2007, **58** (2), 246–50.
5. Xiao, Z. Q. Experimental research on new grouting materials of acidic water glass-calcium carbonate. *J. Rock Soil Mech.*, 2010, **31** (9), 2829–34, 2839.
6. Lv, Q. F. Mechanism of temperature-modification silicification grouted loess. *J. Rock Soil Mech.*, 2013, **34** (5), 1923–8.
7. Wu, G. D. *Research on the Acidic Soluble Glass Material and the Grouting Technology Design*. Master's Thesis, Central South University, Changsha, China, 2008.
8. Hu, Z. F. Research and application of non-alkaline water glass grouting material. *J. Yangtze River Water Res. Edu.*, 1989, (4), 120–34.

# How the United Nations International Year of Glass 2022 arrived and what happens now

*Alicia Duran & John Parker*

Tuesday 18th May 2021 is a date that will be etched forever in our memories; it being the point in time when the United Nations finally approved a resolution promoting 2022 as a ***United Nations International Year of Glass***. This carefully crafted document was submitted by the Spanish Ambassador with 19 co-sponsoring countries. The originators were the International Commission on Glass (ICG), the Community of Glass Associations (CGA) and ICOM-Glass, who built on a concept from the US that we were living in a 'Glass Age'. Its aims were to underline the scientific, economic and cultural roles of glass throughout the world in the context of the UN 2030 goals and to celebrate multiple key anniversaries. Glass supports many vital technologies, facilitates sustainability and a green world; it enriches our lives, yet so often goes unnoticed.

*United Nations building in glass*

The journey began in 2018. Along the way 1500 universities and research centres, societies and associations, museums, artists, educators, manufacturers and supporting companies in 80 countries on five continents added their support. Having successfully negotiated the disruption caused by a pandemic, a draft Resolution outlining our ambitions was created by the Missions of several UN countries during April and passed the UN silent process on 11th May. Our heartfelt thanks go to the Spanish Mission, particularly the Spanish ambassador Agustín Santos Maraver and Ana Alonso, who guided this process through the difficult twists and turns of diplomacy in stressful times. Our thanks also go to all who responded to our vision, especially the many experts and colleagues that collaborated to create splendid videos and documents justifying our project; they were always ready and overflowing with ideas and support.

## The tasks ahead

The task of stimulation, diffusion and coordination of thousands of activities across the planet now begins. Congresses, scientific and technical seminars, industrial fairs and glass schools will coexist with artistic exhibitions, books and technical publications, articles in the press, on the radio and TV and in the social media.

Event planning will rely on grass roots input and a network of volunteers; delegation will be indispensable. Individual efforts will be guided and supported by national and/or regional committees which are currently being set up throughout the world. A list of contacts will be published through our web site (IYOG2022.org). The world-wide web will be used for inputting and accessing advice, sharing best practice and creating lists of planned activities, sortable by geography, date and type.

Among the activities already

agreed are: the printing of dedicated issues of international journals; exhibitions in museums, public and private glass collections; and the preparation of educational materials for universal dissemination. For example, the Spanish Research Council, CSIC, is committed to publishing a celebratory book and organizing exhibitions on: (a) IYOG objectives related to the UN 2030 Agenda and (b) creating a Circular Economy based on recycling and glass containers. English and Spanish versions of exhibition materials will be offered to all supporting countries with translation into other languages also possible.

Many international events are already confirmed. A two-day celebratory Opening Conference is planned in Geneva in February. Several industrial fairs will focus on IYOG beginning with VITRUM (October 2021). First in 2022 will be Mumbai (March), followed by the great China Glass Expo (April), then Glasstec in Düsseldorf (October). All will run satellite events to celebrate glass. The most important scientific event will be the ICG Congress in Berlin, marking the 100th anniversary of the DGG. A conference in Egypt will celebrate the centenary anniversary of the discovery of the tomb of pharaoh Tutankhamun. A US Glass Day in Washington DC, or several congresses will focus on art and museums. The number of proposed activities grows daily.

National annual events are already being re-imagined to integrate IYOG celebrations. Our hope is that local groups will dream big dreams, consider new possibilities and then solidify their ideas of what they can organise, how they will be financed and advertised. Centrally we are working on creating communication channels to share thoughts and examples of best practice, firstly via local committees, then an International Council and finally on into the public domain.

Another task is fundraising, particularly to finance the opening event in Geneva. The IYOG web site will shortly have a tool for accepting smaller donations but if you are able to give more or wish to have your name attached to a specific event please contact the ICG treasurer directly through the same site. Similar fund-raising teams will also be needed to support local events.

IYOG2022 is a dream come true for so many of us in the Glass World, one we scarcely dared to anticipate. We are preparing for the challenges ahead, limited only by our imaginations. But now is the time for you too to act, dream you own dreams, create new contacts, imagine new possibilities, unravel hard problems, and help build a greener, more equal and transparent future.

## UN developmental goals and glass

The UN has set a series of development goals with a completion date of 2030. How does glass help address these targets?

### *GOAL 3: Good health and well-being*

Biocompatible and bioactive glasses have been universally life changing for patients. Specialty glasses can bond broken bones without rejection. Deep, persistent wounds, especially in diabetic patients, are hard to treat but new glasses show a remarkable restorative capacity.

Porous hollow glass microspheres (HGMs) can encapsulate fragile drugs but reject undesired biological agents. Radioactive yttrium-90 delivered in HGMs has treated liver cancer.

Chemically inert glass vials, cartridges, syringes and ampoules prevent interaction with their contents, particularly active formulations. The EpiPen® auto-injector cartridge treats severe allergic reactions - at its heart is chemically strengthened glass; millions have been made. Glass has multiple roles in orthodontics and even toothpaste.

### *GOAL 4: Quality education*

A quality education underpins sustainable development. An inclusive education delivers the tools to create innovative solutions for the world's pressing problems. A well-rounded education offers insights into how society copes with change.

Education is at the heart of the International Commission on Glass (ICG); it links experts in science, technology, art, history and education and in the last decade has organized 20 schools in Europe, China, India, North America and South America. The book *Teaching Glass Better* celebrates these Schools, capturing their development and summarizing content. Sharing staff across schools propagates best practice. An ICG Youth Outreach team arranges events and mentors to attract and retain future talent.

Beyond ICG University Glass Art programs are transforming craftspeople into sculptors. Brazil and India already run educational glass programs for younger students and technicians. ICG actors represent many international teaching organizations - our IYoG goals are to share aspirations,

stimulate fresh ideas and seed new courses.

### GOAL 5: Gender equality

Gender equality is a fundamental human right, a foundation for a peaceful, prosperous and sustainable world. A must for the IYoG, it will be achieved by recruiting top glass-women for plenary and invited talks, as well as CEO and Management Board positions. The IYoG board will propose and support committees chaired by women and, whether organising a congress or a smaller event, a balanced constituency. Educating industry and institutions to manage diversity, making it an engine for innovation and creativity, is the best path to a brighter future. Gender matters. Women are half of the world; they must become half the glass world.

### GOAL 6: Clean water and sanitation

In the last century billions have experienced an unprecedented rise in living standards, but many still live in poverty with little access to clean water. Sufficient fresh water exists but damaging economics or poor infrastructure cause millions to die annually from diseases linked to inadequate supplies, sanitation and hygiene. Similar issues impact adversely on food security, life choices and educational opportunities.

Industrial discharges, excess agrochemicals and domestic waste landfill contaminate surface and groundwater. Glass can mimic current water treatment processes. Porous foam glass or phase separated glass filters can aid sanitation (and purify air, another global issue). Sunlight on coated glass immersed in solutions of organic pollutants can oxidize many into non-toxic products and likewise restore drinking water. Most cost-effective is a combination of porous glass filters with titania-coated glass.

### GOAL 7: Affordable and clean energy

Energy epitomises the opportunities and challenges the world faces. Universal access to energy is crucial to build more sustainable and inclusive communities and in turn entails more efficient generation, renewable energy sources and ways to store it. Sunlight is the main carbon-neutral source and brings more energy hourly to the earth than society consumes annually, But renewable energy is just 17% (2018) of global consumption.

Solar energy harvesting uses: photo-voltaic cells, solar thermal energy generation and photobioreactors. Photovoltaics need glass protective covers which are highly transparent and have antireflection coatings. Solar thermal devices have glass mirrors to reflect the sun's rays and heat a fluid inside a glass tube, which powers a generator. In photobioreactors, microorganisms such as green algae grow in glass tubes, converting solar into chemical energy.

Turbine blades made from fibreglass-reinforced composites convert wind energy to electricity. Improved glass design has created stronger composites and given us larger, more efficient and reliable windmills.

Nuclear power is another low carbon energy source but generates radioactive waste that decays over geological time. New glasses will dissolve and immobilise this waste.

*Siemens Gamesa wind turbines in the Phillipines*

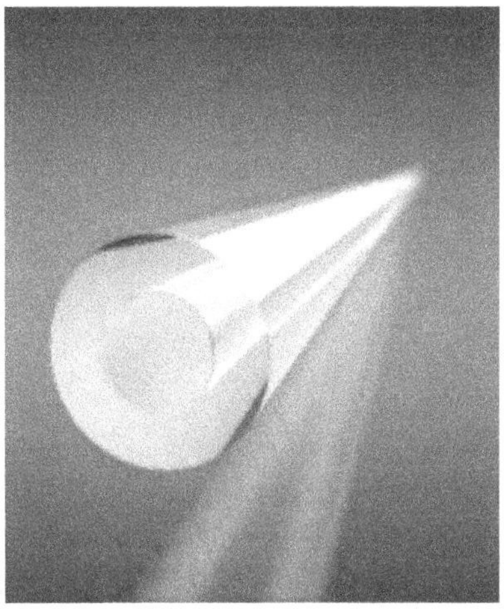

*Glass optical fibre preform*

New glasses are also improving solid-state batteries and hollow glass microspheres may have a role in hydrogen storage for transport.

### GOAL 9: Industry, innovation, infrastructure

Investing in infrastructure such as communication technologies is imperative for sustainable development and increases social cohesion. Low-loss glass optical fibres were the precursor to the Internet and catalysed a paradigm shift in global communications. They are indispensable in our knowledge-based society.

Product development is being driven by the demand to send more data further; on the horizon are photonic crystal fibres (made from glass) and communication via quantumly entangled photons.

But photonics is more than fibres: optical communications industries manage information streams with circuitry fabricated wholly or partly in glass such as spherical lenses, prisms and beam splitters.

Recent advances concern signals such as 5G carried over fibre-optic cables. Fibre lasers exploit rare earth doped glass and are now standard for many applications. Their use in surgery and therapy is expanding and they are furthering studies in nonlinear optics.

LEDs based on light emission

*Schott glass-to-metal seals for medical grade LED light sources can be disinfected in the autoclave*

from crystalline semiconductors, require a phosphor, often a doped glass, to produce white light. One-dimensional photonic structures, which can manipulate light, are made by depositing glass layers with different refractive indices. 2D and 3D photonic crystals require self-assembling glass nanoparticles and are attracting interest as sensors.

Television began with glass cathode ray tubes before moving to glass flat panel displays. As resolution improves and pixel size shrinks, thin film deposition display electronics need panels with exceptionally stable dimensions. Ultra-thin glasses for bendable, even foldable displays are being developed. Integrated optical circuits in films on ultra-thin glasses may soon lead to a breakthrough like that in flexible electronics. Glasses to visualize information through augmented and virtual reality devices offer another revolution.

Glasses have transformed data storage. Rewritable storage is realized by toggling regions between glassy and crystalline states. Magnetic memory disks are high-strength, high-stiffness glasses for faster reading and higher densities. Glass holographic memories promise exceptional capacity.

Sol-gel has been a developing process technology for 50 years. Low temperature and low cost, it works well for coatings and membranes and offers: mechanical and corrosion protection; anti-reflectivity; hydrophobicity; photocatalytic self-cleaning; with optical and optoelectronic functions such as filters, switches, waveguides and integrated optical circuits for solar cells, solid-state lighting and communications.

## *GOAL 11: Sustainable cities & communities*

Cities are hubs for ideas, commerce, culture, science, production, social development and more. At their best, cities have enabled social and economic advances. But by 2030 cities will house 5 billion people and require efficient management. Issues include: congestion, underfunded services, a lack of adequate housing, solid waste management, ageing infrastructure and air pollution.

In transport, glazing allows unimpaired vision and contributes to safety and security, as well as style. So, airplane cockpit windshields are chemically strengthened. Innovative designs offer thermal comfort; improve fuel efficiency by light-weighting; and integrate display features which expand entertainment and connectivity options.

Contemporary architectural designs use larger windows with more energy-efficient coated panes and new double/triple glazing formats. Buildings may soon be energy-neutral or even contribute to the energy grid.

A TNO study showed that new glazing could reduce energy consumption across the EU by 30% in 2030, saving $CO_2$ emissions of 94 million tons. Savings are potentially greater using glazing-integrated photovoltaics, switchable/electrochromic glazing and other novel technologies. The EU aims by 2050 to have the first climate neutral economy.

Glass containers lessen solid urban waste (SUW). The *"Reduce, Reuse, Recycle"* philosophy is vital to a sustainable lifestyle and waste management. 60% of SUW is packaging, mostly single use, often non-renewable or scarce materials. '*Reduce*' means using durable goods such as glass kitchenware. Some hotels and restaurants use returnable glass bottles for '*Reuse*' and glass jars are often *reused* for storage. Supermarkets though prefer single-use containers leading to the third R, '*recycling*', the transformation of an object into a raw material and back. Glass is the only truly recyclable container: one glass bottle produces another, a perfect "circular economy" expressed as "from cradle to cradle". Glass containers are the only ones with the GRASS and Food Safety stamp in Europe and US.

Culture is essential to city life, enhancing the well-being of its inhabitants. Museums have a strategic role, preserving the heritage of humanity for scholarship and enjoyment. In museums with glass collections art, science, archaeology, history and social sciences meet. Glass objects spanning history, works of art and glass for everyday use, are displayed with specialized glassware for industry and science. They describe raw materials and production, telling

*Every 10% increase in cullet usage results in up to 3% energy savings*

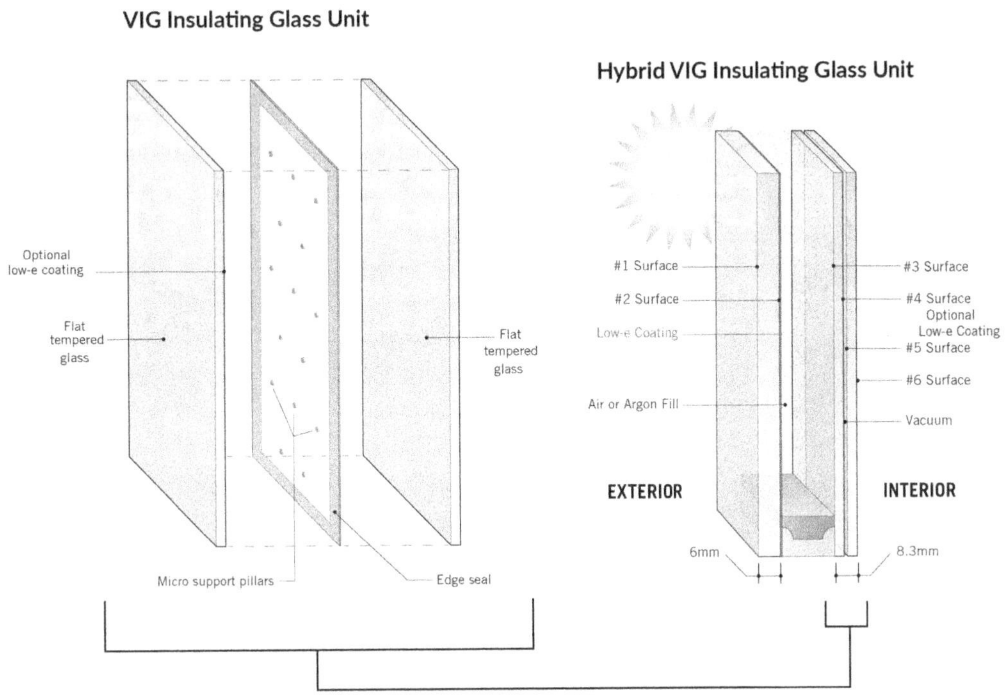


*Adding a vacuum insulated unit to a glazing unit insulates like a wall with the transpancy of glass*

the stories of those who made and used them. Glass museums, particularly ones with hot and cold workshops, promote lifelong learning through programmes for young and old (Goal 4).

## GOAL 12: Responsible consumption & production

Sustainability means "doing more and better with less". It concerns efficiency throughout the life cycle; infrastructure, supply chains and basic services; green and decent jobs; and a better quality of life for all.

Glass firms are addressing these environmental challenges and ICG organizes international benchmarking exercises to stimulate improvement. Glass is innately environmentally friendly. Most glasses are made from safe, readily available raw materials. Recycling rates are high and industry actively seeks more energy efficient melting technologies and alternative glass compositions to reduce its carbon footprint. It has a rich history of responding to challenge, a valuable example for the future.

Educating consumers on sustainable lifestyles depends on distributing information through standards, labels and advertising. The concept of a "circular economy" needs to be understood so all can commit to the challenges of global change; ways to maintain lifestyle without damaging the planet need explaining. Many organizations already do this and an IYoG can help to disseminate best practice.

## GOAL 13: Climate Action

Climate change is affecting every country on every continent, disrupting national economies and individual lives, costing people, communities and even countries.

Glass recycling and product lightweighting have increased the energy efficiency of glass melting substantially; affordable pathways to decarbonization are under development. Glass windows let light into homes and offices while protecting from harsh weather. Double-glazed units increasingly have vacuum insulated glazing, a more effective new technology. Laminated glasses improve acoustic damping to reduce "noise pollution". Architects are using more glass for its functionalities and appealing aesthetics.

Energy saving glass products compensate during their service life several times over for the energy used to make them. So, replacing single with double-glazed windows can give a 5 month energy payback.

## GOAL 14: Life below water

The world's oceans and rivers – their temperature, chemistry, currents and life – drive global systems that make Earth habitable. Careful management is vital for a sustainable future but plastic waste is polluting them. Marine animals ingest it causing health issues for humans consuming seafood: immune disorders, birth defects, some cancers. Glass packaging is infinitely recyclable and a safe, clean alternative made of natural, plentiful materials.

## GOAL 17: Partnerships to achieve goals

A successful sustainable development agenda requires partnerships between governments, the private sector and civil society built on principles and values, a shared vision and goals with people and the planet at the center; partnerships are needed at global, regional, national and local levels. An IYoG will underline the varied roles of glass and stimulate, mobilize and redirect such partnerships to unlock their resources and deliver sustainable development.

A £30m investment that will secure the long-term high quality glass supply in the UK is being made at **Saint-Gobain**'s glass manufacturing factory in Eggborough, East Yorkshire.

This is a major milestone in the company's history and represents the largest industrial investment in the UK by the Saint-Gobain Group since the Eggborough plant was built in 2000. It is also one of the largest single investments made in manufacturing capability in the UK in the glass and window industries in 2021.

The complex engineering and construction programme will see the complete re-build and installation of a state-of the-art float glass furnace. The upgrade will provide greater automation and enhanced control systems, with the use of the very latest equipment and integrated digital technologies, while minimising both the energy used for glass production and emissions from the process. The enlarged capacity furnace will replace the existing furnace which has reached the end of its design life, after manufacturing a 'ribbon of glass' long enough to stretch the length of the UK from Lands End to John O'Groats more than 130 times.

Due to be completed by late summer, the investment programme also includes upgrades to other areas of the factory. The project will result in securing the long term supply of high performance glass that meets the stringent quality requirements of customers in the UK.

The project has been several years in development and planning to ensure continuity of supply for customers throughout the construction process. This has involved a high level of stock build-up in advance and imports of glass from Saint-Gobain's sister plants in Europe which will help to fulfil orders throughout the project.

In addition, the glass coating line and lamination line at Eggborough are continuing to operate as normal to produce the market leading PLANITHERM® range of products.

Supply of solar controlled glass such as the SKN and XTREME ranges, as well as patterned glass, mirrored products, self-cleaning glass, tinted products, and lower-iron glass are also unaffected as these are already manufactured at Saint-Gobain Group plants across Europe.

The new furnace will use less energy and will help the company to optimise manufacturing efficiency, increase production capacity, ensure consistent high quality glass manufacturing and further enhance Saint-Gobain's already market-leading environmental and sustainability credentials.

This will contribute to the Saint-Gobain Group's commitment to achieving 'net-zero carbon' of its global business operations, and deliver its global strategic purpose which the Group has recently defined as 'Making the World a Better Home', in recognition of the need to protect the planet and invent a more sustainable future.

The new facility is strategically important to support and serve the UK window industry with domestically manufactured high-performance products developed specifically to meet the needs of the UK window industry. It demonstrates Saint-Gobain's commitment to the long term future of glass manufacturing in the UK and confidence in the wider construction industry.



The German solution provider **4ET** has installed several new versions of its PEARL laser glass cutting systems for international customers. The innovations go beyond the current typical scope of laser glass cutting which is mostly used in the manufacturing of small to medium sized display cover glasses as used in consumer and automotive electronics. 4JET's next generation solution enables new use cases:

The new PEARL M-UTG is used for cutting of ultra-thin flexible glass. The system is integrated in a fully automated processing line and enables to cut glass foils as thin as 30 μm with excellent breaking strength at highest yield. Potential applications for flexible glass in growth markets such as consumer electronics, medical or semiconductor applications are vast.

Several PEARL systems have also been installed recently to cut very large glass sheets or glass stacks. The Pearl series is capable of cutting glass with a size of up to 2·2 m width and more than 3 m length. For the first time it is possible to use the advantages of laser glass filamentation on large mother glass sheets for e.g. smart windows.

The PEARL systems are based on a granite machine platform, which offers inherent accuracy and long-term stability. Cutting geometries are easily imported from CAD drawings. The specific tuning of the axis controllers not only result in a high cutting speed, but also in excellent path accuracy to achieve high-precision outer contours in the end product. The PEARL DYNAMIC FOCUS ensures that the focus is always within the process window without increasing the process time.

**Prince International**, a portfolio company of American Securities, has acquired **Ferro** in a transaction valued at approximately $2.1 billion, including the assumption of debt, net of cash. The agreement was been unanimously approved by Ferro's Board of Directors.

In conjunction with the closing of the transaction, the company will combine with **Chromaflo Technologies**, a provider of colorant technology solutions and also an American Securities portfolio company. The combination of Prince, Ferro and Chromaflo will result in a global diversified business with sales of approximately $2 billion.

"We are thrilled about combining Prince, Ferro, and Chromaflo, each a leader in their respective markets, to form a true global technology leader in colour solutions, functional coatings and specialty minerals," said D. Michael Wilson, Prince CEO. "The combination will result in a company with unmatched competencies in particle engineering, glass science and color technology. It will enhance our ability, on a global scale, to deliver state-of-the art products and innovative solutions that enrich everyday life."

Scottish engineering company **Woollard & Henry** has acquired the assets of the **Lipex Engineering** from **Eniclay Group** and formed a new company called **WH LIPEX**. This company as a part of Woollard & Henry will continue to serve international glass and basalt fibre industry.

**LIPEX Anlagentechnik und Handel** was originally founded in 1975 and has been a preferred supplier of machinery, equipment and technology for the production of industrial glass fibres. LIPEX is a specialized engineering service provider that conceptually plans both single production lines and complete manufacturing facilities or plants, also as general contractor for the production of industrial glass fibres and creates the necessary designs and provides technology. LIPEX has also strong experience in glass and sizing chemistry for both basalt and glass fibres. WH LIPEX will continue to provide its worldwide customers high quality service and technology in the field of glass fibre industry.

Woollard & Henry's acquisition of LIPEX, follows the acquisition of a major shareholding in the German machine builders, Pama. The collaboration and new group structure will allow better access to product development and manufacturing facilities enabling Lipex

*Schott Xensation Flex cover glass is one of the Society for Information Display (SID) winners of its 27th Annual Display Industry Awards. The 2021 Award winners embody the profound level of innovation within the industry and exemplify how crucial displays have become to nearly every aspect of daily life.*

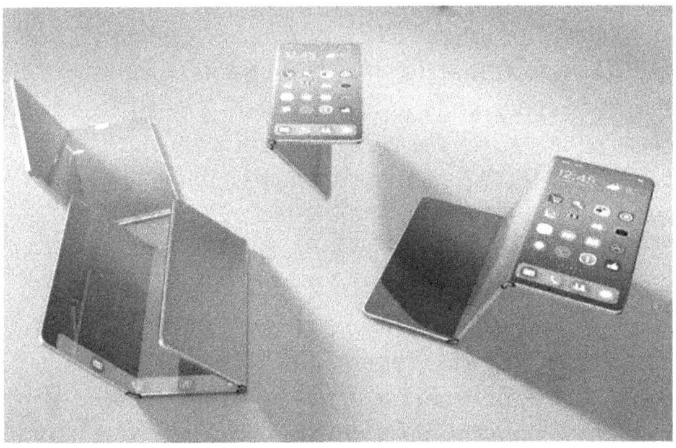

*Xensation Flex cover glass is the first flexible glass that chemically can be toughened for extremely high strength and flexibility, with a bending radius below 2 mm after processing. One of the main challenges Schott faced was adjusting the production process to make aluminosilicate glass types, such as Xensation Flex, extremely thin. They did so by using a proprietary production process called down-draw. During down-draw, a glass ribbon is pulled across various rollers and through the cooling track so that glass of a desired thinness is drawn directly from the melt to achieve stable thinness. The company also developed special handling methods for this fragile material glass. This enabled it to make ultrathin glass suitable for mass production, allowing their clients to develop superior new devices based on ultrathin glass. All told, it points to a better future for foldables and a better user experience for consumers.*

customers to take advantage of the greater synergies while improving their business, which is one of the core values of the new company.

Woollard & Henry is Scottish based employee owned company founded in 1873. Main business sectors for W&H are paper and energy industry and with the acquisition of LIPEX allows W&H to further diverse it's activities into glass and basalt fibres. With strong experience in specialty paper engineering W&H can help LIPEX improve their nonwoven tissue and chopped strand mat technology for the future.

**Samsung Display** has been an investor in **Corning** since 2014, when it received preferred shares convertible after seven years and equivalent to an approximately 7·5% ownership stake as part of a series of strategic and financial agreements that resulted in Corning acquiring full ownership of Samsung Corning Precision Materials. The agreements benefitted both companies' shareholders. Financial and strategic benefits to Corning included full control of its global fusion-glass manufacturing platform, greater flexibility in asset use, greater operational efficiencies, and increased options to pursue precision-glass opportunities.

Following the seven-year anniversary of the agreements, and with the Corning common dividend exceeding the preferred dividend, Samsung Display will convert all of its preferred shares to common, which represents approximately 13% of Corning's common shares on an as-converted basis.

"Corning is a great partner and we are pleased to have an opportunity to strengthen and extend our nearly 50-year relationship," Joo-Sun Choi, president and chief executive officer of Samsung Display, said. "Our latest seven-year commitment reaffirms Samsung's confidence in the value of Corning's capabilities, our ongoing technology collaborations, and our combined innovation leadership."

Based on Samsung Display's

*A new type of unique BIPV façade was installed at **Pilkington Austria** office building in Bischofshofen in cooperation with **Techno-Z**. The BIPV glass panels cover the entire vertical and south-facing (SE 160°) façade surface, both window and spandrel areas.*

*The product Pilkington Sunplus™ BIPV Vision, which is partially covered with PV cells and still offers a good view to the outside, is suitable for the window areas. Sunplus BIPV Spandrel was installed in the spandrel area. It is a product similar to a façade spandrel panel, which is opaque, but it is fully covered with PV cells.*

*The total façade area at Pilkington Austria is 66·5 m², consisting of 30·4 m² Vision glass and 36·1 m² Spandrel glass. In total, the façade has a maximum output of 7·31 kWp, where 2·30 kWp comes from the window glazing and 5·01 kWp from the spandrel area.*

*Photography: www.johannesfelsch.com*

continued confidence in Corning, and the strong performance of its original investment, the two companies will enter into a new agreement for another seven years.

Samsung Display will convert all of its preferred shares to 115 million common shares.

Corning will repurchase 35 million of those common shares, resulting in Samsung Display having an ownership stake of approximately 9%.

The initial repurchase of common shares is expected to close in April 2021 and will immediately reduce Corning's fully diluted share count by 35 million shares, or about 4%.

Samsung Display will maintain its ownership stake in Corning until at least 2028.

"From the early days of CRT, to LCD, to more recent collaborations on OLED, QD Display, and flexible displays, we're proud to be the key material innovation partner to Samsung," said Wendell P. Weeks, Corning's chairman and CEO. "We are honoured by Samsung Display's vote of confidence to remain a significant shareholder for another seven years."

**Borosil Renewables**, the Indian manufacturer of solar panel glass, has announced plans to double furnace capacity to 900 tonnes/day with an investment of Rs 500 crore. The renewable venture of the Borosil Group currently has a 450 tonne per day solar panel glass manufacturing capacity at its Baruch plant in Gujarat, which is enough to power 2·5 GW of solar power plants.

With the brownfield expansion at the Barauch plant in Gujarat, the capacity will jump to 900 tonnes per day or 5 GW of the installable capacity of solar power plants. The new plant, at an investment of Rs 500 crore, should be up and running by July 2022, and this is the second doubling of its capacity in the past five years, Shreevar Kheruka, the managing director of Borosil Group was quoted by local reporters.

The **HEINZ-GLAS Group** is one of the leading manufacturers of glass flacons and caps for the perfume and cosmetics industry. The company produces attractive, individually designed flacons and jars made of crystal-white transparent glass, white opal glass and feeder-coloured glass.

In February, the glass melting technology specialist **HORN** received an order to repair furnace 7 at the Piesau plant in Germany. The order comprised a 49 m² end fired furnace with six forehearths for the production of flacons. In the future, the plant will produce 135 tonnes of flint glass per day.

Buildings are responsible for 40% of primary energy consumption and 36% of total $CO_2$ emissions. And, as we know, $CO_2$ emissions trigger global warming, sea level rise, and profound changes in ocean ecosystems. Substituting the inefficient glazing areas of buildings with energy efficient smart glazing windows has great potential to decrease energy consumption for lighting and temperature control.

*SEM micrograph of vertically standing, flat micromirror array with an inset of magnified area. Credit: Hillmer et al*

Harmut Hillmer and colleagues of the **University of Kassel** in Germany have demonstrated the potential of "MOEMS micromirror arrays in smart windows for daylight steering" in a paper published recently in the inaugural issue of the *Journal of Optical Microsystems.*

"Our smart glazing is based on millions of micromirrors, invisible to the bare eye, and reflects incoming sunlight according to user actions, sun positions, daytime, and seasons, providing a personalized light steering inside the building," Hillmer said.

The micromirror array is invulnerable to wind, window cleaning, or any weather conditions because it is located in the space between the windowpanes filled with noble gas such as argon or krypton. The glazing provides free solar heat in winter and overheating prevention in summer, and it enables healthy natural daylight, huge energy savings (up to 35%), massive $CO_2$ reduction (up to 30%), and a reduction of 10% steel and concrete in high-rise buildings. The smart glass can also reduce reliance on artificial lighting by optimizing natural daylight in a room.

The researchers' MEMS micromirror arrays are integrated inside insulation glazing and are operated by an electronic control system. The orientation of mirrors is controlled by the voltage between respective electrodes. Motion sensors in the room detect the number, position, and movement of users in the room.

The results include much higher actuation speed in the sub-ms range, 40-times lower power consumption than electrochromic or liquid crystal concepts, reflection instead of absorption, and colour neutrality. Rapid ageing tests of the micromirror structure were performed to study reliability and revealed sustainability, robustness, and long lifetimes of the micromirror arrays.

Full open access research article: Harmut Hillmer, Mustaqim Siddi Que Iskhandar, Muhammad Kamrul Hasan, Sapida Akhundzada, Basim Al-Qargholi, Andreas Tatzel, MOEMS micromirror arrays in smart windows for daylight steering, *J. Optical Microsyst.* 2021, **1**(1), 014502. DOI: 10.1117/1.JOM.1.1.014502

---

The scope of supply consists of the design of the melting tank, distributor and the forehearths as well as the steel structure. Furthermore, HORN will supply the equipment for heating, measuring and control, boosting, over the reversing system up to the batch charger HVR® 400 as well as further furnace equipment.

A collaboration between academia and industry is to focus on the recycling of glass fibre products, in a move that could eventually help to reduce the waste produced by wind turbine blades. The **University of Strathclyde** has signed a memorandum of understanding with **Aker Offshore Wind** and **Aker Horizons**. Among other things, the trio will work together to scale-up and commercialise a process developed in the laboratory which centres around recycling glass-reinforced polymer composites used in wind turbine blades.

According to the university, the system focuses on the "thermal recovery and post-treatment of glass fibres" from glass-reinforced polymer composite scrap, with the end result "near-virgin quality glass fibres." The idea is that, using this system, the composite waste could be re-used.

"This is a challenge not only for the wind power industry, but for all industries reliant on GRP materials in their production and manufacturing," Liu Yang, who is head of the Advanced Composites Group at the University of Strathclyde, said in statement. "Retaining and redeploying the embodied energy in the fibres is essential as we move to a more circular economy."

A new series of independent tests carried out to discover the best way to seal glass jars has put Selig's GlassFuze™ on top. The research was commissioned by a peanut butter producer following complaints from customers relating to product spoilage. A new solution had to be found to seal the jars, as the company's brand reputation was being put at risk.

High oil content products,

*Driven to meet its climate commitments, China continues to focus on its solar glass production. The country became the world's biggest manufacturer of photovoltaic (PV) products aiming to reduce coal usage and increase renewable energy sources.*

*With PV glass demand currently outpacing production capacity, local glass manufacturers have rushed to expand existing capacity and build new installations. In addition, a new market trend has developed to replace traditional backboards of solar modules with another layer of PV glass to extend panel lifetime.*

*Following the burst of the solar glass production,* ***Fives*** *has supplied more than 60 annealing lehrs in China over the last 12 months. A new contract for a 6m width lehr - the widest lehr currently in China and internationally - with a production capacity of 600 tonnes per day has recently been signed by Fives Stein Metallurgical Technology (Shanghai), a Fives' subsidiary in China, with Hebei Yingxin Glass Group. Yingxin is a private glass maker located in Shahe City, Hebei Province, which is known as the "China Glass City".*

*"We have been leading technical innovations in annealing lehrs for automotive glass, ultra-thin and PV glass over the last ten years, which allowed us to have a great amount of the local market supply, including to Xinyi Glass, Kibin and now Yinxin. Increasing the lehr for the glass ribbon width from the current 5.2m to 6m can save energy consumption up to 10%, which leads to a decrease in production costs by up to 5%," says Yun Pan, Managing Director of Fives Stein Metallurgical Technology (Shanghai), a subsidiary of Fives in China.*

such as peanut butter, can be challenging to package as the oil makes it difficult to achieve a 100% hermetic seal between the liner and the jar lid. With this in mind, a packaging engineer tested five sealing methods on glass jars, using a 4-minute -300 mBar pressure test on a laboratory Pack-Vac leak detector.

The sealing solutions tested on 100 g glass jars included a plastic screw cap and an induction seal liner, a plastic screw cap and a foam liner, a metal screw cap and a flowed-in seal and a metal twist cap and a flowed-in seal.

Overall, the screw caps were shown to be superior to the twist caps in terms of leakage prevention, but the leak test results showed that only one solution provided consistent, secure sealing that showed no leakage. This was the jar sealed using Selig's induction heat seal liner with GlassFuze heat seal.

Darren Dodd, marketing and service director at Selig says, "With more than two years of research, we're proud that the capabilities of GlassFuze have received independent endorsement. Oily products are notoriously difficult to seal securely, and this small solution can make a big difference to product reliability and a brand's reputation."

**Nipro PharmaPackaging** has strengthening its position as a leading global supplier of glass primary packaging for the pharmaceutical industry with the successful acquisition of **Piramida**, a leading Croatian manufacturer of glass pharmaceutical packaging.

Under the ownership of **Blue Sea Capital**, Piramida enjoyed a modernisation of its facilities, welcomed eight new production lines, and witnessed a strong development of its ampoules and vials business. Moreover, production capacity as well as the revenue doubled. This led the company being ranked among the top five largest ampoule manufacturers for the pharmaceutical industry in the European market.

The **Furnace for the Future** (F4F) has been selected out of 311 projects to progress to the second phase of the EU Innovation Fund. The Fund provides €10 billion of support over 2020–2030 for the commercial demonstration of innovative low-carbon technologies, aiming to bring industrial solutions to decarbonise Europe to the market, and support its transition to climate neutrality.

The F4F project is a breakthrough technology, which will enable the industry to switch to renewable electricity and cut $CO_2$ emissions by up to 60% in the furnace (50% across the whole factory). Container glass production currently uses a mix of roughly 80% gas and 20% electricity. By inverting this to have a mix of roughly 20% gas and 80% renewable electricity, the carbon footprint of glass packaging will be dramatically reduced.

"We are honoured to be among the 70 projects considered by the European Commission as strategic to a climate-neutral Circular Economy," said John Sadlier, Chief Sustainability Officer at Ardagh Group, "The Furnace for the Future represents a key step towards a sustainable future for glass packaging. Decarbonisation is a key priority for our business, but it is also critical for our customers in achieving their sustainability goals."

Electric melting already exists but is limited to small-scale furnaces for clear glass with limited recycled glass content. The F4F will make electric melting work in a large-scale, industrial hybrid oxy-fuel furnace, running on 80%

renewable electricity, for all glass colours, using high amounts of recycled glass.

The F4F project is technically and financially supported by 19 container glass companies who have signed a binding agreement and together represent over 90% of production in Europe. Ardagh Group has been selected to build the furnace at their Obernkirchen site in Germany and will make the application as the beneficiary, strongly supported by the industry-wide consortium.

Each of the industry partners will co-fund the project and in return will have access to the new technology and know-how. Working as a sector has many advantages for the subsequent roll out and scalability of the technology.

"We are proud to be taking the next step forward in our industry transition to a resource-efficient and low-carbon economy," remarked Martin Petersson, CEO Glass Europe at Ardagh Group. "We welcome the support from our partners, our customers and all industry stakeholders in making this important decarbonisation project a reality."

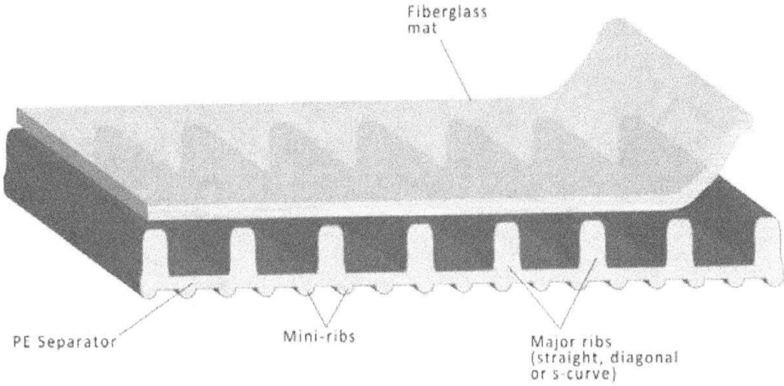


*A glass fibre mat makes up an essential component of modern lead/acid batteries, helping reduce weight, NSG and ENTEK tie up*

**Nippon Sheet Glass** and **ENTEK Technology Holdings** have reached an agreement on the acquisition of a majority stake in NSG's battery separator business. Both companies have executed the relevant agreements and plan to complete the transfer of BSS in August 2021. NSG Group will continue to cooperate with ENTEK in managing and developing battery separator technology by holding a minority stake in and appointing a director to the board of ENTEK's subsidiary in Japan as a commitment to retain and grow trust from the customers.

NSG Group has been supplying battery separators for lead-acid batteries mainly in Asia, including Japan, for decades with manufacturing footprints in Japan and China. ENTEK is a global leader with over 35 years of history in the battery separator business, supplying separators for both lead-acid and lithium batteries to customers globally. ENTEK and NSG Group are partners in a battery separator joint venture in Indonesia and the companies have established a good relationship over the years.

The new company will own all battery separator manufacturing facilities in Japan and China for both polyethylene (PE) separator and absorbent glass mat (AGM). This acquisition combines ENTEK's core competencies in material science and equipment engineering and design along with capital to fund future growth with NSG Group's leadership, customer focus and commitment to quality.

With this strategic acquisition, ENTEK becomes the only separator manufacturer in the world producing all three primary separator technologies (PE, AGM, and lithium separators) for the global energy storage industry.

ENTEK's commitment to innovation and technology has availed the company with unique opportunities to support our customers' growth globally while serving them with regional supply and service. Already recognized as the global market leader for PE separators in the automotive battery industry, this investment provides ENTEK with both capacity and expanded product offerings to help meet broader customer demands as well as enabling entry into new markets.

As joint venture partners in the ENTEK NSG Separindo PE manufacturing plant in Indonesia, NSG Group and ENTEK benefitted from the synergies of combined cultures and the strengths of each company to make that investment a success. This deepening partnership builds on those strengths to bring more products, more innovation and more combined capacity to market for our customers.

"This investment is about taking care of customers and creating additional value by leveraging the strengths of two outstanding companies. It underscores our commitment to the future of the global energy storage industry" said Larry Keith, CEO of ENTEK. "We are coming out of the global pandemic stronger than ever, executing on our strategy of growth, excellent customer service, and quality product expansion globally."

ENTEK's previously announced expansion to double capacity at the Indonesia site continues, with new buildings already erected and equipment installation in progress. ENTEK's acquisition of a majority stake in BSS expands our manufacturing footprint and technical offerings in the region providing an industry-leading service and product platform for customers throughout Asia.

NSG Group will continue to contribute in developing BSS by its minority shareholding and representation at the board of directors.

The **Glass Packaging Institute** (GPI) has released an important resource for municipal recycling leaders, "Questions Municipal Recycling Leaders Should Ask About Glass Recycling." The flyer educates local governments on glass recycling options and helps determine the best direction for satisfying residents' expectations to recycle glass. It also includes

***HEINZ-GLAS Group*** *have commissioned a new glass melting furnace at its Kleintettau site. The project involved investments of around €15 million in a new furnace capable of reacting flexibly to a wide variety of market and customer requirements.*

*The latest generation electric furnace can produce high-quality perfumery glass, clear glass with a high PCR cullet content as well as opal glass. For more than 15 years, the opal glass in the group of companies was produced exclusively at the Dzialdowo, Poland site. The fact that this is now also possible again at one of the German locations opens up further flexible possibilities for HEINZ-GLAS Group to serve its customers.*

*"Technical progress in particular helps to overcome crises and create entrepreneurial perspectives. A certain amount of courage to try something new and then the necessary bit of luck are of course also part of it." says owner and CEO Carletta Heinz.*

*As early as 1971, HEINZ-GLAS was the first company in the glass industry to put an electrically heated melting tank into operation at its Kleintettau location. At the time, the company's competitors, who continued to fire the furnace with gas, were critical, but this investment paid off and, on paper, brought energy savings of around 50% compared to the gas furnace. Together with electricity from renewable sources and the new, highly flexible furnace technology, almost 50 years later, the company has become one of the most sustainable glass manufacturers in the world.*

*The investment secures more than 120 jobs at HEINZ-GLAS and about the same number of jobs at suppliers in the region.*

---

links to support research and urges early collaboration with glass end-markets that are often closer than local leaders are led to believe.

GPI's outreach to local leaders builds off the recently announced initiative, "A Circular Future for Glass," a ten-year plan to raise the U.S.'s glass recycling rate to 50 percent and boost the average use of recycled content in domestically made glass containers.

"Glass bottles and jars are 100% and endlessly recyclable, and there are many options for municipalities to keep glass in their programs," said GPI President Scott DeFife. "Before any decisions are made that will impact the community, it is important for local leaders to understand the variety of avenues and options for glass recycling that may work for their residents. This new flyer is the first of many resources that GPI will provide to municipal leaders to clarify the misinformation and limited options frequently presented to them during contract negotiations, which leave the impression that eliminating glass is the only way to sustain their recycling programs."

GPI looks forward to working closely with communities across the country to improve glass recycling systems. You can explore other case studies, funding opportunities, and end markets by exploring GPI's partner sites, the **Glass Recycling Coalition** (GRC) and the **Glass Recycling Foundation** (GRF).

**Ahlstrom-Munksjö** is investing €58 million in a new glass fibre tissue production line in the US to support the flooring industry in North America. The new line will take in raw glass fibre supplies to produce a full range of glass fibre tissue nonwovens with a main focus on luxury vinyl tiles and vinyl sheet materials. Customer deliveries from the new line are expected to start in mid-2023.

"The investment in this new state-of-the-art foam forming glass fibre in the US is a major step in Ahlstrom-Munksjö's global growth strategy in glass fibre tissue market. In combination with our existing plants in Finland and Russia, this new plant will further strengthen our leadership in this field by consolidating our global position in flooring applications but also by giving us a unique platform to expand into other glass fibre tissue applications," said Hans Sohlström, President &

CEO of Ahlstrom-Munksjö.

The new line will be installed on the Madisonville site, Kentucky, where Ahlstrom-Munksjö already has an established plant developing and producing filtration materials for transportation and industrial applications. The necessary land area has been granted by the local authorities in Kentucky.

"I want to thank Ahlstrom-Munksjö for this significant commitment to Madisonville, Hopkins County and the commonwealth. This facility has long been an important part of the community, and the company's continued growth is incredible news for residents throughout the region. The creation of more than 50 high quality jobs promises a brighter future for families in the area, and positions the company for even more success in the years ahead," comments Kentucky Governor Andy Beshear.

The North American demand for flooring materials is estimated to grow annually by 5%, and specifically by 11% for luxury vinyl tile. "We will use Ahlstrom-Munksjö's extensive know-how and experience in glass fibre tissue manufacturing in responding to the North American flooring market needs. This investment will provide us additional capacity and allow us to better serve our customers in this region from a local site with high quality and cost competitive glass fibre tissue materials," adds Pekka Helynranta, Vice President, Glass Fiber Tissue business. Nonwoven glass fibre tissue also finds uses in reinforced plastics, wall liners, plasterboard and acoustic ceilings.

**RATH**'s Competence Centre for Hot Gas Filtration in Meißen, Germany was recently equipped with further advanced test facilities and a second production line. In future, it will be possible to produce up to 80,000 ceramic hot gas filter elements annually at the research and production site.

For many years, RATH has been developing innovative ceramic hot gas filter elements which make it possible to filter particle-laden industrial flue gases at temperatures up to 1000°C and to reduce emissions of nitrogen oxides. Since 2016, all resources, capabilities and know-how in hot gas filtration have been brought together in Meißen and this new RATH research and production site is home to cutting-edge facilities for the manufacture of FILTRATH hot gas filter cartridges and FILTRATH CAT catalytic filter elements. The latter were developed for pollutant emission control: These rigid yet highly porous and catalytically coated ceramic filter elements are used for multiple pollutant control of particulates, acid gases, dioxins and nitrogen oxides in gas temperatures of 250–420°C and can easily replace conventional textile filters.

*RATH ceramic hot gas filter cartridges operate at up to 1000°C, the catalytic filter version for NOX at up to 420°C*

With around 35 employees, the Hot Gas Filtration Competence Centre is responsible for all research and development, including the laboratory and technical centre, as well as for quality assurance. Additionally, the technical department, which deals mainly with the design of filter systems

*The Engineering Heartspace at the **University of Sheffield** is a central atrium created under a spectacular curved roof linking two of the oldest buildings at the university. White steel 'tree' columns support the approximately 1400 $m^2$ roof, which is made up of interlocking triangular glass and spans the space with sweeping waves. It forms a new entrance to Portobello Street, connecting the faculty with other facilities on campus. The roof not only creates a clear distinction between old and new, its high-tech construction also improves the energy and $CO_2$ balance of the renovated existing buildings by enclosing their interior façades.*

*The more than 900 glazing panels of the roof with white serigraphy of small dots were supplied by **Saint-Gobain Glassolutions**. The Climaplus Cool-Lite Xtreme 50/22 II solar control glass combines functionality and elegance in equal measure. Its properties guarantee ideal solar control and excellent insulation to gain maximum comfort. The high-performance glass contributes significantly to lower energy consumption and provides at the same time an extremely bright and luminous atmosphere under the glass roof at all times of the year. The main entrance used 600 $m^2$ of the high selectivity solar control glass Cool-Lite Xtreme 60/28 II.*

*The eye-catching project made it to the finalist stage of the British Construction Industry Award 2020 in the category Social Infrastructure Project of the Year. (John Kees Photography)*

## Glass blowing becomes expression of Spanish culture

The Council of Ministers, at the proposal of the Minister of Culture and Sports, has approved the Royal Decree declaring the technique of blown glass in Spain as a Representative Manifestation of Intangible Cultural Heritage. For the processing of the Royal Decree, the reports of the Historical Heritage Council and the CSIC have been collected as consultative institutions on the matter.

In the content of this Royal Decree, reference is made to the description of this technique, establishing that its processes, knowledge, products, instruments and associated machinery, as well as architectural spaces present a set of historical, immaterial, technological and artistic values that deserve preservation. , as it is an activity linked to Spanish culture, with large production centers dating back more than three centuries, such as the National Glass Center of the Farm in Segovia and the Gordiola Glass in Mallorca.

In addition, individual experiences with this type of technique coexist throughout the Spanish territory, in the form of small artisan workshops, in many cases linked to museums, which seek to revitalize and give visibility to characteristic productions of centers that are already extinct. At the same time, the variety of production, which currently maintains historical models, along with new designs and a great variety of types, is of great relevance in terms of heritage recognition.

Despite this historical and sociocultural importance, at present the viability of this practice is threatened by several factors such as the competitiveness of the industrial production of glass and other materials compared to this traditional technique. In addition, there is a lack of interest from the new generations, which causes a dangerous situation due to the decrease in the number of holders of knowledge.

### *Brief history of the technique of blown glass in Spain*

The blown glass technique can be traced in the peninsula from Antiquity, coming from the Eastern Mediterranean, where it is usually dated to the middle of the 1st century BC. The exact date of introduction of blown glass in Spain is unknown, but Pliny gives an account of the manufacture of glass in Hispania around the third quarter of the 1st century AD.

Glass blowing was seen in the Islamic art of the peninsula, flourishing workshops that will have their continuation and influence in the Modern Age of Christian context, mainly highlighting the workshops of the Crown of Aragon. Throughout the entire geography, decorative application and filigree techniques will be perfected, as well as types of pieces that will be consecrated with the passage of time.

---

and the entire process technology, is also located in Meißen.

"Where R&D is concerned, there are strong synergy effects between the site in Meißen and other production sites that produce and develop high-temperature insulation wool and vacuum-formed parts," says Manfred Salinger, Director Advanced Ceramic Filtration Solutions. In recent years, RATH's Meißen site has developed into an important hub for hot gas filtration: "It's at our competence center that the entire filtration expertise is bundled and the further development of filter elements is initiated. RATH specialists from a wide variety of fields around the world draw on the knowledge that's accumulated here."

In Meißen, great emphasis continues to be placed on customer consulting, says Mr Salinger, "Because for many prospective customers, there's a need for clarification about operating parameters and the process optimization possibilities which hot gas filtration opens up for them. From time to time, filtration tests are also required to determine the appropriate parameters."

**Apple** is awarding $45 million from its Advanced Manufacturing Fund to **Corning**. The funding will expand Corning's manufacturing capacity in the US and drive research and development into innovative new technologies that support durability and long-lasting product life.

Corning has already received $450 million from Apple's $5 billion Advanced Manufacturing Fund over the last four years. Apple's investment helps support more than 1,000 jobs across Corning's US operations in Kentucky and other facilities. The investment has also helped facilitate research and development into state-of-the-art glass processes, which led to the creation of Ceramic Shield, a new material that is tougher than any smartphone glass.

# CONFERENCE DIARY

**19–21 July**
NGA Glass Conference, Chicago, USA.
*National Glass Association NGA, 8200 Greensboro Drive, Suite 302, VA 22102 McLean, USA. Email attend@glass.org Web www.glass.org*

**1–6 August**
23rd International Conference on Composite Materials (ICCM23), Belfast, Northern Ireland.
*Prof Brian G. Falzon, School of Mechanical and Aerospace Engineering, Queen's University Belfast, Ashby Building, Stranmillis Road, Belfast BT9 5AH, Northern Ireland. Email b.falzon@qub.ac.uk Web iccm23.org/*

**4–8 August**
World Glasstech-MEXICO, Guadalajara, Mexico
*Glasstech Mexico, Av. Santa Fe 170, 01210 Lomas de Santa Fe, Mexico. Email info@glasstechmexico.com Web www.glasstechmexico.com*

**9–13 August** (online only)
Nonlinear Optics, USA.
*The Optical Society, 2010 Massachusetts Ave, NW, Washington, DC 20036 USA. Email custserv@osa.org Web www.osa.org*

***16–20 August* (postponed to 2023)**
Tenth International Conference on Borate Glasses, Crystals and Melts and Third International Conference on Phosphate Materials, Corning, UK.
*Dr Randall Youngman, Corning Incorporated Email youngmanre@corning.com Web boratephosphate2020.org/*

**22–27 August**
Sol-Gel 2021, International Sol-Gel Society Conference, Lyon, France.
*Université Claude Bernard Lyon 1, Cellule Congrès - SOLGEL 2021, 43, bd du 11 Novembre 1918, 69622 Villeurbanne, France. Email solgel2021@univ-lyon1.fr Web www.solgel2021.fr*

**27–30 August**
International Festival of Glass, Stourbridge, UK.
*Ruskin Glass Centre, Wollaston Road, Amblecote, Stourbridge DY8 4HF, UK. Email ifg@rmlt.org.uk Web www.ifg.org.uk*

**29 August–2 September**
XVII ECerS Conference, Dresden, Germany.
*Fraunhofer Institute for Ceramic Technologies and Systems IKTS, Winterbergstrasse 28, 01277 Dresden, Germany. Web www.ecers2021.org/*

**September**
Society of Glass Technology Annual Meeting, online option.
*Christine Brown, Society of Glass Technology, 9 Churchill Way, Chapeltown, Sheffield S35 2PY, UK. Email christine@sgt.org Web www.sgt.org*

**1–4 September**
Glass South America, Sao Paulo, Brazil.
*NürnbergMesse Brasil, R. Verbo Divino, 1547 - 7th floor, Chacara Santo Antonio, 04719-002 Sao Paulo, Brazil. Email glass@nm-brasil.com.br Web www.nm-brasil.com.br*

**~~5–8 September~~ - now 5–8 December**
International Commission on Glass (ICG) Annual Meeting, Incheon, South Korea.

**7 September**
GlassTrend webinar on glass production transition to renewable fuels
*Web www.glasstrend.nl/events/*

**13–17 September**
22nd Congress of the Association Internationale pour l'Histoire du Verre, Lisbon, Portugal.
*Web aihv.org/congress/*

**13–15 September**
GlassBuild America 2021, Atlanta, Georgia, USA.
*National Glass Association, 1945 Old Gallows Road, Suite 750, Vienna, VA 22182, USA. Web www.glassbuildamerica.com/*

**14–17 September**
XXth Biennial Worldwide Congress Unified International Technical Conference on Refractories (UNITECR2021), Chicago, USA.
*The American Ceramic Society, 550 Polaris Parkway, Suite 510, Westerville, OH 43082, USA. Email customerservice@ceramics.org Web www.ceramics.org*

**20–23 September**
Resilient Materials 4 Life (RM4L2021), Cambridge, UK.
*Email info@rm4l.com Web rm4l.com/rm4l2020*

**~~20–23 September~~ Postponed**
16th Conference on Furnace Design: Operation and Process Simulation, Velke Karlovice, Czech Republic.
*Glass Service, Rokytnice 60, 755 01 Vsetin, Czech Republic. Email info@gsl.cz Web www.gsl.cz*

**26–28 September (new dates)**
FIT Show, Birmingham, UK.
*Freya Humphries, Account Manager, FIT Events Ltd, 9 Manchester Square, London, W1U 3PL, UK. Web www.fitshow.co.uk*

**27 September–1 October**
CelSian training course - general glass technology training, Eindhoven, The Netherlands.
*CelSian Glass BV Email infodesk@celsian.nl Web www.celsian.nl/training-and-rd/*

**4–6 October**
Sixth International Conference on Competitive Materials and Technology Processes (ic-cmtp6), Miskolc-Lillafured, Hungary.
*Email cmtp.conf@gmail.com Web www.ic-rmm1.eu/ic-cmtp/*

**5–8 October**
Vitrum 2021, Milan, Italy.
*Fiera Milano, Via Carlo Ilarione Petitti, 16, 20149 Milano MI, Italy. Email vitrum@vitrum-milano.it Web www.vitrum-milano.com*

**6–8 October**
Intersolar 2021, Munich, Germany.
*Web www.intersolar.de/en/home*

**12–15 October**
FESPA Global Print Expo, Amsterdam, The Netherlands.
*Web www.fespaglobalprintexpo.com/*

**12 October**
CelSian training course: heat transfer in glass melting furnaces, Eindhoven, The Netherlands.
*CelSian Glass BV Email infodesk@celsian.nl Web www.celsian.nl/training-and-rd/*

**13–14 October**
CelSian training course: burners, combustion and NOx emissions, Sheffield, UK.

*CelSian Glass BV Email infodesk@celsian.nl Web www.celsian.nl/training-and-rd/*

**16–17 October**
Icon Bath 2021 Conference: Ceramics, Glass and Stained Glass, Bath UK.
*The International Institute for Conservation of Historic and Artistic Works, 3 Birdcage Walk, Westminster, London, W1H 9JJ, UK. Email IconBath2021@gmail.com Web www.iiconservation.org*

**17–21 October**
ACerS 123rd Annual Meeting with Materials Science & Technology 2021, Pittsburgh, USA.
*The American Ceramic Society, 550 Polaris Parkway, Suite 510, Westerville, OH 43082, USA. Email customerservice@ceramics.org Web www.ceramics.org*

**20–22 October**
Glass Performance Days, Tampere, Finland.
*Brown Onduso, GPD, Vehmaistenkatu 5, 33730 Tampere, Finland Email brown.onduso@gpd.fi Web gpd.fi/*

**26–27 October**
CelSian training course: refractory selection, maintenance and related defects, Eindhoven, The Netherlands.
*CelSian Glass BV Email infodesk@celsian.nl Web www.celsian.nl/training-and-rd/*

**27–30 October**
Seventh International Conference on Multifunctinal, Hybrid and Nanomaterials, Genoa, Italy.
*Web www.elsevier.com/events/conferences/international-conference-on-multifunctional-hybrid-and-nanomaterials*

**28–29 October**
CelSian training course - flat glass melting and forming, Eindhoven, The Netherlands.
*CelSian Glass BV Email infodesk@celsian.nl Web www.celsian.nl/training-and-rd/*

**30 October–5 November**
Sixth Wuhan ICG Winter School, Wuhan, China.
*Prof. Kai Xu: SMART, Wuhan University of Technology, 1040 Heping Ave, Wuchang Qu, Wuhan Shi, Hubei Sheng, China. Email kaixu@whut.edu.cn Web www.icglass.org*

**30 October–5 November**
82nd Conference on Glass Problems, Columbus, USA.
*Donna Banks, GMIC, 550 Polaris Parkway, Suite 510, Westerville, OH 43082, USA. Email dbanks@gmic.org Web glassproblemsconference.org*

**~~16–18 November~~ postponed to 26–28 October 2022**
Glasstech Asia 2021, Bangkok, Thailand.
*Web www.glasstechasia.com.sg/*

**17–18 November**
Glass and Modern Technologies - XXI, Moscow, Russia.
*Organizing Committee Email forum@steklosouz.ru Web www.steklosouz.ru*

**18 November**
Glass Focus Awards 2021, UK.
*British Glass, 9 Churchill Way, Chapeltown, Sheffield, South Yorkshire S35 2PY, UK. Web www.britglass.org.uk/glass-focus-2021*

**23–25 November**
Celsian training: glass melting and process improvement for Indian glass industry, Mumbai, India.
*CelSian Glass BV Email infodesk@celsian.nl Web www.celsian.nl/training-and-rd/*

**5–8 December**
International Commission on Glass (ICG) Annual Meeting, Incheon, South Korea.
*Secretariat, Room 403, Meolijae Building, 76 Bangbae-ro, Seochu-gu, Seoul, Republic of Korea. Email secretariat@icg2021.org Web www.icg2021.org/*

**12–17 December (new dates)**
14th Pacific Rim Conference on Ceramic and Glass Technology (PACRIM 14) including Glass & Optical Materials Division 2021 Annual Meeting (GOMD 2021), Vancouver, BC, Canada.
*Andrea Ross, The American Ceramic Society, 550 Polaris Parkway, Suite 510, Westerville, OH 43082, USA. Email aross@ceramics.org Web www.ceramics.org*

**2022 UN International Year of Glass**
**Postponed to 2022 no date announced**
13th International Symposium on Crystallization in Glasses and Liquid, Paris, France.
*Laurent Cormier, CNRS, Paris, France. Email crystallization@sciencesconf.org Web crystallization.sciencesconf.org/*
ACerS Pan American Ceramics Conference, PACC-FMAs, Panama City, Panama.
*Andrea Ross, The American Ceramic Society, 550 Polaris Parkway, Suite 510, Westerville, OH 43082, USA. Email aross@ceramics.org Web www.ceramics.org*

**18–19 February**
Glassman Asia, Seoul, Korea.
*Ken Clark, Quartz Business Media, Quartz House, 20 Clarendon Road, Redhill, Surrey RH1 1QX, UK Email kenclark@quartzltd.com Web www.glassmanevents.com*

**3–5 March**
glasspex INDIA 2022, Mumbai, India.
*Vivek Bohra, Senior Project Manage Email bohrav@md-india.com Web www.glasspex.com*

**6–10 March**
Optical Fiber Communications Conference and Exhibition (OFC), San Diego, California USA.
*The Optical Society, USA. Email custserv@osa.org Web www.ofcconference.org*

**15–18 March**
Unified International Technical Conference on Refractories (UNITECR 2022) 17th Biennial Worldwide Congress on Refractories, Chicago, USA.
*The American Ceramic Society, 550 Polaris Parkway, Suite 510, Westerville, OH 43082, USA. Email aross@ceramics.org Web www.ceramics.org*

**2–4 April**
Deco '22, The Society of Glass and Ceramic Decorating Products annual conference, Columbus, OH, USA.
*Myra Warne, Society of Glass and Ceramic Decorating Products, PO Box 2489, Zanesville, OH 43702, USA. Email info@sgcd.org Web www.sgcd.org*

**3–6 April**
TMS World Congress on Artificial Intelligence in Materials and Manufacturing (AIM 2022), Pittsburgh, Pennsylvania, USA.
*TMS Meeting Services, 5700 Corporate Drive Suite 750, Pittsburgh, PA 15237, USA. Web www.tms.org/AIM2022*

**13–16 April**
China Glass 2022, Shanghai, China.
*Web www.chinaglass-expo.com*

**18–20 April**
Int. Conf. on Glass Science and Technology from the Golden Era of the Pharaohs to a Sustainable Future, Egypt.

**11–12 May**
Glassman Latin America, Monterrey, Mexico.
*Ken Clark, Quartz Business Media, Quartz House, 20 Clarendon Road, Redhill, Surrey RH1 1QX, UK Email ken-clark@quartzltd.com Web www.glassmanevents.com*

**18–20 May**
REFRA PRAGUE 21, 21st Conference on modern refractory materials and key achievements in high temperature technologies, Prague, Czech Republic.
*Ing. Jan K. Luxa, Silikátová společnost České republiky, Novotného lávka 5, 110 01 Praha 1, Czech Republic. Email hluxa@seznam.cz Web www.silikaty.cz*

**18–21 May**
GAS 2022, Glass Art Society conference, Tacoma, WA, USA.
*Glass Art Society, 2208 NW Market Street, Suite 200, Seattle, Washington 98107, USA. Email info@glassart.org Web www.glassart.org*

**22–26 May**
2022 Glass and Optical Materials Division Annual Meeting (GOMD2022), Baltimore, USA.
*Andrea Ross, The American Ceramic Society, 550 Polaris Parkway, Suite 510, Westerville, OH 43082, USA. Email aross@ceramics.org Web www.ceramics.org*

**6–9 June**
Mir Stekla 2022, Moscow, Russia.
*Ekaterina Razumeitseva, Project Head, EXPOCENTRE, Moscow, Krasnopresnenskaya nab., 14, 123100, Russia. Email re@expocentr.ru Web www.mirstekla-expo.ru/en/*

**16–17 June**
XXXIV ATIV Int. Conf., Where Glass Science, Art and Technology Meet Together, Parma, Italy.
*ATIV Secretariat, MV Congressi spa Via Marchesi 26 d 43126 Parma, Italy. ativ@ativ-online.it Web www.ativ.eu*

**21–24 June**
Techtextil 2022, Frankfurt, Germany.
*Email techtextil@messefrankfurt.com Web techtextil.messe-frankfurt.com*

**23–24 June**
Challenging Glass Conference 8: International Conference on the Architectural and Structural Application of Glass, Ghent, Belgium.
*Email cgc8@challengingglass.com Web www.challenging-glass.com*

**3–8 July**
26th International Congress on Glass, Berlin, Germany. Hosted by the DGG in their centenary year.
*Annette Doms, Deutsche Glastechnische Gesellschaft, Siemenstrasse 45, D-63071 Offenbach, Germany. Email dgg@hvg-dgg.de Web www.hvg-dgg.de*

**4–8 July**
Ninth International Conference on Optical, Optoelectronic and Photonic Materials and Applications (ICOOPMA) and EuroDIM: 14th Europhysical Conference on Defects in Insulating Materials, Ghent, Belgium.
*Web icoopma-eurodim.ugent.be*

**10–15 July**
Physics of Non-Crystalline Solids, now also including Society of Glass Technology Annual Meeting Canterbury, UK.
*Christine Brown, Society of Glass Technology, 9 Churchill Way, Chapeltown, Sheffield S35 2PY, UK. Email christine@sgt.org Web www.sgt.org*

**10–14 July**
Ninth Int. Congress on Ceramics (ICC9), Krakow, Poland.
*European Ceramic Society AISBL, Av. Gouverneur Cornez 4, B-7000 Mons, Belgium. Email ecers@bcrc.be Web ecers.org/*

**24–29 July**
Sol-Gel 2021, International Sol-Gel Society Conference, Lyon, France.
*Université Claude Bernard Lyon 1, Cellule Congrès - SOLGEL 2021, 43, bd du 11 Novembre 1918, 69622 Villeurbanne, France. Email solgel2021@univ-lyon1.fr Web www.solgel2021.fr*

**20–23 September**
Glastec 2022, Düsseldorf, Germany.
*Anke Seltmann, glastec, D-40474 Düsseldorf, Am Staad (Stockumer Höfe), Germany. Email seltmanna@messe-duesseldorf.de Web www.glasstec-online.com/*

**9–13 October**
ACerS 124th Annual Meeting with Materials Science & Technology 2022, Pittsburgh, USA.
*The American Ceramic Society, 550 Polaris Parkway, Suite 510, Westerville, OH 43082, USA. Email customerservice@ceramics.org Web www.ceramics.org*

**11–13 October**
GlassBuild America 2022, Las Vegas, USA.
*National Glass Association, 1945 Old Gallows Road, Suite 750, Vienna, VA 22182, USA. Web www.glassbuildamerica.com/*

**26–28 October 2022**
Glasstech Asia 2022, Singapore.
*Web www.glasstechasia.com.sg/*

**2023**
**5–9 March**
Optical Fiber Communications Conference and Exhibition (OFC), San Diego, California USA.
*The Optical Society, USA. Email custserv@osa.org Web www.ofcconference.org*

**4–9 June**
2023 Glass and Optical Materials Division Annual Meeting (GOMD 2023), New Orleans, USA.
*The American Ceramic Society, 550 Polaris Parkway, Suite 510, Westerville, OH 43082, USA. Email customerservice@ceramics.org Web www.ceramics.org*

**10–14 July**
Nonlinear Optics, Honolulu, Hawaii, USA.
*The Optical Society, 2010 Massachusetts Ave, NW, Washington, DC 20036 USA. Email custserv@osa.org Web www.osa.org*

**17–21 July**
Tenth International Conference on Borate Glasses, Crystals and Melts and Third International Conference on Phosphate Materials, Corning, USA.
*Dr Randall Youngman, Corning Incorporated Email youngmanre@corning.com Web boratephosphate2020.org/*

**August**
NCM15, Structure of Non-Crystalline Materials 15, combined with the SGT Annual Meeting, UK.
*Christine Brown, Society of Glass Technology, 9 Churchill Way, Chapeltown, Sheffield S35 2PY, UK. Email christine@sgt.org Web www.sgt.org*

**2024**
**June**
Tenth International Conference on Optical, Optoelectronic and Photonic Materials and Applications (ICOOPMA), Pardubice, Czech Republic.
*Center of Materials and Nanotechnology and Department of General and Inorganic Chemistry, Faculty of Chemical*

*Technology, University of Pardubice, Cs. Legii sq. 565, 532 10 Pardubice, Czech Republic. Email info@icoopma.com Web www.icoopma.com*

**2–6 July**
XVIII ECerS Conference, Lyon, France.
*European Ceramic Society AISBL, Av. Gouverneur Cornez 4, B-7000 Mons, Belgium. Email ecers@bcrc.be Web ecers.org/*

**14–19 July**
Tenth Int. Congress on Ceramics (ICC10), Montreal, Canada.
*The American Ceramic Society, 550 Polaris Parkway, Suite 510, Westerville, OH 43082, USA. Email customerservice@ceramics.org Web www.ceramics.org*

**20–23 August**
Materials Challenges in Alternative & Renewable Energy 2023 combined with the Sixth Annual Energy Harvesting Society Meeting (EHS 2023), Bellevue, Washington, USA.
*The American Ceramic Society, 550 Polaris Parkway, Suite 510, Westerville, OH 43082, USA. Email customerservice@ceramics.org Web www.ceramics.org*

The Glass Manufacturers' Educational Trust is the primary UK charity in relation to education within the field of glass. Applications for funding are now open to all parties interested in glass education.

The funding is available to both individuals and organisations as long as educational outcomes can be identified. In the past, the trust has given grants for glass related training courses, literature about the glass industry, glass events, and to develop courses that enable students to learn about glass.

The trust is always interested to hear about your glass projects. When applying for funds, individuals and organisations should provide a brief description of the reason for the application, and the amount of money they would like to receive.

For more information about the grants programme please contact Louise Mackrell: l.mackrell@britglass.co.uk

## Impossible Objects enters joint development agreement with Owens Corning

3D printing machine developer Impossible Objects has established a composite-based additive manufacturing (CBAM) process to be used with Owens Corning fibre glass to offer composites with high strength-to-weight ratios and excellent chemical resistance. Also boasting low costs, the materials are intended as a viable alternative to metals such as aluminium, enabling high-performance 3D printed parts.

"Owens Corning is committed to the development of composite materials and their applications," said Dr. Chris Skinner, vice president of strategic marketing, composites, Owens Corning. "We seek to be at the forefront of new processing and new applications involving composites."

### *Composite-based additive manufacturing*

The CBAM process operates on a high-speed inkjet-based 3D printing method. First, long fibre sheets of glass are fed into the CBAM system, where layers of a part are deposited onto the sheets using clear fluid and thermal inkjet technology. The binding polymer powder is then used to coat the fibre sheet, where it selectively adheres to the sections with the printing fluid. When the excess powder is removed, a powdered layer in the shape of the printed fluid is left behind.

Once the process is repeated for all layers in a part, the individual long fibre sheets are stacked on top of each other and pressed at high temperatures. This melts the polymer and compresses the consoli-

dated part into its final height. Finally, the unbonded portions of the fibre sheets are mechanically or chemically removed, leaving nothing but the final 3D printed part.

Impossible Objects claims its 3D printing technology can produce parts up to ten times faster than FFF, and is compatible with high-performance polymers such as polyamide and PEEK.

### *A composite collaboration*

By combining the CBAM process with Owens Corning's glass non-woven manufacturing expertise, the partnership is expected to scale CBAM 3D printing for high strength, high volume applications. The end goal is for the technology to eventually compete with processes such as injection moulding, except CBAM would also eliminate long lead times and tooling costs, all while enabling mass part customisation.

The partners also see great value in lowering material costs when it comes to additive manufacturing adoption, as conventional 3D printing materials can cost up to eight times more than traditional manufacturing materials.

Glass Technol.: Eur. J. Glass Sci. Technol. A, 2021, 62 (2), 65–74

# Experimental and numerical study on abrasive water jet machining of laminated glass

*Ajitanshu Vedrtnam,*[a,b,1] *Dheeraj Gunwant,*[a,2] *Shashkant Chaturvedi,*[a] *Dheeraj Sagar*[a] *& S. J. Pawar*[b]

[a] *Department of Mechanical Engineering, Invertis University, Bareilly, UP, India-243001*
[b] *Department of Applied Mechanics, Motilal Nehru National Institute of Technology Allahabad, Prayagraj, UP, India-211004*

Manuscript received 25 January 2020
Revision received 15 July 2020
Manuscript accepted 14 December 2020

*The use of laminated glass (LG) is growing in structural applications. The conventional methods used to cut LG involve cutting glasses and interlayer separately and, thus, are less productive. The present work reports the use of abrasive water jet machining (AWJM) for LG cutting for the first time in the open literature. The statistical methods have determined the optimum process parameters of AWJM for the LG cutting. The boundaries of LG sample cut using conventional method and AWJM are examined using atomic force microscopy (AFM). Three-point bending tests have highlighted the superiority in the load-bearing ability and fracture pattern for LG samples cut by AWJM. A finite element (FE) model formed using an explicit dynamic module of ANSYS has explored the experimental findings. The FE model has predicted the stresses and strain for the glass and polyvinyl butyral (PVB) interlayer due to the strike of single SiC abrasive particles. The findings should be useful for the practical guidance to both the academic community and industry.*

## 1. Introduction

The laminated glass (LG) consists of glass and polymer layers sandwiched between glasses. The LG is widely used nowadays in structural glazing, automobiles, photovoltaic panels, specialty markets and blast resistance applications. The traditional method of cutting LG includes first the cutting of glasses using a diamond cutter and then the interlayer using a blade. This method is unproductive, and the cut LG requires further finishing as the cutter traces remain on the cut surfaces. Thus, an alternative efficient cutting process of LG is required. The present work reports on the use of abrasive water jet machining (AWJM) for the cutting of LG samples, which is not cited in the scientific literature. The AJWM uses a slurry having abrasive grains suspended in water.[(1)] The material is removed due to the shearing of workpiece material by abrasive particles under the high pressure stream of water. AWJM offers high machining versatility and no thermal distortion on the workpiece. Thus, the AWJM is frequently used for machining difficult to machine materials like ceramics, composites, and superalloys, where conventional machining is often technically or economically not feasible.[(2)] The AWJM is used for milling,[(3)] drilling,[(4)] polishing[(5)] and for cutting otherwise challenging materials.[(6)] The AWJM process produces an excellent surface finish even for hard to cut materials.[(7–10)]

The AWJM is associated with erosion,[(11,12)] which can be modelled using analytical and computational fluid dynamics (CFD) models.[(13–15)] Balamurugan *et al* studied the effects of pressure, stand-off distance, and traverse speed on material removal rate (*MRR*), kerf angle and surface finish through mathematical modeling.[(16)] Mm *et al* have reported machining of hybrid carbon and glass fibre reinforced composites by AWJM and concluded that stand-off distance was dominant for minimising kerf ratio.[(17)] Armagam *et al* studied the effect of performance parameters on the roughness and kerf width in AWJM for the vinyl ester matrix reinforced with glass fibres. The standoff distance has been reported as the most crucial parameter for surface roughness; although, the kerf width was also affected by traverse speed to some extent.[(18)] Chakraborty *et al* used the grey wolf optimiser algorithm based on the hunting pattern of the grey wolves for determining the optimal parameters in AWJM.[(19)] Kumar *et al* used the response surface method-box Behnken design for minimising the surface roughness for AWJM of Inconel 718 material.[(20)] Selvam *et al* reported the use of AWJM for machining of E-glass and carbon fabric epoxy hybrid composites.[(21)] Prabhu *et al* used AWJM for banana fibre reinforced polyester composites.[(22)] Ponselvan *et al* reported the effect of various cutting parameters on the depth of cut in phosphate glass and observed a direct relationship between the input and output variables studied. The depth of cut was observed to be directly proportional to jet pressure and inversely proportional to traverse speed.[(23)] The optimisation of AWJM process is discussed widely for different grades of steel, composites, copper, aluminium and their alloys, carbides, alumina ceramic, concretes, stones and other alloys

[1] Corresponding author. Email ajitanshu.m@invertis.org
[2] Now at: GET Group of Institutions, Bazpur, Uttarakhand, India
DOI: 10.13036/17533546.62.2.002

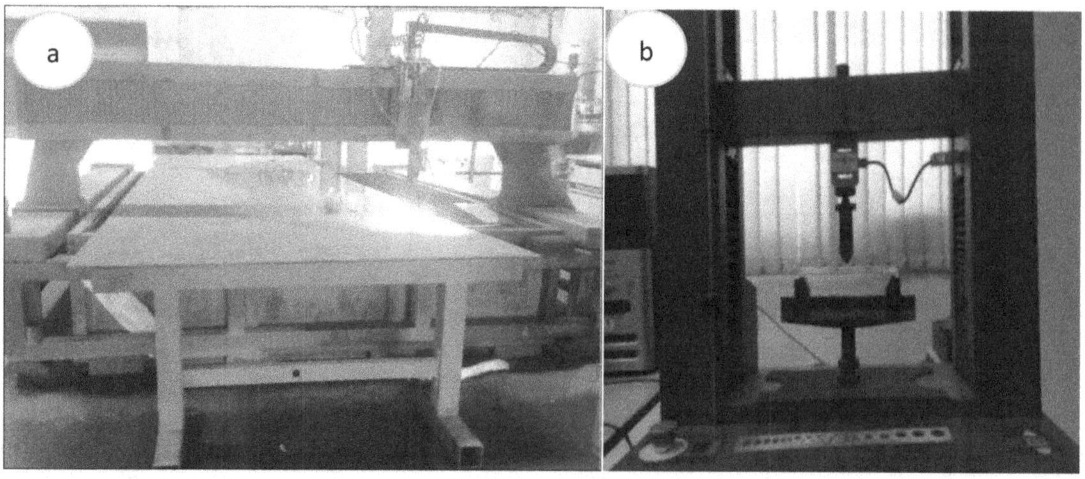

*Figure 1. (a) AWJM machine test section with a test piece (b) The bending test machine coupled with Tinius Olsen Q-set software*

using Taguchi method, response surface methodology, Grey relation analysis, and soft computing methods.[7–16] Cutler has used AWJM for glass plate cutting in the artistic glass work.[24]

The US patents (US1731820A, US4471895A, and US1966353A) reported methods of machining LG that involve heating with cutting, but they are time consuming and less productive processes. The flex diamond saw, use of flammable liquids, and other commercially available methods do not have the generalised application and also have limitations with the finishing of the cut surface of LG. Many of them require finishing operation at the boundaries after cutting of LG. A detailed review of the cutting of glass, ceramics, and silicon through laser ablation has been reported in Rihakova & Chmelickova.[25] The average laser power required is 100 W to cut the glasses by ablation using continuous infrared or visible light waves. The excessive laser-fluency of the high power laser causes a heat affected zone on the glass surface.[26,27] A laser-induced thermal crack propagation method was proposed for cutting the sandwich structure of a glass–silicon–glass wafer.[28] The laser-assisted machining of LG is considered to be a costly affair in which heat affected surface may remain a problem and the effect of heat may also cause the interlayer to melt.

The present work suggests a new application of AWJM, as the cited literature includes machining and finite element studies on glass, glass–polymer composites, different grades of steel, copper, aluminium, 87% alumina ceramics, Ti6Al4V, different types of stone using AWJM but not LG.[29–32] The optimisation of AWJM for cutting of LG is performed using the Taguchi method in the first section. A mathematical model for evaluating the *MRR* of LG samples is developed. Further, the LG samples were prepared by the conventional method and AWJM to report the difference in the surface finish of cut LG samples. The literature review on the load bearing ability of LG is reported by the authors in their earlier work.[33–35] The influence of boundaries on the load bearing ability of LG is reported. Finally, the FE model has reported the stress and strains produced in the glasses and the interlayer during cutting.

## 2. Materials and methods

The LG samples are cut by abrasive water jet machine (Model: OMAX 2652) shown in Figure 1(a). The water pressure (KSI), traverse speed (mm/min), and standoff distance (mm) are considered as control and *MRR* as the responses parameter. *MRR* is defined as the volume of material removed (mm$^3$) divided by the machining time in minutes and measured on jet entrance face on the workpiece. Different levels of the machining parameters are selected based on machine specifications, and experimentation was planned as per L9 Taguchi orthogonal array. The material removal rate (*MRR*) is measured for selected levels of the standoff distance, traverse speed, and water pressure. The parameters were ranked based on their contribution to the *MRR*. The levels of the input parameters are selected based on the availability of the machine setting (Table 1).

The L9 orthogonal array is chosen based on the number of independent design variables and levels as the experimental design and corresponding nine experiments are conducted according to the orthogonal array settings. The values of response are then considered for model building and optimisation. The cutting of LG is achieved in the single pass with the irregular abrasive particles of silicon carbide (SiC). The SiC particles (10% volumetric concentration) are

*Table 1. Parameters and their levels*

| *Process parameter* | *Level 1* | *Level 2* | *Level 3* |
|---|---|---|---|
| Stand-off distance ($S_d$) (mm) | 0·8 | 1·5 | 2·5 |
| Traverse speed ($T_s$) (mm/min) | 473·57 | 254·79 | 141·60 |
| Water pressure ($W_p$) (ksi) | 10·3 | 18·2 | 26·4 |

available with the AWJM. While cutting the LG, a tapered mask was applied such that abrasive particles remained directed towards the intended cutting zone. The impingement angle was 90° during cutting. The cutting operation was performed in a closed space to minimise the effect of external disturbances during cutting. The LG-PVB samples were having two annealed glass beams of 5 mm thickness, each with an interlayer of PVB of 0·76 mm thickness in between. Figure 1(b) shows a Tinius Olsen machine on which three-point bending test was performed. The process and standard followed for bending test are the same as in author's previous work.[34,35] The statistical analysis is performed for an insight into the experimental findings using MINITAB.

The effect of traverse speed and an inertia force on the stress distribution in the glass and PVB interlayer were analysed using FEM. The glass and PVB interlayer were separately modelled as the single-particle of abrasive (SiC) should have the impacted separately on the glasses and PVB interlayer during cutting. The dynamic nature of problem was duly considered using the explicit dynamics module of the ANSYS WB. The properties used in simulation are given in Table 2.[33–36]

The dimensions of SiC particle (3 μm diameter), glass (50×50×5 μm), and PVB layers (50×50×5 μm) were decided based on experiments, computational efficiency, and the accuracy of results after several simulations run. The 33 000 3D elements comprising of 44 531 nodes were considered during meshing. The impact angle was taken 90° to maintain the sphericity of crater equal to 1 irrespective of the impact velocity.[37] The time at the end of simulation was kept $2\times10^{-6}$ seconds to make the simulation computationally efficient without compromising with accuracy. The linear elastic material is assumed for all the materials used during experimentation.[33,38]

## 3. Results and discussion

A mathematical model is constituted using experimental data and MATLAB platform for predicting *MRR* of the LG samples. The model is used for obtaining the optimal working conditions of the AWJM process. The *MRR* of LG is considered as a function of water pressure ($W_p$), traverse speed ($T_s$), and standoff distance ($S_d$) as given in Equation (1), other control parameters are neglected, and it is assumed that considered parameters act independently

$$MRR=f(S_d,T_s,W_p) \quad (1)$$

Table 2. Material properties used for the FE analysis[33–36]

| S. No. | Constituent | Young's modulus (MPa) | Poisson's ratio | Density (kg/m$^3$) |
|---|---|---|---|---|
| 1 | Glass | 70000 | 0·23 | 2500 |
| 2 | SiC | 410000 | 0·14 | 3200 |
| 3 | PVB | 220 | 0·495 | 1100 |

Power law function is used for the model development for *MRR* of LG (Equation (2))

$$MRR=a_0S_d^{a_1}T_s^{a_2}W_P^{a_3} \quad (2)$$

Here $a_1$, $a_2$ and $a_3$ are constants; the power-law function taking natural logarithm is given in Equation (3)

$$\ln(MRR)=\ln(a_0)+a_1(\ln(S_d)+a_2\ln(T_s)+a_3\ln(W_p)... \quad (3)$$

The regression analysis is performed based on the linear model stated in Equation (3) and nine experimental runs for calculating the unknown coefficients. The expression thus developed is given in Equation (4)

$$\ln(MRR)=-3{\cdot}9475+0{\cdot}0277\ln(S_d)+0{\cdot}1397\ln(T_s) -0{\cdot}0242\ln(W_p) \quad (4)$$

Further, Equation (5) shows the *MRR* for LG using AWJM.

$$MMR=e^{-3{\cdot}9475}S_d^{0{\cdot}0277}T_s^{0{\cdot}1397}W_p^{-0{\cdot}0242} \quad (5)$$

The validity of Equation (5) is examined by follow up experimental runs. The actual mean *MRR* was 482·5 mm$^3$/min for five experimental runs with the same input variables, whereas predicted *MRR* from the Equation (5) was 499·56 mm$^3$/min. The value of mean absolute percentage error 3·64% suggested that the model is adequate for calculating *MRR* of LG using AWJM.

The *MRR* of LG using AWJM for optimum setting of control parameters is maximised using Equation (5) as the objective function. As was reported in an earlier study that silicon carbide (SiC) abrasive particles are harder and produce better results than $Al_2O_3$ and garnet abrasives during glass cutting,[29] the SiC abrasive particles of irregular shape were used. The response for means (raw data) and signal/noise (S/N) data at the three-parameter levels are calculated. These results have been plotted as shown in Figure 2 using MINITAB 17.

It is reflected in Figure 2(a) that $S_d$ does not influence much, but for level 2 S/N ratio is comparatively higher than level 1 and level 3. $T_s$ have the most significant effect on *MRR*; S/N ratio is least for level 1, and the maximum for level 3, whereas $W_p$ also has a significant influence on *MRR*; for $W_p$ level 3 has the maximum S/N ratio. The larger-the-better rule of the Taguchi method is utilised for maximising the *MRR*. Table 3 shows the optimal level of process parameters. Analysis of variance (ANOVA) is conducted based on the predicted values obtained through the estimated model given by Equation (5). The ANOVA depicts the contributive effect of the factors and their interactions

Table 3. Optimal level of process parameters

| Process parameter | Optimum level | Value |
|---|---|---|
| $S_d$ | 2 | 1·5 mm |
| $T_s$ | 3 | 473·57 mm/min |
| $W_p$ | 3 | 26·4 Ksi |

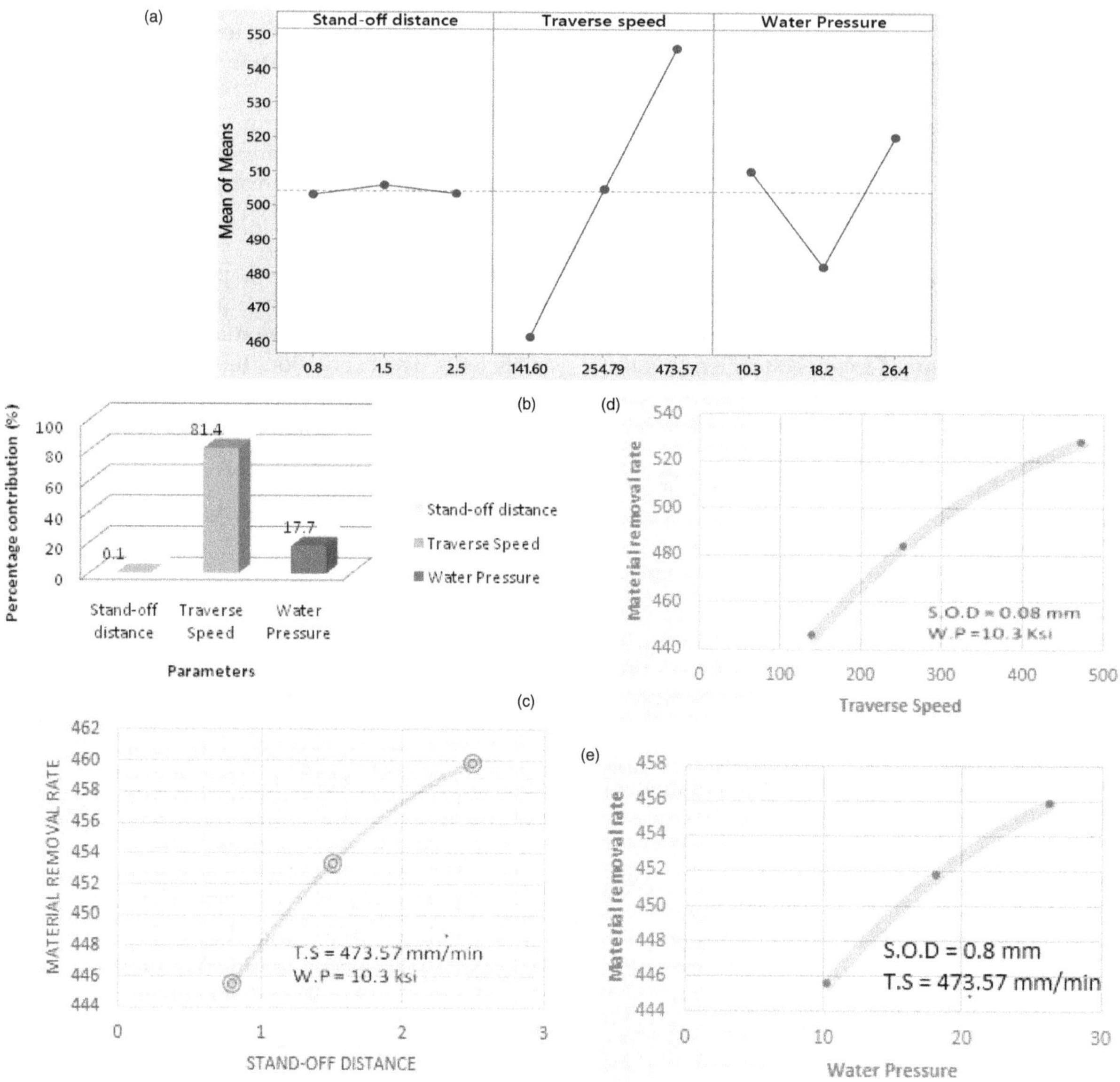


*Figure 2. (a) Main effects plot for means (MRR) (b) Percentage contribution for MRR. (c) Effect of $S_d$ on MRR (mm$^3$/min) (d) Effect of $T_s$ on MRR (mm$^3$/min) (e) Effect of $W_p$ on MRR (mm$^3$/min)*

on *MRR* (Figure 2(b)).

Two of the control parameters-water pressure and traverse speed have a significant contribution to the depth of cut. The effect of $S_d$ on *MRR* of LG is given in Figure 2(c). The $S_d$ did not show a significant influence on *MRR*; however, *MRR* slightly increases with the increase in $S_d$ while keeping $T_s$ and $W_p$ to be constant. One might expect a reduction in *MRR* due to the reduction in kinetic energy of the jet at impingement, but a slight increase in *MRR* is reported as the area of LG at which jet strikes increase when stand-off distance is increased from 0·8 mm to 1·5 mm. The higher $S_d$ allows the jet to expand before impingement, which increases vulnerability to external drag from the surrounding environment. Therefore, increase in the stand-off distance results an increased jet diameter as cutting is initiated and in turn, reduces the kinetic energy of the jet at impingement. So the little increment in *MRR* is observed with an increase in $S_d$.

Figure 2(d) shows the effect of $T_s$ on *MRR*; for decreasing the machining costs, the higher feed

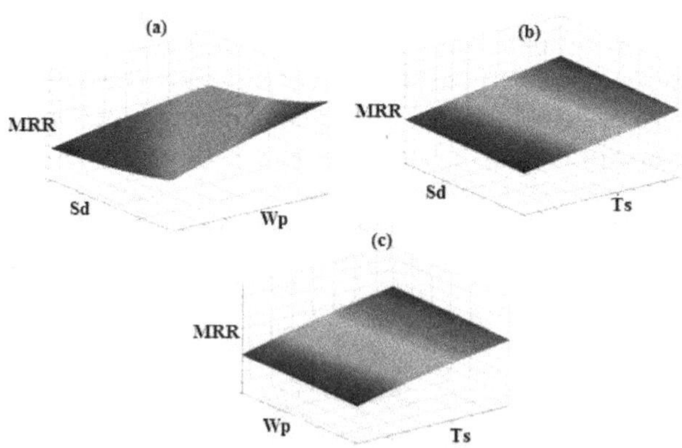


*Figure 3. (a) MRR versus $S_d$, $W_p$ keeping $T_s$ constant (b) MRR versus $S_d$, $T_s$ keeping $W_p$ constant (c) MRR versus $W_p$, $T_s$ keeping $S_d$ constant*

rates are preferred, but increasing the $T_s$ may result in inaccuracy. However, *MRR* may increase as the result of increasing $T_s$, but at the same time, less number of abrasive particles pass through a unit area. Therefore, less number of impacts and cutting edges are available per unit area for material removal. Thus,

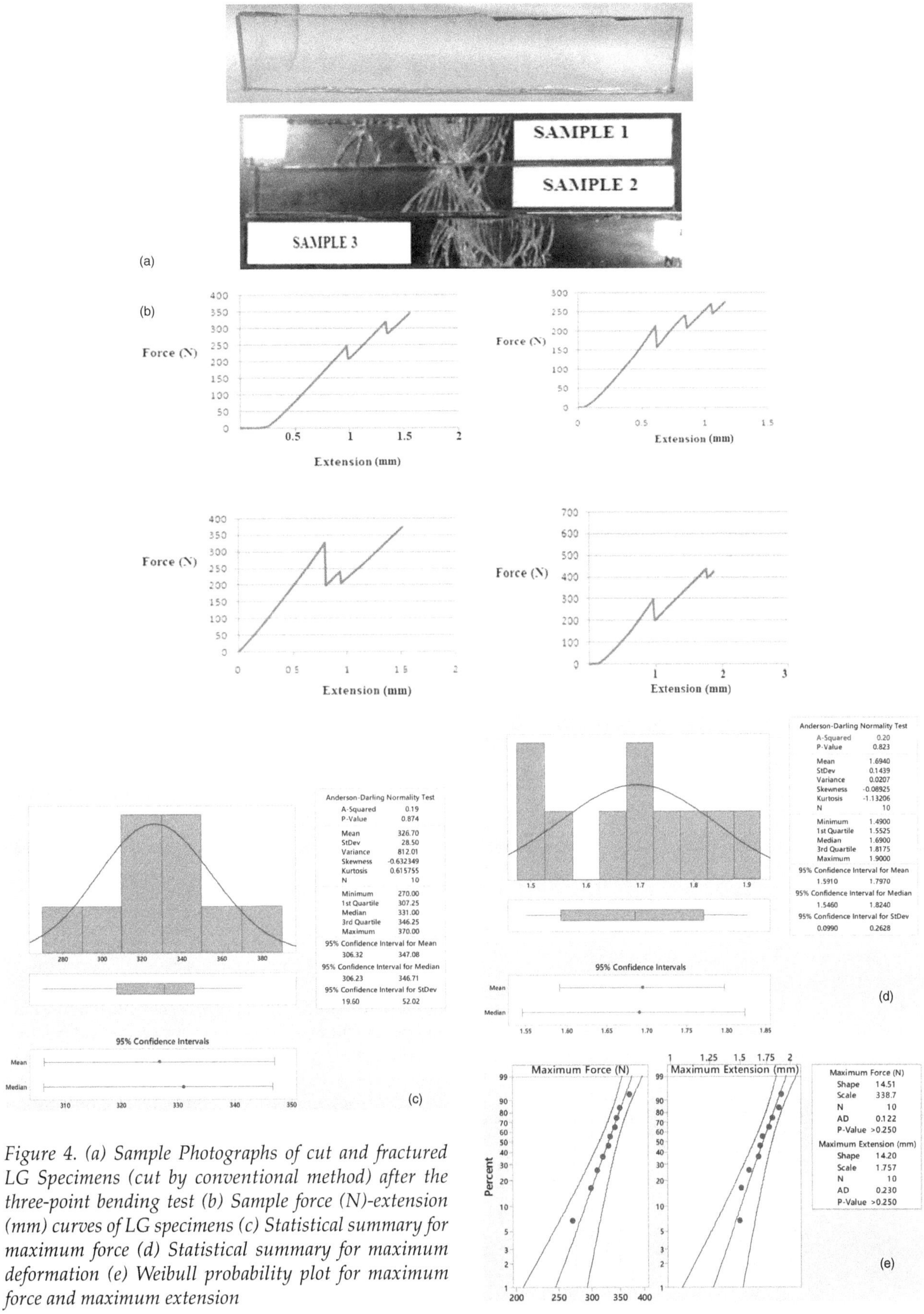


*Figure 4. (a) Sample Photographs of cut and fractured LG Specimens (cut by conventional method) after the three-point bending test (b) Sample force (N)-extension (mm) curves of LG specimens (c) Statistical summary for maximum force (d) Statistical summary for maximum deformation (e) Weibull probability plot for maximum force and maximum extension*

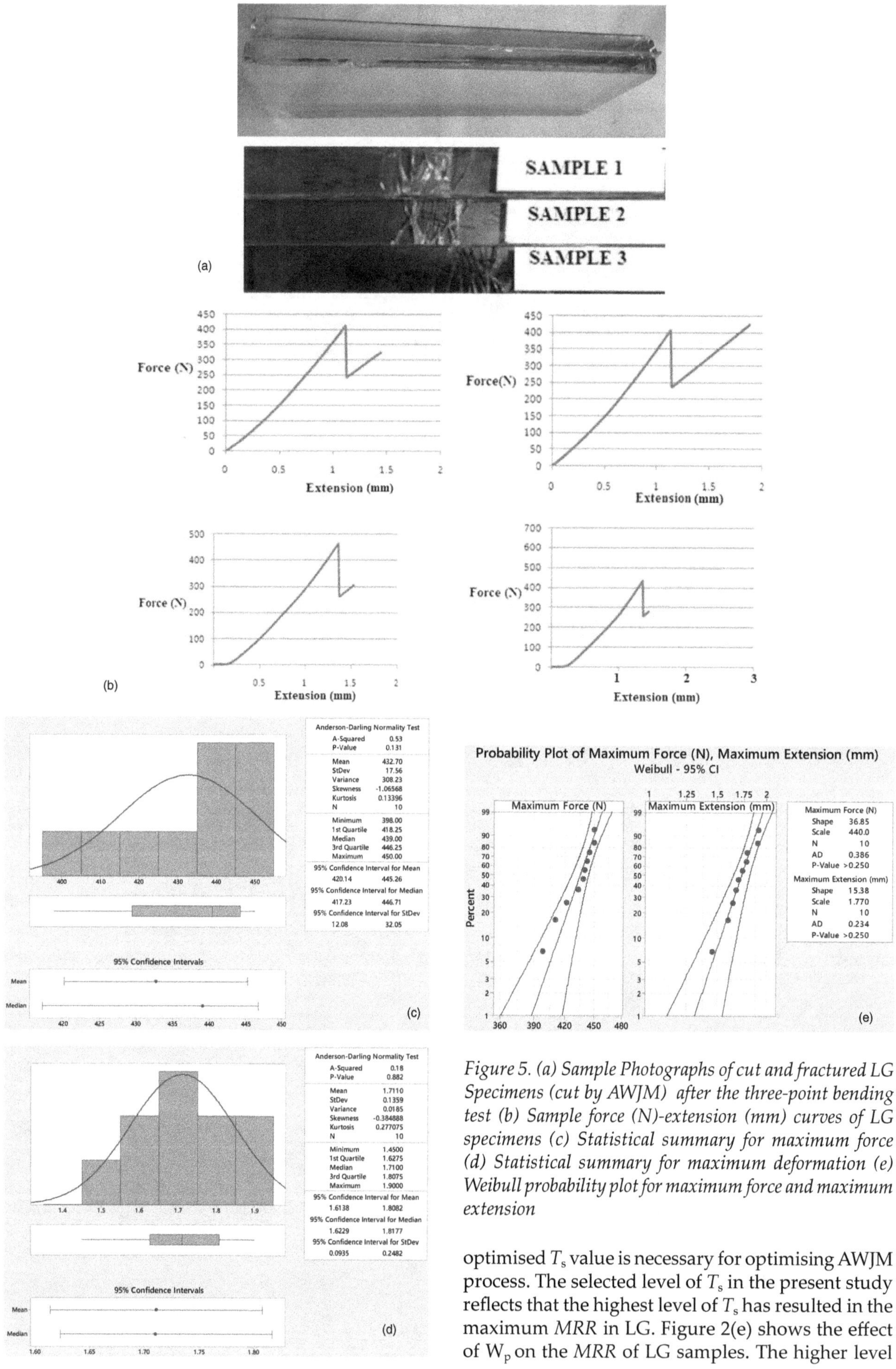


*Figure 5. (a) Sample Photographs of cut and fractured LG Specimens (cut by AWJM) after the three-point bending test (b) Sample force (N)-extension (mm) curves of LG specimens (c) Statistical summary for maximum force (d) Statistical summary for maximum deformation (e) Weibull probability plot for maximum force and maximum extension*

optimised $T_s$ value is necessary for optimising AWJM process. The selected level of $T_s$ in the present study reflects that the highest level of $T_s$ has resulted in the maximum *MRR* in LG. Figure 2(e) shows the effect of $W_p$ on the *MRR* of LG samples. The higher level of jet pressure increases the *MRR*. With the increase in the jet pressure, brittle abrasives break down (as observed after the experiment) into smaller ones, and as a result, LG samples experience quicker removal of glass and polymeric interlayer.

Figure 3 (a) shows the surface plot for *MRR* versus $S_d$, and $W_p$ obtained using Equation (5). It reflects that $S_d$ has less influence than $W_p$ on *MRR*, at the higher $W_p$ *MRR* is the maximum. Figure 3(b) shows the surface plot for *MRR* versus $S_d$, $T_s$. The higher values of $T_s$ results in the higher *MRR*; the influence of $S_d$ is negligible. Figure 3(c) shows the surface plot for *MRR* versus $W_d$, $T_s$; the higher level of both parameters results in the higher *MRR*. Figure 3(c) also displays the maximum *MRR* at the higher level of $T_s$ and $W_p$. The surface plots represent an excellent match to the results of the Taguchi analysis. The optimal level suggested by the Taguchi method was verified satisfactorily in a verification experiment as the *MRR* with optimal values was 16·66% higher than the maximum values of *MRR* during earlier experiments. However, a further intensive study is required to calculate the difference in cutting time for glass and interlayer. Such a study will be useful for determining further efficient process parameter variation.

The boundaries of glass cut by conventional method and AWJM were examined to observe the visible difference on the cut surfaces. Three-point bending was conducted for comparing the effect of machining on the fracture pattern of LG. Figure 4 shows a photograph of fractured samples (cut by conventional method) after a three-point bending test and force-extension curve of all LG samples and results of statistical analysis. The photograph of sample 1 reflects multiple fractures nearby the region where the load is applied. It was observed that the first crack is generated across the width in the lower glass plate near the mid-span region followed by multiple nearby fractures and then fracture of the upper glass plate that finally fractures. The multiple fractures and severe damage have noticed in the sample around the mid-span region. The load-extension curve of sample 2 indicates the same pattern as followed by sample 1, but the fracture pattern reported is slightly different. The fracture zone at one side around the mid-span of LG is extended while the other sides, lines of fracture, are getting concentrated. A little insight into the sample reflects that the sides cut by the diamond cutter are having extended zone while another side of the LG sample is having the surface finish as initially supplied by the manufacturer. Sample 3 has an even wider zone of fracture and the associated load–extension curve report multiple fractures. Both sides of the samples were machined by the conventional cutting method.

Figure 5 shows the photograph of cut LG samples and fractured samples (cut by AWJM) after the test, the force-extension curve of all LG samples. The fracture of sample 1 is having a shorter fracture zone than all conventionally machined samples. The fracture of sample 2 is similar to sample 1, but the load–extension curve has indicated some resistance offered by the interlayer after the fracture of both the glass plates. Sample 3 also has a similar fracture pattern and load–extension curve as sample 1 and sample 2 are having, but the maximum load at fracture is different. The results of the statistical analysis given in Figures 4 and 5 demonstrate that the LG samples cut by AWJM have above 32 % higher average bending strength than LG samples cut by conventional method. The difference in maximum deformation is insignificant. The higher p-value indicates that variability exists, and data may not represent the population completely, but the Weibull probability plots support the conclusion drawn on bending strength and deformation. The variation of the maximum load at fracture is due to the non-uniform behaviour of glass due to surface (crack) and edge defect, non-uniformity of the chemical composition, the duration of the loading, characteristics of glass manufacturing techniques (including annealing, environmental parameters), the geometry production, cutting, storage, and transportation of the sample, which requires further intensive investigation.[33,34] It can be safely concluded based on the observations, as the edges of samples cut by AWJM were smoother than a conventional method; they were having lesser edge defects, which in turn resulted in fewer crack initiation and further propagation from the boundary and thus the lesser damage. Figure 6(a) shows the surface finish of boundaries of glass cut by the conventional method, and Figure 6(b) shows the surface finish (boundary) of glass cut by AWJM. It is clearly reflected that conventionally cut material has visible marks of diamond cutter traces that are lesser in AWJM cut glass pieces. Thus, AWJM has clear superiority over other methods presented in the cited literature. Improved surfaces

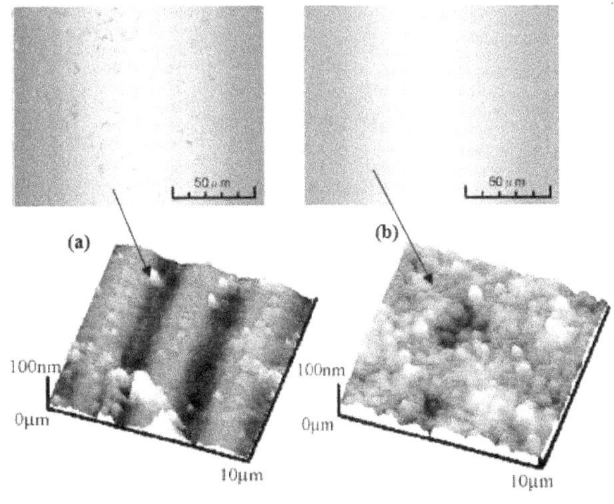


*Figure 6. (a) Surface finish (boundary) of LG cut by the conventional method, (b) Surface finish (boundary) of LG cut by AWJM*

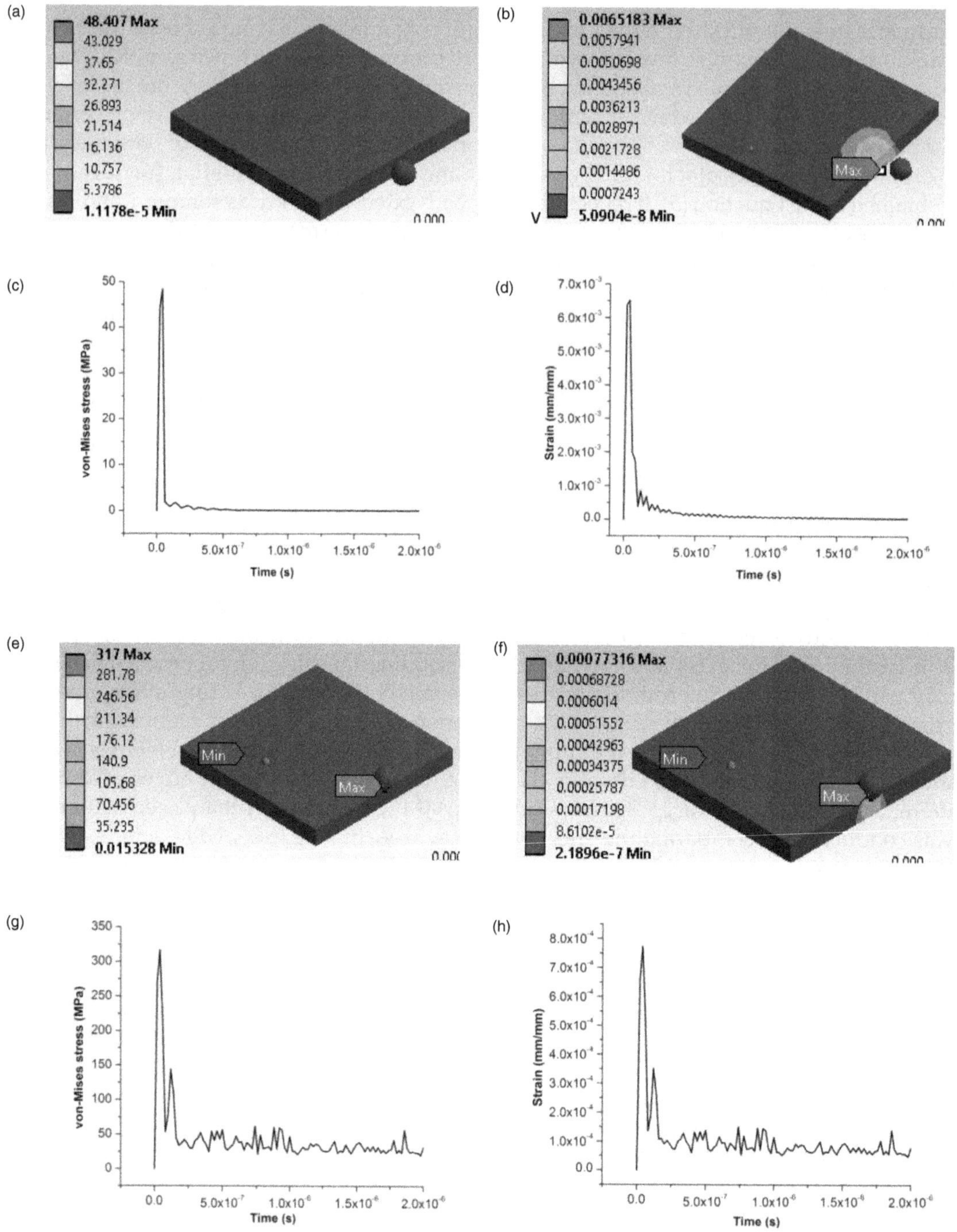

*Figure 7. Von Mises stress contours and corresponding plots for (a)–(d) PVB layer and (e)–(g) glass layer*

finish increase load bearing ability, fatigue strength, and impact behaviour of materials.[33,34,37] The *MRR* depends on abrasive particle size in AWJM. It was reported that the coarse abrasive particle increases the *MRR*, while the fine abrasive particle results in minimum kerf angle and good surface finish for selected machining conditions.[38–42] The effect of variation in abrasive particle size on LG cutting can be considered as future work.

Figure 7 shows the sample simulation results, including the variation of stress and strain with time for SiC particle interaction with the PVB and glasses in the form of contour and time–quantity plots. The complete results are documented in Table 4. The maximum von Mises equivalent stresses and strains were observed at a time of $4\times10^{-8}$ s. Figure 8 shows the variation of the stresses and strains with the traverse speed. The maximum von Mises stress in the case of the PVB layer for the traverse speed of 141·60 mm/min

*Table 4. Explicit dynamics analysis of AWJM process*

| *S. No.* | *Traverse speed (mm/min)* | *von-Mises stress (PVB) (MPa)* | *von-Mises strain (PVB)* | *von-Mises stress (Glass) (MPa)* | *von-Mises strain (Glass)* |
|---|---|---|---|---|---|
| 1 | 141·60 | 48·41 | 0·006518 | 317 | 0·0007732 |
| 2 | 254·79 | 78·65 | 0·01056 | 570·60 | 0·0016236 |
| 3 | 473·57 | 106·50 | 0·013688 | 732·27 | 0·0036369 |

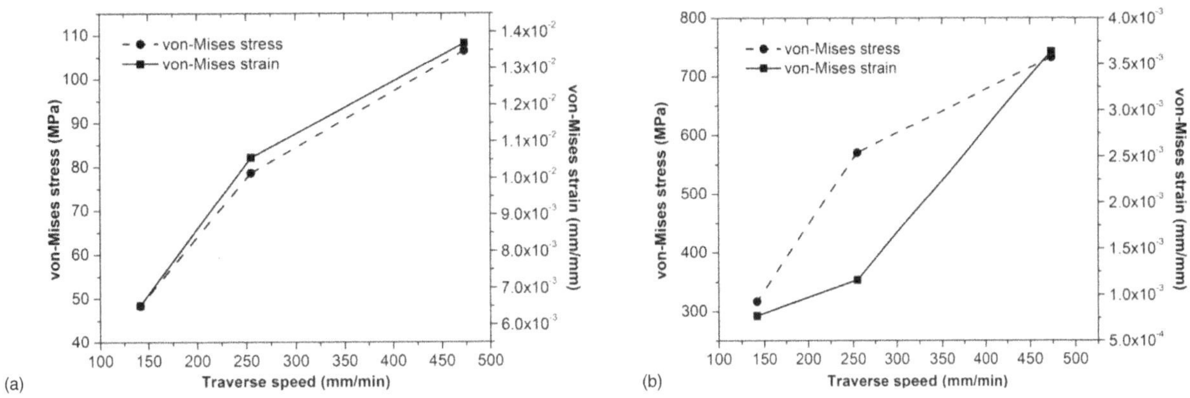


*Figure 8. Variation of von Mises stress for (a) PVB and (b) glass layer with a traverse speed*

was 48·41 MPa. The von Mises stress was increased to 78·65 MPa, and 106·50 MPa (67·45% and 119·92% higher) as the velocity was increased to 254·79, and 473·57 mm/min from 141·60 mm/min whereas the von Mises strain in the PVB layer was increased to 0·01056 and 0·013688 (62 and 110% higher). The von Mises stress and strain in the glass layer were 317 MPa and 0·0007732 for a traverse speed of 141·60 mm/min. The von Mises stress in the glass layer increased by 80 and 131% to 570·6 MPa and 732·27 MPa with a corresponding increase in von Mises strain to 0·0016236 and 0·036369 (50 and 236%) as the traverse speed was increased to 254·79 and 473·57 mm/min, respectively.

The simulation results reflect that the glasses have higher stress but lower strain than PVB interlayer for all travel speeds of abrasive particles. The higher travel speed has resulted in maximum *MRR* in the experiment, and the same is evident in the simulation results. This model can be used to predict the cutting time for different glass and interlayer thicknesses while cutting LG-PVB using AWJM.

## 4. Conclusions

The use of AWJM for LG cutting in a single pass is successfully demonstrated in the present work. AWJM offers higher productivity than the conventional cutting of LG. AWJM overcomes the problem of intermittent cutting due to interlayers in the conventional process and produces lesser surface imperfections. A mathematical model based on experimental results was developed for predicting the *MRR*. For three selected levels of machine parameters, it was found that *MRR* is affected most by travel speed (81·4701%) followed by water pressure (17·757%), and standoff distance has negligible (0·10959%) effect. The optimum level of all process parameters is found. However, further optimisation of the process variable is possible if the exact time of cutting of first glass plate, polymeric interlayer, and the second glass plate is found, and process parameters are set accordingly for glass and the interlayer. It can be concluded that the cutting of LG has a significant role in the fracture of LG due to the difference in the surface finish at the boundaries. AWJM should be considered for the cutting of LG to avoid boundary defects be a primary reason for failure. AWJM machined LG samples were having lesser cutter traces, smoother surface, and lesser defects. During the testing of LG samples by convention cutting mostly, the fracture was originated from the boundary of samples, which is not the case while AWJM cut the LG samples. For the crack-free surface, the abrasive particles should strike the LG surface at trivial angles and should move through the cutting line at a high velocity, a mechanical model based on mechanics of cutting followed by CFD analysis could be considered as future work for further optimisation of AWJM method.

A three-point bending test has revealed a smoother fracture of LG cut by AWJM in comparison to the conventional method due to lesser irregular and defected boundaries. The statistical analysis results have confirmed a higher bending strength of LG samples cut by AWJM than the conventional method. The prominent reason was the lesser boundary defects in LG samples cut by AWJM. The FE analysis of AWJM process using explicit dynamics simulation displayed increase in the equivalent stresses and strains with traverse speed. The proposed FE model can be used to predict the cutting time of LG-PVB using AWJM. It would be interesting to find the relation for how the increase in depth of cut reduces the *MRR* for LG. The AWJM can be examined for surface polishing of LG at boundaries for the superior surface finish.

## References


1. Liu, H. T. '7M' advantage of abrasive waterjet for machining advanced materials. *J. Manuf. Mater. Proc.*, 2017, 1 (1), 11. DOI:10.3390/jmmp1010011
2. Liu, X., Liang, Z., Wen, G. & Yuan X. Waterjet machining and research developments: a review. *Int. J. Manuf. Tech.*, 2019, **102** (5–8), 1257–335. DOI: 10.1007/s00170-018-3094-3
3. Hashish, M. An investigation of milling with abrasive-waterjets. *J. Eng. Ind.*, 1989, **111** (2), 158–66. DOI: 10.1115/1.3188745
4. Ruiz-Garcia, R., Mayuet, Ares, P., Vazquez-Martinez, J. & Salguero, Gómez, J. Influence of abrasive waterjet parameters on the cutting and drilling of CFRP/UNS A97075 and UNS A97075/CFRP stacks. *Mater.*, 2019, **12** (1). DOI: 107. 10.3390/ma12010107
5. Chen, F., Miao, X., Tang, Y. & Yin, S. A review on recent advances in machining methods based on abrasive jet polishing (AJP). *Int. J.*

*Manuf. Tech.*, 2017, **90** (1–4), 785–99. DOI: 10.1007/s00170-016-9405-7

6. Momber A. W. & Kovacevic, R. *Principles of Abrasive Water Jet Machining*. Springer Science & Business Media, 2012.
7. Górka, J. & Kotarska, A. The quality of water jet cutting of selected construction materials. In IOP Conference Series: *Mater. Sci. Eng.*, 2018, **400**, 022031. DOI: 10.1088/1757-899X/400/2/022031
8. Tiwari, T., Sourabh, S., Nag, A., Dixit, A. R., Mandal, A., Das, A. K., Mandal, N. & Srivastava, A. K. Parametric investigation on abrasive waterjet machining of alumina ceramic using response surface methodology. In IOP Conference Series: *Mater. Sci. Eng.*, 2018, **377** (1), 012005. IOP Publishing. DOI: 10.1088/1757-899X/377/1/012005
9. Siddiqui, T. U., Shukla, M. & Tambe, P. B. Optimisation of surface finish in abrasive water jet cutting of Kevlar composites using hybrid Taguchi and response surface method. *Int. J. Mach. Machinab. Mater.*, 2008, **3** (3–4), 382–402. DOI: 10.1504/IJMMM.2008.020970
10. Singh, D. & Chaturvedi, V. Investigation of optimal processing condition for abrasive water jet machining for stainless steel AISI 304 using grey relational analysis coupled with S/N ratio. *Appl. Mech. Mater.*, 2014, **592**, 438–43. DOI: 10.4028/www.scientific.net/AMM.592-594.438
11. Hashish, M. Optimisation factors in abrasive-waterjet machining. *J. Eng. Ind.*, 1991, **113** (1), 29–37. DOI: 10.1115/1.2899619
12. Wilkins, R. J. & Graham, E. E. An erosion model for waterjet cutting. *J. Eng. Ind.*, 1993, **115** (1), 57–61. DOI: 10.1115/1.2901639
13. Dhanawade, A. & Kumar, S. Experimental study of delamination and kerf geometry of carbon epoxy composite machined by abrasive water jet. *J. Compos. Mater.*, 2017, 51 (24), 3373–90. DOI: 10.1177/0021998316688950
14. Williams, R. E. & Rajurkar, K. P. Stochastic modeling and analysis of abrasive flow machining. *J. Eng. Ind.*, 1992, **114** (1), 74–81. DOI: 10.1115/1.2899761
15. Kamarudin, N. H., Rao, A. P. & Azhari, A. CFD based erosion modelling of abrasive waterjet nozzle using discrete phase method. In IOP Conference Series: *Mater. Sci. Eng.*, 2016, **114**, 012016. IOP Publishing. DOI: 10.1088/1757-899X/114/1/012016
16. Balamurugan, K., Uthayakumar, M., Sankar, S., Hareesh, U. S. & Warrier, K. G. Mathematical modelling on multiple variables in machining $LaPO_4/Y_2O_3$ composite by abrasive waterjet. *Int. J. Mach. Machinab. Mater.*, 2017, **19** (5), 426–39. DOI: 10.1504/IJMMM.2017.087616
17. MM, I. W., Azmi, A. I., Lee, C. C. & Mansor, A. F. Kerf taper and delamination damage minimisation of FRP hybrid composites under abrasive water-jet machining. *Int. J. Manuf. Tech.*, 2018, **94** (5), 1727–44. DOI: 10.1007/s00170-016-9669-y
18. Armağan, M. & Arici, A. A. Cutting performance of glass-vinyl ester composite by abrasive water jet. *Mater. Manuf. Process.*, 2017, **32** (15), 1715–22. DOI: 10.1080/10426914.2016.1269919
19. Chakraborty, S. & Mitra, A. Parametric optimization of abrasive waterjet machining processes using grey wolf optimizer. *Mater. Manuf. Process.*, 2018, **33** (13), 1471–82. DOI: 10.1080/10426914.2018.1453158
20. Kumar, A., Singh, H. & Kumar, V. Study the parametric effect of abrasive water jet machining on surface roughness of Inconel 718 using RSM-BBD techniques. *Mater. Manuf. Process.*, 2018, **33** (13), 1483–90. DOI: 10.1080/10426914.2017.1401727
21. Selvam, R., Karunamoorthy, L. & Arunkumar, N. Investigation on performance of abrasive water jet in machining hybrid composites. *Mater. Manuf. Process.*, 2017, **32** (6), 700–6. DOI: 10.1080/10426914.2016.1198039
22. Prabu, V. A., Kumaran, S. T. & Uthayakumar, M. Performance evaluation of abrasive water jet machining on banana fiber reinforced polyester composite. *J. Nat. Fibers.*, 2017, **14** (3), 450–7. DOI: 10.1080/15440478.2016.1212768
23. Selvan, M. C., Sampath, S. S., Madara, S. R. & Raj, N. S. Effects of process parameters on depth of cut in abrasive waterjet cutting of phosphate glass. In *Advances in Science and Engineering Technology International Conferences (ASET)*, 2018, Feb 6, pp. 1–6. IEEE. DOI: 10.1109/ICASET.2018.8376868
24. Cutler, V. *New Technologies in Glass*. A&C Black, 2012. ISBN: 9781408139547
25. Rihakova, L. & Chmelickova, H. Laser micromachining of glass, silicon, and ceramics. *Adv. Mater. Sci. Eng.*, 2015. DOI: 10.1155/2015/584952
26. Plat, K., von Witzendorff, P., Suttmann, O. & Overmeyer, L. Process strategy for drilling of chemically strengthened glass with picosecond laser radiation. *J. Laser Appl.*, 2016, **28** (2), 022201. DOI: 10.2351/1.4944508
27. Contuzzi, N., Campanelli, S. L. & Ludovico, A. D. 3d finite element analysis in the selective laser melting process. *Int. J. Simul. Model.*, 2011, **10** (3), 113–21. DOI: 10.2507/IJSIMM10(3)1.169
28. Cai, Y., Wang, M., Zhang, H., Yang, L., Fu, X. & Wang, Y. Laser cutting sandwich structure glass–silicon–glass wafer with laser induced thermal–crack propagation. *Opt. Laser. Technol.*, 2017, **93**, 49–59. DOI: 10.1016/j.optlastec.2017.01.028
29. Zhimalov, A. B., Solinov, V. F., Kondratenko, V. S. & Kaplina, T. V. Laser cutting of float glass during production. *Glass Ceram.*, 2006, **63** (9–10), 319–21. DOI: 10.1007/s10717-006-0112-y
30. Perec, A. Experimental research into alternative abrasive material for the abrasive water-jet cutting of titanium. *Int. J. Manuf. Tech.* 2018, **97**, 1529–40. DOI: 10.1007/s00170-018-1957-2
31. Vigneshwaran, S., Uthayakumar, M. & Arumugaprabu, V. Abrasive water jet machining of fiber-reinforced composite materials. *J. Reinf. Plast. Comp.*, 2018, **37** (4), 230–7. DOI: 10.1177/0731684417740771
32. Wang, T., Hou, R. & Lv, Z. Experimental investigation on the material removal of the ultrasonic vibration assisted abrasive water jet machining ceramics. *Adv. Mater. Sci. Eng.*, 2017. DOI: https://doi.org/10.1155/2017/1365786
33. Vedrtnam, A. & Pawar, S. J. Experimental and simulation studies on fatigue behaviour of laminated glass having polyvinyl butyral and ethyl vinyl acetate interlayers. *Fatigue Fract. Eng. M.*, 2018, **41** (6), 1437–46. DOI: 10.1111/ffe.12788
34. Vedrtnam, A. & Pawar, S. J. Experimental and simulation studies on flexural strength of laminated glass using ring-on-ring and three-point bending test. *P. I. Mech. Eng. C-J Mec.*, 2018, **232** (21), 3930–41. DOI: 10.1177/0954406217744815
35. Vedrtnam, A. & Pawar, S. J. Experimental and simulation studies on bending behavior of laminated glass with polyvinyl butyral and ethyl vinyl acetate inter-layers of different critical thicknesses. *J. Sandwich Struct. Mater.*, 2019, **2**. https://doi.org/10.1177/1099636219830143
36. Chen, X. C., Deng, S, S., Chen, M., Hua, W. X., Liao, S. & Yao, S. Simulation analysis of striation phenomena in abrasive water jet cutting (AWJC) process of AISI 304 stainless steel. *Metalurgija.*, 2018, **57** (1–2), 114–16.
37. Taylor, D. & Clancy, O. M. The fatigue performance of machined surfaces. *Fatigue Fract. Eng. M.*, 1991, **14** (2–3), 329–36. DOI: 10.1111/j.1460-2695.1991.tb00662.x
38. Vedrtnam, A. Experimental and simulation studies on delamination strength of laminated glass composites having polyvinyl butyral and ethyl vinyl acetate inter-layers of different critical thicknesses. *Defence Technol.*, 2018, **14** (4), 313–17. DOI: 10.1016/j.dt.2018.02.002
39. Thirumalai Kumaran, S., Uthayakumar, M., Mathiyazhagan, P., Krishna Kumar, K., & Muthu Kumar, P. Effect of abrasive grain size of the awjm performance on AA(6351)-SiC-B4C hybrid composite. *Appl. Mech. Mater.*, 2015, **766–767**, 324–329.
40. Paul, S., Hoogstrate, A. M. & van Praag, R. Abrasive water jet machining of glass fibre metal laminates. Proc. Inst. Mech. Eng. B: J. Eng. Manuf., November 1, 2002 https://doi.org/10.1243/095440502320783396
41. Azmir, M. A. A study of abrasive water jet machining process on glass/epoxy composite laminate. *J. Mater. Proc. Technol.*, 2009, **209** (20), 6168–73.
42. Abarca, A. M. *Tests Performed On Laminated Glass Bonded With A New Interlayer Composition*. Thesis, Graduate Faculty of Texas Tech University, USA, 2003.

# TEACHING GLASS BETTER

The ICG, a platform for global cooperation in the world of glass, has as part of its mission the promotion and stimulation of understanding and interaction between glass experts in science and technology, art, history and education. To revitalize the role that education plays a committee was created, using a European Grant, to support and develop teaching programs on glass fundamentals in 2005. The target audience was those new to the subject, whether in industry or academia, groups perceived as the lifeblood of research and development and the key to the ongoing success of the material in the marketplace. Based on this background, the first ICG summer school took place in Montpellier in 2009. It has been repeated annually ever since and at the 2017 school everyone agreed that 2018 should be a celebratory event. One aim was to prepare a book which summarized the course content by which also captured the historical development and the philosophy of the schools, explaining the lessons learned and offering a framework for others wishing to follow.

Preparation and publication was completed to coincide with the Tenth School in July 2018 and to high standard using full colour. The early chapters review the history and the evolutionary processes involved in creating the school's present structure. The major part of the book presents the core of the course – spanning glass science lectures on: structural analysis using the latest analytical techniques and atomistic simulation; optical, chemical and mechanical properties; thermodynamics and transport phenomena; and phase separation, glass ceramics and glass surfaces. Over the last five years technical streams on: glass melting technology; glass production; glass surfaces and thin films; and numerical modeling, have run alongside the more fundamental courses. The chapters are written in a tutorial style by the lecturers – all recognized glass scientists and technologists. The appendices include more personal and less formal descriptions of each event, an example of questionnaire responses from the students, biographies of the contributors and last, but not least an index.

The book provides an invaluable introduction to those starting a career in glass. It offers a starting point for the next generation of glass experts who will in turn be able to influence the health of academic glass research and stimulate an efficient and productive glass industry in the face of global concerns such as sustainability, global warming, health issues, energy shortages and population growth. The International Commission of Glass appreciates from the success of these schools that this future is in good hands.

A key feature of the courses is to encourage networking, links that can and do continue well beyond the close of each school. While a book such as this encapsulates what is taught, the only way to experience the value of the contacts made between all the participants, whether teachers or students, is to participate. The dates of future schools and how to join can be found on the ICG web site (icglass.org).

**Teaching Glass Better:** Tenth Anniversary of the ICG Summer School
Edited by: Akira Takada, John Parker, Alicia Durán, Klaus Bange
ISBN 978-84-17528-04-1
Weight: 884g;
Pages: 416;
Size: 240×170×20 $mm^3$
Printed in full colour.
Cost: € 50/ £45 plus postage

The Society of Glass Technology: http://store.sgt.org/index.php?app=gbu0&ns=prodshow&ref=TeachingGlass

Corning, NY

# Borate X

**July 17–21, 2023**

# Phosphate III

## Revised Conference Dates

The 10th International Conference on
**BORATE GLASSES, CRYSTALS AND MELTS**
**July 17–19, 2023**

This marks the 10th in the series of conferences on borate glasses, crystals and melts, and will return to New York, USA close to where the first Borate Conference was held (Alfred, 1977). The most recent meeting took place in Oxford, UK (2017). The Borate Conference is a forum for the discussion of current scientific research on borate and B-containing materials. The conference will be dedicated to Dr. E.I. (Stratos) Kamitsos from the National Hellenic Research Foundation (Athens), to honor his lifetime of work in the field of borate glasses.

The 3rd International Conference on
**PHOSPHATE GLASSES, CRYSTALS AND MELTS**
**July 19–21, 2023**

This conference continues the tradition established with the first two conferences held in Pardubice, Czech Republic (2014) and Oxford, UK (2017). It will cover all aspects of phosphates and P-containing materials, including P-O, P-S, P-Se and P-O-N networks. Topics of interest will cover basic science, applications and technologies. The Phosphate Materials conference will be dedicated to Professor Ladislav Koudelka, University of Pardubice, Czech Republic, recognizing his career in phosphate and borophosphate glasses.

## Conference Themes

This joint meeting, to be held at the Corning Radisson Hotel, is comprised of two conferences that will run sequentially. On the third day of the meeting there will be a combined day dedicated to borophosphate glasses followed by a conference dinner to mark the close the Borate Conference and the start of the Phosphate Conference. Topics will encompass anything related to borate and phosphate materials, including:

- **Energy**
- **Healthcare**
- **Optical and data storage materials**
- **Property-structure relationships**
- **Fundamentals of the glassy state**
- **Novel glasses, crystals and melts**
- **Technology and Industry**

### Conference Chair

Dr. Randall Youngman
Corning Incorporated
youngmanre@corning.com

### Local Organizing Committee

R. Youngman (chair), J. Wu (Corning), L. Ma (Corning), L. Huang (RPI), A. Goel (Rutgers), D. Möncke (Alfred), A. Clare (Alfred)

***www.boratephosphate2020.org***

INTERNATIONAL YEAR OF GLASS 2022

# 'Glass in our Lives' is a boon says the youth

**-Photography winners announced on the International Youth Day**

*AIGMF Photograph Contest: First Prize, S. Christy Laura aged 15, 10th class student of Sri Akilandeswari Vidyalaya, (Trichy) Tamil Nadu*

Following the United Nations approval of 2022 as the International Year of Glass (IYOG2022). Many glass related activities are being planned worldwide; as a kickstart to these events and a partner organisation in India, the All-India Glass Manufacturer's Federation (AIGMF) announced its First Photography Contest for the age group between 7–24 years on the theme: Glass in our Lives. Entries were invited online by July 31, with results to be announced on the International Youth Day on August 12.

The following distinguished jury members representing a cross section of glass sectors and regions decided the top placing photographs out of ~7000 entries received from educational institutes and youth across India:

- Sanjay Somany, Former President AIGMF and CMD HNG Inds. Ltd
- Sanjay Ganjoo, Former President AIGMF and COO Asahi India Glass Ltd
- Dr Suman Kumari Mishra, FNASc, Director CSIR-Central Glass and Ceramic Research Institute and Adjunct Prof. AcSIR and Ex-Dean Eng. Sc. AcSIR
- Prof. Manoj Choudhary, Former President ICG, Adjunct Prof. Materials Science & Engineering, The Ohio State University and President, MKC Innovations
- Dr Arun K. Varshneya, Emeritus Professor (Glass Science) Alfred University and President Saxon Glass Technologies
- Dave Fordham, Publisher Glass Worldwide (UK).

First Prize (Rs. 20 000) was given to S. Christy Laura aged 15, 10th class student of Sri Akilandeswari Vidyalaya, (Trichy) Tamil Nadu

Second Prize (Rs. 10 000) was given to Medhansh Singhla aged 8, 3rd class student of Delhi Public School (Vasant Vihar) New Delhi

Third Prize (Rs. 5000) was given

*AIGMF Photograph Contest: Second Prize, Medhansh Singhla aged 8, 3rd class student of Delhi Public School (Vasant Vihar) New Delhi*

to Brijesh Jethva aged 22, from (Junagadh) Gujarat.

"The AIGMF's First Photography Contest proved a worthy kick-start to the Federation's planned activities for the International Year of Glass 2022 and *Glass Worldwide* was very pleased to be involved in the process. The jury were presented with a vast selection of wide-ranging photographs to evaluate under the 'Glass in our Lives' theme, many of which were of a very high standard. All entrants should be commended for supporting the initiative and following a thorough judging process, I offer sincere congratulations to the winners", said Dave Fordham, Publisher, *Glass Worldwide.*

"The enthusiasm shown by the Youth through a large participation highlights the importance of glass in our daily lives. Not just photographs but some very good artwork of glass decorations, paintings were also submitted by the school children, emerging out of this interesting project. AIGMF would sincerely like to thank Education Departments of Delhi Government and other States, who helped us connecting with thousands of children on this educative project", commented the Secretary AIGMF, Vinit Kapur.

*AIGMF Photograph Contest: Third Prize, Brijesh Jethva aged 22, from Gujarat*

A formal award ceremony would be held parallel to the virtual Annual General Meeting of the AIGMF on September 10, where an online exhibition of select photos of the event would also be unveiled. This would form a part of a touring exhibition on the theme 'Glass in our Lives'. Detailed coverage and best entries will be published in *Kanch (Glass)* and *Glass Worldwide.*

A lucky 500 entries will be given a specially designed Glass Bottle made out of recycled glass (manufactured by Hindustan National Glass & Industries Ltd., under its CSR initiative), carrying logos of International Year of Glass 2022 and Swachh Bharat Abhiyaan (Clean India Campaign).

www.aigmf.com

# Recycling of lead glass from landfill

Glasriket, the Kingdom of Crystal, in Sweden is the location of a project looking at methods for recycling lead glass from landfill sites across the area. The three year project started in September 2019 and is run by RISE, Sweden's independent, state-of-the-art research institute for future-proof technologies, products and services.

The project provides companies and local authorities with a basis for being able to invest in recycling and innovative remediation. The broad consortium of interested parties is looking at five work packages: project management, sorting, smelting, reuse and implementation. The project is financed by SGI, the County Administrative Boards, the participating companies and the municipalities with a total budget is SEK5.45 million.

The Glasriket has about 40 landfill sites to be decontaminated by recycling the lead glass. This has the potential to remove pollutants from the environment as well as reduce the extraction of new raw materials. Methods for recycling lead glass are being explored and verified on an industrial scale. This provides companies and public actors with a basis for being able to invest in waste recovery and innovative remediation of the Kingdom of Crystal.

The Kingdom of Crystal flourished during the 19th century until the middle of the 20th. Many of the glassworks produced lead crystal glass for some of this period. The typical solid crystal contains 30% by weight of lead and usually about 1% by weight of arsenic or antimony. In a geographical area today containing a total of 14 glassworks in the municipalities of Emmaboda, Nybro, Uppvidinge, and Lessebo in southern Sweden there are currently around 40 glassworks properties and landfills that are classified by the county administrative boards as very high risk. These must be decontaminated sooner or later. All landfills are unique with different amounts of glass and glass qualities. How the glass is distributed over the area and how accessible also varies. What was waste could now become a resource of glass and constituent metals, will it be possible to exploit this?

In a previous project surveying glass landfills, the project manager RISE showed that it was possible to separate lead and glass on a small scale. In the succeeding project, the technology is being scaled up to recycle glass from the landfills and investigate how it can be implemented across all of Glasriket. To do this it is necessary to:

- identify crystal glass-dense areas
- separate the glass into different glass qualities
- find suitable melting and scaling factors
- develop a melt separation technology for industrial scale
- provide examples of what you can do with the glass
- set up the guidance for a contractor to implement the results.

By verifying the technology at full scale, a contractor can be guided on investing in a suitable facility. Reusing the glass in Glasriket can then become a reality. The metals can also be purified and reused instead of extracting from new minerals.

The project has been extended by three months and will run until November 30, 2022. The delay is a direct effect of the covid-19 pandemic. In order for the project to continue, The Glass Factory

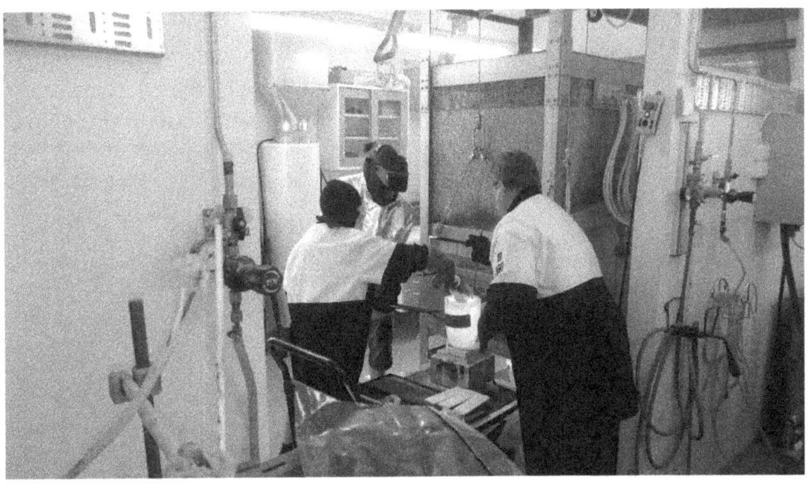

*Reduction experiments where glass and metal are separated*

has become a new project partner. They took over the first of the large-scale melts that would have been made in Målerås Glasbruk.

## Sampling and sorting

The first preparatory sorting of samples from Strömbergshyttan have been carried out. The sorting is done at Ragnsell's facility in Heljestorp through, among other things, eddy current, magnetism and screening. The next step is to send the prepared material to a demonstration facility in Germany. There the crystal glass will be sorted by measuring the lead content (XRF). Automatic camera sorting will be used to remove potential contaminants. Colour sorting can, for example, remove glass that contains cadmium. The goal is to obtain four samples of 100 kg of excavated crystal glass. This will be used for test melting and must meet certain criteria in order for it to be used.

*Scaling up the waste recovery provided cast tiles for analysis*

## Melting at The Glass Factory

During Easter 2021 the first large scale melt was achieved. This was an important milestone for in the development of the method that will hopefully lead to a recycling industry in the Kingdom of Glass. The first melts in a large crucible with a volume of 100 litres.

In the first melt, the glass was scooped out about 24 hours after a reducing agent was added to the almost 1400°C melt. Among other things, tiles were cast.

The second melt was treated in the same way as in the previous experiments with small crucibles. The whole oven was lowered in temperature after the reducing agent had worked for 24 h.

The large-scale trials exceeded expectations. The crucible held the molten metal and there was clear separation of the lead from the glass mass.

After evaluation of the first two large-scale experiments, some experiments on a lab scale will be made before the next large-scale experiment at Målerås Glasbruk.

## Reuse with opportunities

The prototyping of the potential product is now entering its second phase. This means that the possibilities and properties of the remelted glass are tested to see how dramatic shapes the material tolerates. This will determine the overall design of the final product.

## Implementation

At the same time as the experiments at The Glass Factory an effort was made to document all the work by filming. The idea is that the material can be used for different purposes to communicate this important work. The experiments at The Glass Factory also created media attention and were followed up by both television and national radio.

Communication is important in the work. During the spring, the County Administrative Board of Kronoberg had led work in compiling a communication plan that will guide communication with different target groups, both locally, regionally and nationally. Equipping the research group with the right tools to be able to maintain public interest, at the same time as it also creates the conditions for further development funds.

A new element in the collaboration between the county administrative boards, SGU and the municipalities is a regular planning meeting to keep all parties updated on which grants can be applied for and what the status is in various projects. The planning meetings will be a good addition to ensure that the remediation work continues to be successful.

The project management has had conversations with the business leaders in the municipalities to discuss how we can create an interest in an industrial establishment in Glasriket.

*Coverage on national television for the waste recovery project*

# "24-hour turnaround" for Mersen UK

## High precision products facilitated by Teesside University digitised manufacturing KTP

Mersen UK Teesside, a world-leading manufacturing facility specialising in machined graphite products, anti-corrosion and hot glass handling equipment (Cerberite®), is now offering a 24 hour turnaround time for low volume, high precision engineering products following the introduction of a digitised manufacturing process developed in a Knowledge Transfer Partnership (KTP) with Teesside University.

Mersen is a global expert in electrical specialties and advanced materials for high-tech industries. With more than 50 industrial sites and 16 R&D centres in 35 countries, the company develops customised solutions and delivers key products to its clients to meet the new technological challenges shaping tomorrow's world. For Mersen UK Teesside, the KTP fed directly into the company's overall aim to increase revenue from low volume, high value sales, maximising manufacturing operations profitability and a growing market for highly variable precision graphite parts in short lead times.

A KTP is an Innovate UK-funded programme that helps businesses to improve their competitiveness and productivity through the better use of knowledge, technology and skills that reside within universities. Mersen UK Teesside sought the University's expertise through a two-year project, overseen by Dr David Hughes (below), a Senior Lecturer and Enterprise Fellow (Materials Engineering) in the institution's School of Computing, Engineering & Digital Technologies.

"Due to the nature of products that Mersen manufacture, the part volume is generally low but high value, with small batch (less than 10) and one-off production of specialist parts (known as 'Strangers') generating a majority of their current sales revenue," Dr Hughes said. "Mersen's objective is to expand its high precision engineering capabilities to meet growing demands from specialist customers it cannot currently serve. Such clients have regular low volume, high value requirements, stringent focus on accuracy, tolerance, traceability and short lead times that were not previously achievable within Mersen's (or most of their competitors) operations. To react quickly to these unique part requirements Mersen recognised the need for more intelligent manufacturing systems and processes to ensure sufficient flexibility, automation and integration of the Strangers pipeline."

In response, the KTP project introduced an Electrical Discharge Machining (EDM) webstore offering customers a one-day turnaround time, a significant improvement compared to previous lead times. Activities from the KTP are aiming to increase current market share over the coming years, whilst enhanced data collection and analysis offer significant opportunities for optimisation and a cultural shift to welcome digitisation. This included the addition of Robotic Process Automation (RPA). The project, led by Knowledge Transfer Partnership associate Daniel Pybus – who has subsequently enrolled on a PhD at the University – has addressed several key opportunities, Dr Hughes explained.

"Developing knowledge and tools to collect, measure and monitor production data has been an important development which has allowed Mersen to identify opportunities for optimisation by analysing accurate, real-time data," he said. "The management and display of such data has highlighted several key areas for continuous improvement which will directly contribute to its strategy to meet growing demands from specialist customers. Whilst traditional manufacturing processes are reactive and disruptive, the KTP project has built the tools required to become more predictive to enable a more flexible, dynamic manufacturing process with less disruption. This enables it to minimise industry lead times toward a market-leading position, maximising opportunities in existing UK and export markets and new specialist markets such as electrical discharge machining electrodes."

Additionally, stock cutting was traditionally done by eye, risking lower yields and higher waste through offcuts. Mersen needed a solution to challenge this practice and replace it with a system that can analyse inventory and suggest cutting techniques for higher yield and lower waste. The KTP project has addressed this by developing an evolutionary algorithm which suggests optimised cutting order. This enables Mersen to improve profit margins by maximising raw material utilisation and has shown savings of 20% in current trials. In order for the required opportunities to be addressed, a culture shift needed to occur to ensure the team embraced Industry 4.0 solutions. The KTP assisted this shift through training, communication, and giving responsibility and accountability to the team in the developments undertaken.

*Mersen UK Teesside Managing Director Andrew Freer*

"The KTP partnership has provided the company with the means to identify and solve complex problems to improve efficiency and accelerate growth, as well as helping Mersen to expand its industrial network and giving the opportunity to support digitisation and development in the local region," said Mersen UK Teesside Managing Director Andrew Freer. "As digitisation continues, significant savings are expected through automation and smarter, leaner working practices. It is expected that more value-adding tasks will be carried out, improving company efficiency, financial performance and customer satisfaction. New working practices will enable Mersen to offer more competitive bids for work, and further research and development has the scope to increase the range of products offered. Critically, all these capabilities aim to provide a competitive advantage that will secure Mersen's place in the European market post-Brexit. Without this advantage, new trading arrangements could otherwise mean that our products and lead times are less attractive in Europe."

Following on from the learning developed in this KTP, the University has launched the Industrial Digitalisation Technology Centre (IDTC) to support SMEs in the Tees Valley explore the opportunities Industry 4.0 technologies bring, such as Internet of Things (IoT), big data analytics, modelling, simulation and sensor technology – creating an essential competitive edge that generates real value for businesses. Funded through the European Regional Development Fund (ERDF), the IDTC offers consultancy and mentoring for digitalisation projects that typically last from six to 12 months, giving businesses a practical opportunity to investigate the digital tools and services available to help plan their digitalisation journey.

Mersen is supporting the IDTC and, as members of the University's School of Computing, Engineering & Digital Technologies industrial advisory board, feed into teaching design and the implementation of digitalisation across its engineering degrees.

**Noelle + von Campe** processes up to 550 tons of glass to approximately 1·8 million glass containers daily. The family-run company is located in Boffzen, Germany, and supplies more than 35 countries.

Following an expansion initiative, **Zippe Industrieanlagen** has received an order for an extension of the existing batch house and a completely new cullet return system. Cullet addition is 50%. The Zippe scope of supply includes the engineering, the equipment, the silos, as well as the complete mechanical and electrical installation including the supervision of the works.

The batch house work will feed a new furnace with a maximum of 420 tons daily in addition to the existing melting furnace with 320 tons melting capacity. **CM. Project. Ing** was responsible for the complete coordination of the construction activities including detailed planning of the buildings and supply media.

Installation is planned to start in summer 2021, commissioning at the end of 2021.

Following a competitive funding call, the **UKRI-ISCF Transforming Foundation Industries Network+** has awarded over £270,000 in funding to six collaborative projects supporting energy efficient manufacturing across the foundation industry sectors.

Two of the projects are related to the glass industry: Improved Energy Efficiency of Float Glass Production – Dr Peter Green (**University of Liverpool**) in collaboration with **NSG Group**; and Using Blast Furnace Waste Heat to Convert Various Wastes into New Raw Materials for Low Energy Glass – Dr Zushu Li (**University of Warwick**) in collaboration with **British Glass, Glass Futures, Calumite** and **Materials Processing Institute**. A third project looks at the cold sintering of silica based shaped refractories led by Dr Linhao Li (**University of Sheffield**) in collaboration with **Vesuvius UK**.

"The TFI Network+ has remit to help the foundation industries get ready for net-zero 2050. These projects are in key research areas that could radically alter the carbon footprint of several foundation industry sectors. We look forward to reporting their outcomes and to funding more high-quality projects in the future" said Professor Ian Reaney, Director of the Transforming Foundation Industries Network+.

All projects will begin in September, with project descriptions available on the TFI Network+ website. In July, the Network launched a second funding call, inviting proposals under the themes of resource efficiency and industrial symbiosis.



The **NSG Group** has successfully developed a new ultra-thin type of SELFOC Micro Lens with the diameter of 125 μm, which is as fine as optical fibres.

In the "Beyond 5G" era, where the number of network-connected devices as well as network traffic is expected to increase explosively, opticalisation of network infrastructure is being considered a solution to address such challenges as ensuring high speed and large capacity communication and, at the same time, drastically reducing electricity consumption.

The newly developed ultra-thin microlens will be able to make various devices that connect optical fibres and various terminals smaller, simpler and more cost competitive. Thus realising the all-optical network, which could reduce power usage significantly as optical signal is used instead of electrical one. The Group also aims to contribute to the mitigation of climate change impacts.

The SELFOC Micro Lens, the unique cylindrical and flat-end-

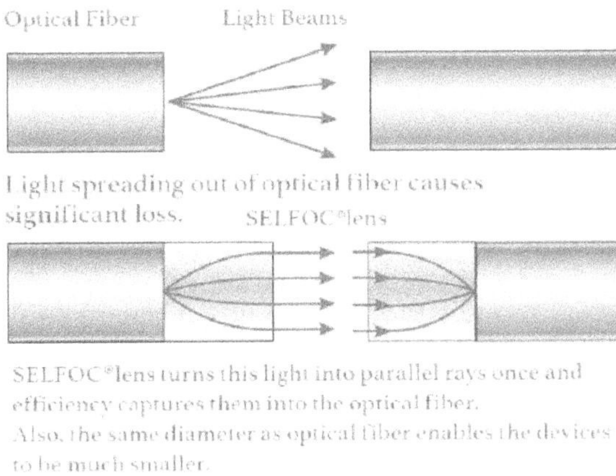


*NSG's SELFOC Micro Lens with the diameter of 125 μm, as fine as optical fibres. Offering the opportunity to develop all-optical devices*

shaped glass lens with a gradient-index, was developed by NSG in 1968 and has been used in a wide range of applications such as coupling and collimating lens for telecommunication devices, imaging lens for various endoscopes and microscopes used in the medical, life science and industrial areas because of its easy-to-handle shape and minimal transmission loss.

While the diameter of conventional lens ranged from 1·0 to 4·0 mm for telecommunication devices and from 0·25 to 2·70 mm for imaging devices, the new product's diameter is much smaller at 0·125mm. Being finer and having the same diameter as optical fibers, optical devices can be designed smaller and simpler. The NSG Group is now preparing for its commercial production and, going forward, plans to develop new applications in the imaging area as well.

**Ciner Glass** has launched a public consultation on plans for a container glass manufacturing and recycling facility in Ebbw Vale, Wales. The proposed Dragon Glass Bottle Manufacturing Facility at the Rassau Industrial Estate will create around 400 jobs initially. The aim is to satisfy the demand for glass in the local Wales market.

The company is committed to providing training and employment opportunities for local people. Its preferred strategy is to employ young people, with the workforce in Turkey having an average age of 26.

"Ciner Glass intends to make a positive and long-term contribution to the future of Wales. We want to work constructively in partnership with the Welsh and UK Governments as well as other relevant stakeholders in the Valleys and across Wales. The hope is that the size and scale of this project can act as a catalyst to provide further innovation and investment to this area," said Mrs Didem Ciner. "We think it is important to cut the number of miles bottles have to travel to reach the customer. Our vision is to build a centre of engineering excellence in glass technology here in Wales which can compete with the best in the world."

Ciner has headquarters in London and is part of the Ciner Group a Turkish family business with glass container factories in Turkey

*Mrs Didem Ciner*

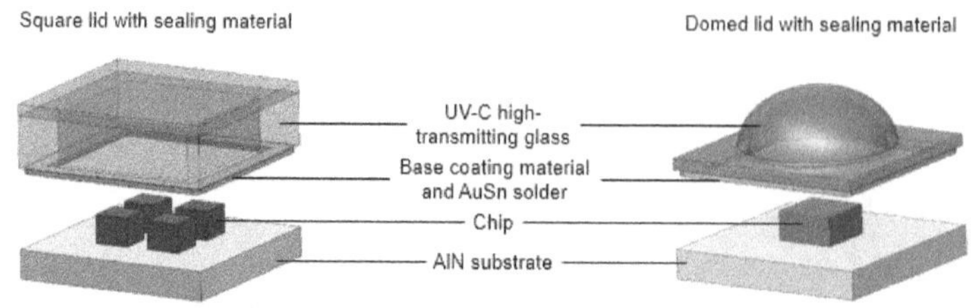


**Nippon Electric Glass** has developed and commercialized two types of cap lids: a square lid (box-shaped lid) for UV-C LEDs with the world's highest light extraction efficiency and a domed lid (bowl-shaped lid) for UV-C LEDs with the world's highest light extraction efficiency and a wider light distribution angle.

The square lid developed is integrated with gold–tin solder for sealing a substrate, to which the necessary anti-reflection coating is applied. It achieved a light extraction efficiency of 96%, which exceeds that of conventional quartz glass cap lids. The domed lid is a cap lid that is made by processing UV-C high-transmitting glass into a dome (hemispherical) shape and integrating gold–tin solder for sealing a substrate onto the flange. It achieved a light extraction efficiency of 93% and a wide light distribution angle.

There are high hopes for UV-C LEDs in sterilization and virus inactivation. Many manufacturers have been trying to increase the power of LEDs, but the power remains at less than one-hundredth of that of mercury lamps. For this reason, lids used for UV-C LEDs are required to have improved light extraction efficiency, and lids for high-power LEDs are shifting from a structure to which flat glass is attached to a cap shape with excellent light extraction efficiency.

By supplying the two types of cap lids developed to the market, the company will contribute to the improvement in performance of UV-C LEDs and the realisation of mass production.

---

as well as energy and mining, shipping and media. There are also proposals for glass factories in Belgium and Kazan, Russia.

**Berlin Packaging** has acquired **Elias Valvanis** of Larissa, Greece. All employees and locations for this acquisition will be retained.

With locations in Bulgaria, Romania, and throughout Greece, Elias Valavanis has been a key supplier of glass bottles and jars for olive oil, wine, spirits, soft drinks and water. Elias Valavanis has long-term relationships with its supplier partners and its 1500+ customers, who benefit from the company's value-added services, including package structural design, decoration capabilities and the company's own glass recycling facility.

"Elias Valavanis is the perfect partner for Berlin Packaging as we expand our operations to the Balkans and continue to augment our Mediterranean glass business," said Paolo Recrosio, CEO of Berlin Packaging Europe. "Elias Valavanis shares our passion for glass design and artistry and our commitment to help our customers grow their business by offering industry-leading packaging products and services."

"We take tremendous pride in all aspects of our business, and we know we've found a like-minded partner," said Elias Valavanis, "I am excited about combining our company with Berlin Packaging, as I know together, we can accelerate investments in our products, capabilities, and employees like never before."

This is the 15th acquisition that Berlin Packaging has completed in Europe since 2016, and its fifth acquisition in Europe during 2021.

**Encirc** is set to reduce its carbon output in its Elton, Cheshire facility by more than 15600 tonnes per year by installing a new, intelligent end-to-end process control system.

The company, with bid writing support from **Glass Futures**, has been awarded funding from the **Department for Business, Energy, and Industrial Strategy** to begin its 'DEEP Control Project', ('Deployment of End-to-End Process control for Encirc's Elton site'). The technology will allow the container producer to optimise its furnaces to run at minimum viable energy, delivering huge carbon reduction levels every year, while also improving productivity and product quality.

The DEEP Control project, which will digitally link the furnaces to Encirc's 14 production lines, is being carried out in partnership with Glass Futures, the **Science and Technology Facilities Council** (STFC) Hartree Centre and **Siemens**. The new system will also further increase the value of Encirc's 'Industry 4.0-Ready' glass production line, built at the end of 2020, which uses advanced technologies to ensure that glassmaking is safer, easier and efficient.

*Encirc in both Elton and Derrylin have seen huge decarbonisation initiatives*

Paddy O'Doherty, Encirc's deputy operations director, commented: "Glass is not only a historic material, it's the packaging choice of the future. Glass's sustainability benefits are unparalleled, but we know there is work to be done in further reducing the carbon output of the production process.

"Industry leading projects such as this help us boost production, upskill our workforce and reduce our carbon footprint in line with consumer demands for more sustainable products. This is just one part of glass's decarbonisation challenge. We are also pushing the boundaries when it comes to using more innovative fuels, such as with our recent biofuel project in our Northern Ireland factory in Derrylin, which used sustainable biofuel in place of high carbon natural gas to reduce the carbon footprint of the bottle-making process by up to 90%, creating the world's most sustainable glass bottles to date.

"We're also championing better, more innovative ways to encourage glass recycling once our products are in the market, and we're an active supporter of the implementation of a Digital Deposit Return Scheme (DDRS) across the UK."

Senior R&D Projects Manager at Glass Futures, Dr Ajith Appukuttan said, "Glass Futures has a highly qualified innovation team and the funding bid for this project was an excellent opportunity to offer Encirc support by identifying this project and then leading them through the complex bid writing process."

**SCHOTT** has celebrated the official opening of its new factory in Jinyun, China. The production, which went into operation at the end of 2020, is manufacturing high quality borosilicate glass tubing for vials and syringes for COVID-19 vaccines. With an initial

*Consolidating housing, childcare, classrooms, study spaces, food and student services into one 14-story, 335,000 square-foot building, the Peter George Centre for Living and Learning (PGCLL) at* ***McMaster University****, Hamilton, Ontario, is a true hybrid building.* ***Diamond Schmitt****'s innovative architectural design includes a four-story, skylit atrium; a 650-seat auditorium and connecting helical staircase; and a U-shaped block of daylit residential and study spaces. Solarban 67 and Solarban 70 solar control, low-emissivity (low-e) glasses by* ***Vitro Architectural Glass*** *were chosen to enhance the building's aesthetics and environmental performance, while "Nuance" acid-etched patterns from* ***Walker Glass*** *make the glass curtainwalls more decorative. Solarban 67 glass was used on all the building's vertically installed glass, as it provided the best balance of energy performance and clarity. Solarban 70 glass was selected for the atrium skylight due to its increased energy performance, lower VLT and protection against glare. The high-performance glass reduced the building's heating and cooling loads, which was critical in the south-facing kitchen and lounge spaces, and the west-facing study areas which could have been prone to overheating.*

capacity of 20000 tons and room for further expansion, the plant is functioning as a production and supply hub in the region. Schott has invested €60 million in the new plant as part of a $1 billion investment to expand its global pharmaceutical business.

Building a melting tank for high-quality pharma glass tubing requires extraordinary skills. Despite the pandemic, a team of German and Chinese experts managed to complete the construction of the factory in record-breaking "China speed" of just 15 months. "Of course, our glass stands at the very beginning of an ambitious value chain. But it's one piece of the bigger picture: Turning the vision of "Healthy China 2030" into reality. We're very proud that it also marks the first time that Schott is melting glass in China," says Dr Frank Heinricht, chairman of the management board.

With the help of big data and artificial intelligence, each tube is inspected 100% based on the company's perfeXion process.

The manufacturing site will support the move away from the low borosilicate glass (so called 7.0 glass type) to the higher quality middle borosilicate glass (so called 5.0 glass type), which is already well established globally.

**Nipro PharmaPackaging** is investing over $60 million in its facilities for glass primary packaging in France and the US. The sum addresses the global increase in quality and demand, as well as supports Nipro's growth in its pharmaceutical packaging business. In France, the investment means the construction of a greenfield operation in Aumale in the Normandy region.

Housed in a new building adjacent to the current manufacturing facility, the new lines will be producing NSV borosilicate type I glass. Automation combined with the pure composition of Nipro's glass enables the site to produce improved quality glass tubing for high-end pharmaceutical packaging products.

Mr. Stephan Arnold, CEO commented: "We have come a long way with this site in Aumale. We are celebrating our 150th anniversary this year and have had a long history with many different partners. When we joined Nipro in 2011, we immediately felt a very strong sense of commitment. This commitment resulted in the biggest investment ever in this plant's history. Together with my team here, I look forward to a stronger future in France. I want to thank Nipro for its trust in us, and I want to thank the team for their relentless efforts in getting this project started. Together, we are building the future here!"

**De Dietrich** have opened a new 2000 m$^2$ factory in Wuxi, China less than 10 minutes from the company's first plant. The new plant increases production in glass-lined spare parts for customers, in China and abroad, providing them with the De Dietrich quality and best-in-class products, compliant with all international standards.

With this new factory, the company will have the capacity to produce 20000 items a year (50000 items expected in 2025), and this allows the first Wuxi plant to be dedicated to manufacturing reactors, tanks and columns.

With an increasing demands in the pharmaceutical, chemical, and plant-based sectors, De Dietrich Process Systems is consequently aligned with its markets and customers evolutions.

**AMETEK Land** has enabled **Thai Glass**, one of the leading container glass makers in Thailand, to optimise its production quality with the installation of a high-performance temperature monitoring system.

Traditional CCTV cameras in the glass melt tanks were replaced with a system that would add both visual and temperature measurement to its process control. AMETEK Land's NIR-Borescope-656-Glass thermal imaging solution produces high-definition thermal images and highly accurate temperature measurements in the range of 1000 to 1800°C. The new system allows Thai Glass to obtain a wider variety of temperature measurements in different locations, and can trend measurements at key points, including the crown, ports, and burner blocks.

A spokesperson for Thai Glass commented: "We have seen significant advantages of using AMETEK Land's NIR-B-656-Glass in terms of process optimisation and quality control. The data obtained is invaluable in enabling us to monitor batch flow closely and to ensure optimum efficiency in the operation of the melt tank."

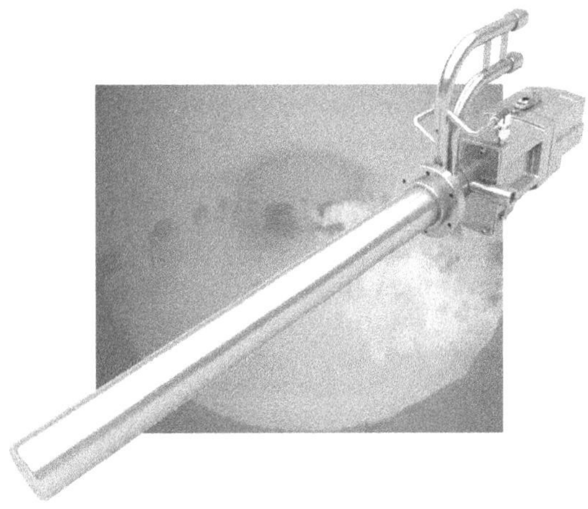

*AMETEK Land's NIR-Borescope-656-Glass thermal imaging*

Phillipe Kerbois, Glass Sector Manager at AMETEK Land, said: "Glass producers worldwide are looking for more robust thermal imaging solutions that provide much more than just CCTV images. Our systems offer a truly unrivalled data-driven solution to the challenge of effective temperature measurement of glass melt tanks and increase production efficiency."

The **International Finance Corporation** (IFC) is investing $40 million in **NSG UK Enterprises**, a long-standing IFC client and a member of the **NSG Group**. The corporate loan will finance the completion and start of operations of the group's greenfield glass production plant in Cardales, Argentina, with an estimated production of 900 tpd.

Cardales is around 70 km north-west of Buenos Aires, the plant will utilise the latest NSG worldwide technology supporting the achievement of industry and sustainability targets. This includes aspects such as application of Industry 4.0 standards and technologies that support the decarbonisation ambition of NSG Group.

IFC's loan to NSG UK, through the company's subsidiary in Argentina, **Vidriería Argentina** (VASA), will positively impact the local economy by substituting imports and generating exports, adding value to domestic raw materials and enhancing value chains. Furthermore, IFC will provide an advisory service to VASA to help identify energy efficiency opportunities.

A working group comprising **Snam**, **RINA**, **Bormioli Luigi**, **Bormioli Rocco**, **STARA GLASS**, **Università degli Studi di Genova**, **Stazione Sperimentale del Vetro**, **IFRF Italia**, **SGRPRO** and **RJC SOFT** has begun a collaboration aimed at reducing emissions in the glass industry through hydrogen.

The manufacture of glass objects, of which Italy is the second-largest producer in Europe with over 5 million tonnes/year, is energy-intensive and difficult to power with electricity. To this end, the 'Divina' project (Decarbonisation of the Glass Industry: Hydrogen and New Equipment), co-ordinated by Snam, RINA and Bormioli, aims to reduce emissions in the glass melting stage, which accounts for more than 50% of total energy consumption throughout the production process.

Snam CEO Marco Alverà said: "Hydrogen will play a key role in decarbonising energy-intensive sectors such as glass production in order to meet domestic and European climate targets. This project complements what we are already doing in the steel, rail transport and ceramics sectors."

Ugo Salerno, President and CEO of RINA, commented: "Following the first test with a mix of natural gas and 30% hydrogen in steel processing that we carried out in May, our expertise and laboratories are also being used for the 'Divina' project, an important milestone towards the decarbonisation of another of the most significant sectors in the Italian economy."

Vincenzo Di Giuseppantonio, CEO of the Bormioli Luigi Group, remarked: "The partnership with leading players in the energy sector, the glass industry and academia makes the Divina project a sound and concrete proposal for a path to green transition and sustainability."

The initiative will make it possible, in the short to medium term, to assess the results of introducing an increasing proportion of hydrogen blended with natural gas into existing melting furnaces operating under regular production conditions. Testing significant quantities of hydrogen on operational furnaces will be an opportunity to evaluate the compatibility of hydrogen combustion with glass material in real industrial production contexts following appropriate testing in laboratories. The project will also define and subsequently optimise the design rules for future furnaces, which can guarantee the best performance even with higher hydrogen percentages up to 100%.

Juergen Maier CBE, former Chief Executive of Siemens UK, has joined Glass Futures as an advisor to the Board.

Mr Maier is passionate about championing manufacturing, improving infrastructure and engineering skills and is enthusiastic about the benefit of industrial digitalisation to encourage advanced manufacturing and investment in digital technologies. He is currently Chair of the Digital Catapult and Co-Chair of the UK's Made Smarter programme.

He said: "I was delighted to be asked to become involved with Glass Futures to encourage their members' to introduce new digital manufacturing technique, increasing production efficiencies and yields."

Glass Futures' Chief Executive, Richard Katz said: "I'm delighted Juergen has agreed to join Glass Futures as an Advisor to the Board. This is a further indication of Glass Futures' acceptance as a growing major industrial innovator and thought leader, as demonstrated a few months ago with the incredibly successful biofuel trial reducing the carbon footprint of some 3 million beer and spirits bottles by some 90%."

"Juergen joining us also provides a tremendous opportunity for Glass Futures' members to benefit from his vast experience on the delivery of innovation and technology across the glass and other foundation industries."

Researchers at **Zhejiang University**, Hangzhou, People's Republic of China have developed a way to create intricate three-dimensional structures out of glass by mimicking origami. The process allows the team to make complex shapes that would be difficult or impossible to achieve with traditional glass-working methods

The technique involves curing a silica nanoparticle-filled liquid precursor into a polymer composite sheet, which is cut into the desired shape. The flexible sheet is then manually folded at room temperature into a 3D structure, just as is done with paper in origami. Once the structure is complete, the material undergoes pyrolysis and sintering to remove the polymer binder and convert the remaining material into glass.

Although this technique currently relies on manual folding of the polymer composite, the team hopes that the process could be automated. As the cutting process is done using a low-power, digitally controlled laser, the technique may be suitable for large-scale manufacturing.

Y Xu *et al*, *Nat. Commun.*, 2021, **12**, 4261 (DOI: 10.1038/s41467-021-24559-x)

**Owens Corning** has launched a new version of its PINK fibre insulation featuring advanced technology that enables up to 23% faster installation compared to existing products.

The PINK Next Gen® Fiberglas creates a tightly woven network of soft, fine fibres to form a super-resilient blanket of insulating micro-pockets. The new product is safe for installers and residents and faster to install than competitive insulation products and spray foam insulation. Because no

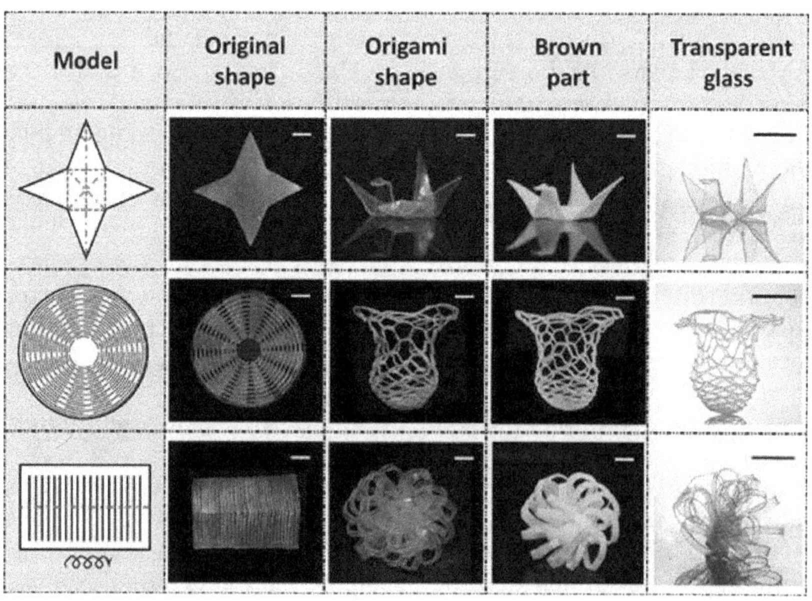


*Fabrication of transparent origami glass DOI: 10.1038/s41467-021-24559-x*

***British Glass*** *has reported the latest figures from the* ***Department for Environment, Food and Rural Affairs*** *show a glass recycling rate of 75·8% for 2020 making glass the second most recycled packaging in the UK. The provisional figure is based on a reported 1 841 000 tonnes recovered/recycled from 2 430 000 tonnes of packaging waste arising*

hazardous chemicals are required to be mixed on the install site, there is less potential for human error, increasing peace of mind for installers and residents.

"Contractors today have high performance expectations and extremely tight timelines," said Joe Wagner, insulation marketing director. "This product evolution is a significant step-change in meeting the needs of contractors, as well as installers, builders and homeowners."

The product is designed to meet building codes and help professionals manage through tight labour timelines and budgets by reducing install time.

"PINK Next Gen insulation sets a new standard for insulation as the right choice for contractors, builders and homeowners," Wagner added. "This is just the beginning of a series of ongoing innovations in 2021 and beyond that will enable Owens Corning to continue exceeding customers' high expectations."

With net zero targets being high on the governments' agendas in not just the UK but worldwide, the **Glass and Glazing Federation** (GGF) is working with **British Glass** to research the impact glass and glazing has in making buildings more energy efficient and carbon reducing. This new research will show how glass types perform in buildings and why the government should be giving the glass and glazing industry more consideration in future initiatives or policies when it comes to energy efficiency and zero carbon.

John Agnew GGF group managing director comments "The GGF has been working closer with British Glass for a few years and this latest collaboration cements our strong working relationship."

Dave Dalton, chief executive of British Glass adds, "British Glass and the GGF have had a long-standing partnership for many years and on several projects and campaigns. It's essential for the industry to collaborate to amplify our voice to government for the benefit of the sector."

Aligned to the research, the GGF will also survey other trade bodies in the glass, glazing and fenestration sectors to explore concerns and find common ground. In doing so the GGF aims to present the importance of energy efficient glass and glazing, to improve the existing and new building stock, whilst making a valuable contribution to reducing carbon emissions.

Mr Agnew summarises, "Since 1977, when trade associations from all sides of the industry came together to form the GGF, the Federation has continuously worked with other trade organisations to advance the interests of Members and in many cases, for the betterment of the entire industry. Our industry collaborations are imperative to increase our voice and progress our shared objectives for the benefit of all in the industry and the consumer."

**Asahi India Glass** (AIS) and **Vishakha Group** have announced a joint venture into solar glass with an in-principle agreement to set up India's largest and state-of-the-art green field solar glass plant at Mundra, Gujarat. With phase-1 glass manufacturing capacity of up to 3 GW of the installable capacity the factory is expected to be commissioned within 18–24 months. AIS will take a minority stake in the solar glass business.

Vishakha Group is one of the largest manufacturers of flexible polymer packaging. It has manufacturing facilities for key components in the entire solar panel value chain: solar backsheets, EVA encapsulant sheets and solar module aluminium frames.

Sanjay Labroo, managing director, AIS, said: "Although a different market segment, solar glass is still an adjacent business to the existing businesses of AIS. We are delighted to partner with Vishakha group in their vision of creating the entire ecosystem and value chain for indigenizing solar energy generation in India."

Jigish Doshi, chairman Vishakha Group, said: "India has been one of the few nations that has accelerated its global commitment towards climate change and we intend to do our part to achieve the renewable energy target. The joint venture into solar glass manufacturing will support the "Atmanirbhar Bharat" initiative of Government of India, contribute in achieving energy independence and lead to massive job creation in the Gujarat region. We are delighted to partner with AIS in this new journey. We plan to compliment the technical expertise & experience of AIS with synergies from our other businesses in the entire solar value chain".

# CONFERENCE DIARY

**16–20 *August* (postponed to 2023)**
Tenth International Conference on Borate Glasses, Crystals and Melts and Third International Conference on Phosphate Materials, Corning, UK.
*Dr Randall Youngman, Corning Incorporated Email youngmanre@corning.com Web boratephosphate2020.org/*

**~~29 August–2 September~~ (postponed to 2022)**
XVII ECerS Conference, Dresden, Germany.
*Fraunhofer Institute for Ceramic Technologies and Systems IKTS, Winterbergstrasse 28, 01277 Dresden, Germany. Web www.ecers2021.org/*

**1–4 September**
Glass South America, Sao Paulo, Brazil.
*NürnbergMesse Brasil, R. Verbo Divino, 1547 - 7th floor, Chacara Santo Antonio, 04719-002 Sao Paulo, Brazil. Email glass@nm-brasil.com.br Web www.nm-brasil.com.br*

**4–12 September**
Venice Glass Week, Venice, Italy and online.
*Camilla Purdon, Festival Coordinator, Venice Glass Week, Istituto Veneto di Scienze, Lettere ed Arti, Palazzo Loredan, Venezia 30124, Italy. Email info@theveniceglassweek.com Web www.theveniceglassweek.com*

**7 September**
GlassTrend webinar on glass production transition to renewable fuels
*Web www.glasstrend.nl/events/*

**13–17 September**
22nd Congress of the Association Internationale pour l'Histoire du Verre, Lisbon, Portugal.
*Web aihv.org/congress/*

**13–15 September**
GlassBuild America 2021, Atlanta, Georgia, USA.
*National Glass Association, 1945 Old Gallows Road, Suite 750, Vienna, VA 22182, USA. Web www.glassbuildamerica.com/*

**14–17 September**
XXth Biennial Worldwide Congress Unified International Technical Conference on Refractories (UNITECR2021), Chicago, USA.
*The American Ceramic Society, 550 Polaris Parkway, Suite 510, Westerville, OH 43082, USA. Email customerservice@ceramics.org Web www.ceramics.org*

**20–23 September**
Resilient Materials 4 Life (RM4L2021), Cambridge, UK.
*Email info@rm4l.com Web rm4l.com/rm4l2020*

**~~20–23 September~~ Postponed**
16th Conference on Furnace Design: Operation and Process Simulation, Velke Karlovice, Czech Republic.
*Glass Service, Rokytnice 60, 755 01 Vsetin, Czech Republic. Email info@gsl.cz Web www.gsl.cz*

**~~26–28 September~~ (new date May 2022)**
FIT Show, Birmingham, UK.
*Freya Humphries, Account Manager, FIT Events Ltd, 9 Manchester Square, London, W1U 3PL, UK. Web www.fitshow.co.uk*

**27 September–1 October**
CelSian training course - general glass technology training, Eindhoven, The Netherlands.
*CelSian Glass BV Email infodesk@celsian.nl Web www.celsian.nl/training-and-rd/*

**4–6 October**
Sixth International Conference on Competitive Materials and Technology Processes (ic-cmtp6), Miskolc-Lillafured, Hungary.
*Email cmtp.conf@gmail.com Web www.ic-rmm1.eu/ic-cmtp/*

**5–8 October**
Vitrum 2021, Milan, Italy.
*Fiera Milano, Via Carlo Ilarione Petitti, 16, 20149 Milano MI, Italy. Email vitrum@vitrum-milano.it Web www.vitrum-milano.com*

**6–8 October**
Intersolar 2021, Munich, Germany.
*Web www.intersolar.de/en/home*

**12–15 October**
FESPA Global Print Expo, Amsterdam, The Netherlands.
*Web www.fespaglobalprintexpo.com/*

**12 October**
CelSian training course: heat transfer in glass melting furnaces, Eindhoven, The Netherlands.
*CelSian Glass BV Email infodesk@celsian.nl Web www.celsian.nl/training-and-rd/*

**13–14 October**
CelSian training course: burners, combustion and NOx emissions, Sheffield, UK.
*CelSian Glass BV Email infodesk@celsian.nl Web www.celsian.nl/training-and-rd/*

**16–17 October**
Icon Bath 2021 Conference: Ceramics, Glass and Stained Glass, Bath UK.
*The International Institute for Conservation of Historic and Artistic Works, 3 Birdcage Walk, Westminster, London, W1H 9JJ, UK. Email IconBath2021@gmail.com Web www.iiconservation.org*

**17–21 October**
ACerS 123rd Annual Meeting with Materials Science & Technology 2021, Pittsburgh, USA.
*The American Ceramic Society, 550 Polaris Parkway, Suite 510, Westerville, OH 43082, USA. Email customerservice@ceramics.org Web www.ceramics.org*

**20–22 October**
Glass Performance Days, Tampere, Finland.
*Brown Onduso, GPD, Vehmaistenkatu 5, 33730 Tampere, Finland Email brown.onduso@gpd.fi Web gpd.fi/*

**26–27 October**
CelSian training course: refractory selection, maintenance and related defects, Eindhoven, The Netherlands.
*CelSian Glass BV Email infodesk@celsian.nl Web www.celsian.nl/training-and-rd/*

**27–30 October**
Seventh International Conference on Multifunctinal, Hybrid and Nanomaterials, Genoa, Italy.
*Web www.elsevier.com/events/conferences/international-conference-on-multifunctional-hybrid-and-nanomaterials*

**28–29 October**
CelSian training course - flat glass melting and forming, Eindhoven, The Netherlands.
*CelSian Glass BV Email infodesk@celsian.nl Web www.celsian.nl/training-and-rd/*

**30 October–5 November**

Sixth Wuhan ICG Winter School, Wuhan, China.
*Prof. Kai Xu: SMART, Wuhan University of Technology, 1040 Heping Ave, Wuchang Qu, Wuhan Shi, Hubei Sheng, China. Email kaixu@whut.edu.cn Web www.icglass.org*

**30 October–5 November**
82nd Conference on Glass Problems, Columbus, USA.
*Donna Banks, GMIC, 550 Polaris Parkway, Suite 510, Westerville, OH 43082, USA. Email dbanks@gmic.org Web glassproblemsconference.org*

**~~16–18 November~~ postponed to 26–28 October 2022**
Glasstech Asia 2021, Bangkok, Thailand.
*Web www.glasstechasia.com.sg/*

**17–18 November**
Glass and Modern Technologies - XXI, Moscow, Russia.
*Organizing Committee Email forum@steklosouz.ru Web www.steklosouz.ru*

**18 November**
Glass Focus Awards 2021, UK.
*British Glass, 9 Churchill Way, Chapeltown, Sheffield, South Yorkshire S35 2PY, UK. Web www.britglass.org.uk/glass-focus-2021*

**23–25 November**
Celsian training: glass melting and process improvement for Indian glass industry, Mumbai, India.
*CelSian Glass BV Email infodesk@celsian.nl Web www.celsian.nl/training-and-rd/*

**5–8 December**
International Commission on Glass (ICG) Annual Meeting, Incheon, South Korea.
*Secretariat, Room 403, Meolijae Building, 76 Bangbae-ro, Seochu-gu, Seoul, Republic of Korea. Email secretariat@icg2021.org Web www.icg2021.org/*

**12–17 December (new dates)**
14th Pacific Rim Conference on Ceramic and Glass Technology (PACRIM 14) including Glass & Optical Materials Division 2021 Annual Meeting (GOMD 2021), Vancouver, BC, Canada.
*Andrea Ross, The American Ceramic Society, 550 Polaris Parkway, Suite 510, Westerville, OH 43082, USA. Email aross@ceramics.org Web www.ceramics.org*

**2022 UN International Year of Glass**
**Postponed to 2022 no date announced**
13th International Symposium on Crystallization in Glasses and Liquid, Paris, France.
*Laurent Cormier, CNRS, Paris, France. Email crystallization@sciencesconf.org Web crystallization.sciencesconf.org/*
ACerS Pan American Ceramics Conference, PACC-FMAs, Panama City, Panama.
*Andrea Ross, The American Ceramic Society, 550 Polaris Parkway, Suite 510, Westerville, OH 43082, USA. Email aross@ceramics.org Web www.ceramics.org*

**10–12 February**
Opening Ceremony of the International Year of Glass 2022, Geneva, Switzerland.
*Web www.IYoG2022.org*

**18–19 February**
Glassman Asia, Seoul, Korea.
*Ken Clark, Quartz Business Media, Quartz House, 20 Clarendon Road, Redhill, Surrey RH1 1QX, UK Email kenclark@quartzltd.com Web www.glassmanevents.com*

**3–5 March**
glasspex INDIA 2022, Mumbai, India.
*Vivek Bohra, Senior Project Manage Email bohrav@md-india.com Web www.glasspex.com*

**6–10 March**
Optical Fiber Communications Conference and Exhibition (OFC), San Diego, California USA.
*The Optical Society, USA. Email custserv@osa.org Web www.ofcconference.org*

**15–18 March**
Unified International Technical Conference on Refractories (UNITECR 2022) 17th Biennial Worldwide Congress on Refractories, Chicago, USA.
*The American Ceramic Society, 550 Polaris Parkway, Suite 510, Westerville, OH 43082, USA. Email aross@ceramics.org Web www.ceramics.org*

**2–4 April**
Deco '22, The Society of Glass and Ceramic Decorating Products annual conference, Columbus, OH, USA.
*Myra Warne, Society of Glass and Ceramic Decorating Products, PO Box 2489, Zanesville, OH 43702, USA. Email info@sgcd.org Web www.sgcd.org*

**3–6 April**
TMS World Congress on Artificial Intelligence in Materials and Manufacturing (AIM 2022), Pittsburgh, Pennsylvania, USA.
*TMS Meeting Services, 5700 Corporate Drive Suite 750, Pittsburgh, PA 15237, USA. Web www.tms.org/AIM2022*

**5–6 April**
GlassPrint 2022, Düsseldorf, Germany.
*Email sales@glassworldwide.co.uk Web www.glassprint.org*

**13–16 April**
China Glass 2022, Shanghai, China.
*Web www.chinaglass-expo.com*

**18–20 April**
Int. Conf. on Glass Science and Technology from the Golden Era of the Pharaohs to a Sustainable Future, Egypt.

**19–22 April**
13th European Conference on Industrial Furnaces and Boilers, Algarve, Portugal.
*Raquel Reis, Cenertec, Rua Gago Coutinho, 185-187, 4435-034 Rio Tinto, Portugal. Email raquel.reis@cenertec.pt Web infub.pt*

**11–12 May**
Glassman Latin America, Monterrey, Mexico.
*Ken Clark, Quartz Business Media, Quartz House, 20 Clarendon Road, Redhill, Surrey RH1 1QX, UK Email kenclark@quartzltd.com Web www.glassmanevents.com*

**18–20 May**
REFRA PRAGUE 21, 21st Conference on modern refractory materials and key achievements in high temperature technologies, Prague, Czech Republic.
*Ing. Jan K. Luxa, Silikátová společnost České republiky, Novotného lávka 5, 110 01 Praha 1, Czech Republic. Email hluxa@seznam.cz Web www.silikaty.cz*

**18–21 May**
GAS 2022, Glass Art Society conference, Tacoma, WA, USA.
*Glass Art Society, 2208 NW Market Street, Suite 200, Seattle, Washington 98107, USA. Email info@glassart.org Web www.glassart.org*

**22–26 May**
2022 Glass and Optical Materials Division Annual Meeting (GOMD2022), Baltimore, USA.
*Andrea Ross, The American Ceramic Society, 550 Polaris Parkway, Suite 510, Westerville, OH 43082, USA. Email*

*aross@ceramics.org Web www.ceramics.org*

**6–9 June**
Mir Stekla 2022, Moscow, Russia.
*Ekaterina Razumeitseva, Project Head, EXPOCENTRE, Moscow, Krasnopresnenskaya nab., 14, 123100, Russia. Email re@expocentr.ru Web www.mirstekla-expo.ru/en/*

**8–9 June**
Furnace Solutions and Training Day, venue TBC, UK.
*Christine Brown, Society of Glass Technology, 9 Churchill Way, Chapeltown, Sheffield S35 2PY, UK. Email christine@sgt.org Web www.sgt.org*

**16–17 June**
XXXIV ATIV Int. Conf., Where Glass Science, Art and Technology Meet Together, Parma, Italy.
*ATIV Secretariat, MV Congressi spa Via Marchesi 26 d 43126 Parma, Italy. ativ@ativ-online.it Web www.ativ.eu*

**21–24 June**
Techtextil 2022, Frankfurt, Germany.
*Email techtextil@messefrankfurt.com Web techtextil.messe-frankfurt.com*

**23–24 June**
Challenging Glass Conference 8: International Conference on the Architectural and Structural Application of Glass, Ghent, Belgium.
*Email cgc8@challengingglass.com Web www.challenging-glass.com*

**3–8 July**
26th International Congress on Glass, Berlin, Germany. Hosted by the DGG in their centenary year.
*Annette Doms, Deutsche Glastechnische Gesellschaft, Siemenstrasse 45, D-63071 Offenbach, Germany. Email dgg@hvg-dgg.de Web www.hvg-dgg.de*

**4–8 July**
Ninth International Conference on Optical, Optoelectronic and Photonic Materials and Applications (ICOOPMA) and EuroDIM: 14th Europhysical Conference on Defects in Insulating Materials, Ghent, Belgium.
*Web icoopma-eurodim.ugent.be*

**10–15 July**
Physics of Non-Crystalline Solids, now also including Society of Glass Technology Annual Meeting Canterbury, UK.
*Christine Brown, Society of Glass Technology, 9 Churchill Way, Chapeltown, Sheffield S35 2PY, UK. Email christine@sgt.org Web www.sgt.org*

**10–14 July**
Ninth Int. Congress on Ceramics (ICC9), Krakow, Poland.
*European Ceramic Society AISBL, Av. Gouverneur Cornez 4, B-7000 Mons, Belgium. Email ecers@bcrc.be Web ecers.org/*

**24–29 July**
Sol-Gel 2021, International Sol-Gel Society Conference, Lyon, France.
*Université Claude Bernard Lyon 1, Cellule Congrès - SOLGEL 2021, 43, bd du 11 Novembre 1918, 69622 Villeurbanne, France. Email solgel2021@univ-lyon1.fr Web www.solgel2021.fr*

**27–30 August**
International Festival of Glass, Stourbridge, UK.
*Ruskin Glass Centre, Wollaston Road, Amblecote, Stourbridge DY8 4HF, UK. Email ifg@rmlt.org.uk Web www.ifg.org.uk*

**20–23 September**
Glastec 2022, Düsseldorf, Germany.
*Anke Seltmann, glastec, D-40474 Düsseldorf, Am Staad (Stockumer Höfe), Germany. Email seltmanna@messe-duesseldorf.de Web www.glasstec-online.com/*

**9–13 October**
ACerS 124th Annual Meeting with Materials Science & Technology 2022, Pittsburgh, USA.
*The American Ceramic Society, 550 Polaris Parkway, Suite 510, Westerville, OH 43082, USA. Email customerservice@ceramics.org Web www.ceramics.org*

**11–13 October**
GlassBuild America 2022, Las Vegas, USA.
*National Glass Association, 1945 Old Gallows Road, Suite 750, Vienna, VA 22182, USA. Web www.glassbuildamerica.com/*

**19–21 October**
XXXIV ATIV Int. Conf., Where Glass Science, Art and Technology Meet Together, Parma, Italy.
*ATIV Secretariat, MV Congressi spa Via Marchesi 26 d 43126 Parma, Italy. ativ@ativ-online.it Web www.ativ.eu*

**26–28 October 2022**
Glasstech Asia 2022, Singapore.
*Web www.glasstechasia.com.sg/*

**2023**

**5–9 March**
Optical Fiber Communications Conference and Exhibition (OFC), San Diego, California USA.
*The Optical Society, USA. Email custserv@osa.org Web www.ofcconference.org*

**4–9 June**
2023 Glass and Optical Materials Division Annual Meeting (GOMD 2023), New Orleans, USA.
*The American Ceramic Society, 550 Polaris Parkway, Suite 510, Westerville, OH 43082, USA. Email customerservice@ceramics.org Web www.ceramics.org*

**10–14 July**
Nonlinear Optics, Honolulu, Hawaii, USA.
*The Optical Society, 2010 Massachusetts Ave, NW, Washington, DC 20036 USA. Email custserv@osa.org Web www.osa.org*

**17–21 July**
Tenth International Conference on Borate Glasses, Crystals and Melts and Third International Conference on Phosphate Materials, Corning, USA.
*Dr Randall Youngman, Corning Incorporated Email youngmanre@corning.com Web boratephosphate2020.org/*

**~~August~~ (moving to 2024)**
NCM15, Structure of Non-Crystalline Materials 15, combined with the SGT Annual Meeting, UK.
*Christine Brown, Society of Glass Technology, 9 Churchill Way, Chapeltown, Sheffield S35 2PY, UK. Email christine@sgt.org Web www.sgt.org*

**2024**

**June**
Tenth International Conference on Optical, Optoelectronic and Photonic Materials and Applications (ICOOPMA), Pardubice, Czech Republic.
*Center of Materials and Nanotechnology and Department of General and Inorganic Chemistry, Faculty of Chemical Technology, University of Pardubice, Cs. Legii sq. 565, 532 10 Pardubice, Czech Republic. Email info@icoopma.com Web www.icoopma.com*

**2–6 July**
XVIII ECerS Conference, Lyon, France.
*European Ceramic Society AISBL, Av. Gouverneur Cornez 4, B-7000 Mons, Belgium. Email ecers@bcrc.be Web ecers.org/*

ERNST-ABBE-FONDS
IM STIFTERVERBAND

# CALL FOR APPLICATIONS

## OTTO SCHOTT RESEARCH AWARD

In 2022 the Ernst-Abbe-Fonds will be presenting 17th Otto Schott Research Award. Since 1991 the prize honours biannually excellent researchers for their scientific and technological achievements in basic and applied research in the fields of glass, glass-ceramic and related advanced materials.

The Otto Schott Research Award is awarded for individual achievements, but also the work done by smaller groups. It is endowed with a total of 25,000 euros.

Scientists from universities, non-university research institutes and industrial research institutions worldwide are eligible. Candidates can be nominated by an internationally renowned scientist or engineer. Moreover, self-applications are allowed, but must be accompanied by a letter of support.

Central criteria for the awarding of the prize are the candidate´s scientific activity in recent years, the quality and originality of his/her original scientific publications, and the extent how he/she has influenced the research field.

The board of trustees of the Ernst-Abbe-Fonds is responsible for selecting the laureates. It will base its decision on the recommendations of a prize jury which is composed of internationally renowned material scientists. The board´s decision is final.

Each application/nomination should include:

» the candidate´s five top ranked and/or most influential publications in the field of glass sciences or related fields;
» one or more letters of support from outstanding scientists acknowledging the candidate´s scientific achievements;
» metrics (h-index) of scientific activity over the last 5 and 10 years;
» a list of other activities that highlight the candidate´s scientific impact (e.g. prizes, peer review, editorships, committee functions);
» the candidate´s curriculum vitae.

Applications should be submitted only via this website until
**3rd October 2021.**

If you have any questions please contact

Klaus Kuli
Ernst-Abbe-Fonds, Deutsches Stiftungszentrum
Postfach 16 44 60 . 45224 Essen
T ++49 201 8401-160 . klaus.kuli@stifterverband.de

The Ernst-Abbe-Fonds is a foundation that is dedicated to scientific research in the fields of fine mechanics, optics, electronics, glass, and ceramic glass. It was founded in 1989 by the Carl Zeiss Foundation in honor of its 100th anniversary. The Carl Zeiss Foundation is the only shareholder of SCHOTT AG, Mainz, Germany, and Carl Zeiss AG, Oberkochen, Germany. For further information on the Ernst-Abbe-Fonds please refer to www.abbe-fonds.de.

STIFTERVERBAND
Bildung. Wissenschaft. Innovation.

Glass Technol.: Eur. J. Glass Sci. Technol. A, 2021, 62 (3), 96–108

# Thermal, optical and mechanical properties of new glass compositions containing fly ash

*Clarissa Justino de Lima,*[a,b,1] *Fred Veer,*[c] *Hongzhi Zhang,*[a] *Fernando França de Mendonça Filho,*[a] *Oguzhan Copuroglu*[a] *& Rob Nijsse*[a,c]

[a] *Department of materials, mechanics, management & design (3Md), Delft University of Technology, the Netherlands*
[b] *American Glass Research, the Netherlands*
[c] *Department of Architectural Engineering & Technology, Delft University of Technology, the Netherlands*

Manuscript received 25 July 2019
Revision received 7 March 2021
Manuscript accepted 16 March 2021

*The investigation of new compositions is crucial for the expansion of possible applications of glass, from the typical applications for building engineering, in the form of cast blocks or float glass, to more advanced technologies, such as 3D-printed glass or glass-to-metal connections. Since high melting temperatures and brittleness are two important drawbacks of glass, this work aims to improve both properties. Characterisation techniques, such as thermal analysis, nano-indentation, and UV/VIS spectroscopy, are used to evaluate the properties of the samples. The modification of the properties is achieved via changes in the composition of the glass, using compounds such as phosphorus pentoxide, aluminium oxide and boron oxide. Then, the choice of different glass formers and modifiers contributes to the development of compositions with lower melting and glass transition temperatures. The reduction of the melting temperature allows a saving of energy during the manufacturing. The structures of the glasses differ from the standard soda–lime–silica and borosilicate glasses, leading to a different mechanical behaviour. Furthermore, these new compositions incorporate up to 35% of fly ash in their formulas. The valorisation of these by-products reduces costs and gas emission.*

## 1. Introduction

The utilisation of industrial waste in manufacturing building materials has increased considerably due to concerns over the amount of waste produced and the significant environmental problems caused by its disposal. Expanding populations, urbanisation, and increased wealth are ramping up the global production of solid waste. Factors like population and per capita gross domestic product (GDP) are used to measure total global municipal solid waste (MSW) production.[1] The waste production rates per capita usually grow with wealth, despite the fact that there is a tendency toward anti-materialism in wealthier countries. When combined, these aspects indicate a scenario in which, over the next decades, global waste generation will go beyond our capacity to control it.[1]

Fly ash is a by-product from burning pulverised coal in electric power generating plants, it can be used in Portland cement concrete (PCC) pavement, paving roads, regular structural concrete, or any other normal concrete, among others. Even with the shift from the energy generation modes from thermal to other renewable sources, the dependence on coal is expected to be significant in the next decades. This culminates in a large amount of fly ash generation and the ways that need to be explored for its better utilisation.

Coal-fired power plants are declining as a result of the growing popularity of renewable energy and cleaner forms of fuel, such as natural gas. As a by-product of the coal combustion process, the fly ash generation is directed by the consumption of coal for electricity generation. The percentage of electric power from coal is predicted to decrease from 39% of total generation in 2013 to 34% in 2033.[2] However, the combination of economic growth and an increasing population will drive increasing demand for electricity. Consequently, coal electric generation is projected to grow by 3·4% from 2013 to 2033, according to the US Energy Information Association.[2] According to a survey by the American Coal Ash Association (ACAA), approximately 37 million tons of fly ash were produced in the US in 2018.[3] Of this, around 20 million tons were used in concrete and other applications. The rest was disposed of in landfills.

Many countries still did not reach an optimal utilisation of fly-ash. For instance, in India and China, the top two countries based on annual fly ash generation, the current percentage utilisation is quite behind the target. As per a 2018 report by the Central Electricity Authority (CEA), about 57% of the total power generated in India is produced by coal-based

[1] Corresponding author. Email cdelima@agrintl.com
DOI: 10.13036/17533546.62.3.007

power plants. Globally, coal provides around 27% of overall energy supplies and generates about 38% of the world's electricity. This percentage is expected to remain stable for the next 30 years.[4]. Even with numerous initiatives, taken at the policy level by various ministries and a large number of technological developments by the various R&D organisations for gainful utilisation, the percent of fly ash utilisation in India was 67·13% in 2018. From this amount of fly ash, approximately 38·67% is utilised in the construction sector for various modes, like 25·60% in cement, 9·01% in bricks and tiles, 3·40% in roads and flyovers, and 0·66% in concrete.[5] In China, the percent of fly ash utilisation was 70% in 2016.[5]

As a considerable amount of coal combustion products (CCPs) is still sent to disposal, pollution control agencies and environmental groups are putting increased pressure on power producers to find more valuable uses for ash. A conservative estimation of the quantity of CCPs in disposal across the US exceeds 2 billion tons.[6] Technologies under development can provide new opportunities to use CCPs, and part of this volume, disposed in landfills and ponds, will be reclaimed for beneficial use. Even with the long list of applications for fly ash utilisation, no individual application can assure 100% use of fly ash. For this reason, a reasonable mix of applications is unavoidable. Cement and concrete have reached the threshold limit and will not accommodate more than 35–40% of overall ash generated in the next years.[5] Therefore, it is understood that the cement and concrete alone would not be able to utilise the total production of fly ash and new alternatives have to be found and promoted.

Despite a significant number of products developed from fly ash, most of them consist of non-transparent materials, partly because it is a challenge to get transparent materials in reasonable temperatures from this by-product. Fly ash contains many elements that are also present in typical glass formulas. For instance, the elements found in higher amounts in the chemical systems of standard silicate glasses are $SiO_2$, $Na_2O$, CaO, $K_2O$, MgO and $Al_2O_3$.[7] All these elements are found in fly ash compositions. Some of these components are highly refractory and their presence in complex compositions means they are highly likely to crystallise and have high working temperatures. However, glass is a material that allows large amounts of various elements in solution, being suitable to assimilate complex compounds in its compositions. Therefore, this work aimed at characterising new glass recipes, which incorporate up to 35% of fly ash in their compositions, while keeping transparency and relatively low working temperatures.

The choice of the element used as the glass former is the first step in designing a glass composition. Phosphate glasses possess relatively low working temperatures, low liquidus viscosity, large ultraviolet transparency, and high solubility for other glass modifiers or intermediaries.[8] For this reason, phosphate was used as a glass former in this work.

The compound $KPO_3$ was chosen as the phosphate source. Like other alkali metal ions, potassium can be incorporated between the chains, each formed from $PO_4$ tetrahedron. Alkali metal ions have a small electric field intensity and a small electronegativity. Consequently, the bonding strength between the alkali metal ion and an oxygen ion is small, while linear chains formed from $PO_4$ tetrahedra can be connected to each other by weak bonding strength through alkali metal ions.[9] The effect of these weak bonds is reflected in the low melting temperature of the glass; however, it also has poor chemical durability.

Since binary alkali phosphate glasses have poor chemical durability, it is necessary to add additional elements into the formula of the glass, namely polyvalent element oxides. Cations of polyvalent elements can inhibit the penetration of water into a glass structure by increasing the bonding strength between the chains of a metaphosphate glass. Among the polyvalent element oxides, there are divalent elements, which can improve water resistance to a certain extent, as they have a relatively small electric field intensity and electronegativity and have a relatively small bonding strength compared to an oxygen ion. Among these oxides, there is MgO and CaO, present in fly ash. Other oxides, such as $Al_2O_3$ and $Fe_2O_3$, are also present in fly ash (Table 1). These oxides contain cations that form a crosslinking structure between those and a phosphorous ion through an oxygen ion.[9] They are able to improve the water resistance more than the divalent elements. Fly ash proved to be a great and inexpensive source of polyvalent element oxides.

The first attempts at melting aimed to obtain samples from a mixture of $KPO_3$ and fly ash, however produced non-transparent materials. The addition of aluminium oxide ($Al_2O_3$) yielded transparent glass samples. Diboron trioxide ($B_2O_3$) was added to some samples in an attempt to decrease the thermal expansion coefficient of the glasses.

The developed samples already had their structures analysed by x-ray diffraction (XRD), which confirmed the amorphous character of the samples with the highest amounts of fly ash.[10] Some samples had their compositions verified using x-ray fluorescence (XRF); the results are shown in Table 2. The concentration of $B_2O_3$ could not be determined as boron cannot be measured by XRF. Thus, the concentration of this element was predicted based on the concentrations

*Table 1. Composition of fly ash (FA) determined by x-ray fluorescence*

| | $SiO_2$ | $Al_2O_3$ | CaO | MgO | $Fe_2O_3$ | $SO_3$ | $Na_2O$ | $K_2O$ | $TiO_2$ | $P_2O_5$ | LOI |
|---|---|---|---|---|---|---|---|---|---|---|---|
| FA | 54·28 | 23·32 | 4·23 | 1·62 | 8·01 | 0·64 | 0·85 | 1·97 | 1·23 | 0·54 | 3·37 |

*Table 2. Compositions of the glass samples containing fly ash, expressed in wt% and mol%*

| *Sample* | *70KPO₃–15FA–15Al₂O₃* | | *60KPO₃–20FA–20Al₂O₃* | | *40KPO₃–25FA–15Al₂O₃–20B₂O₃* | |
|---|---|---|---|---|---|---|
| *concentration:* | *wt%* | *mol%* | *wt%* | *mol%* | *wt%* | *mol%* |
| $SiO_2$ | 8·36 | 14·97 | 10·31 | 17·47 | 13·86 | 21·17 |
| $B_2O_3$ | - | - | - | - | 14·32 | 18·88 |
| $Al_2O_3$ | 19·55 | 20·62 | 24·44 | 24·40 | 21·75 | 19·57 |
| $Na_2O$ | - | - | - | - | 0·18 | 0·26 |
| $K_2O$ | 27·96 | 31·92 | 25·47 | 27·53 | 18·54 | 18·06 |
| MgO | 0·24 | 0·65 | 0·29 | 0·72 | 0·38 | 0·86 |
| CaO | 0·69 | 1·33 | 0·78 | 1·42 | 1·15 | 1·89 |
| SrO | 0·02 | 0·02 | 0·03 | 0·02 | 0·04 | 0·03 |
| BaO | 0·28 | 0·19 | | | 0·10 | 0·06 |
| $ZrO_2$ | 0·01 | 0·01 | 0·18 | 0·15 | 0·11 | 0·08 |
| ZnO | 0·09 | 0·12 | 0·11 | 0·14 | 0·01 | 0·01 |
| $TiO_2$ | 0·17 | 0·23 | 0·23 | 0·30 | 0·34 | 0·39 |
| $Cr_2O_3$ | - | - | - | - | 0·01 | 0·01 |
| MnO | 0·01 | 0·02 | 0·01 | 0·02 | 0·02 | 0·02 |
| $SO_3$ | 0·02 | 0·02 | 0·02 | 0·02 | 0·01 | 0·01 |
| $Fe_2O_3$ | 1·26 | 0·85 | 1·64 | 1·05 | 2·51 | 1·44 |
| $P_2O_5$ | 38·29 | 29·01 | 37·22 | 26·69 | 26·64 | 17·22 |
| CuO | - | - | - | 0·01 | 0·01 | 0·01 |
| $GeO_2$ | - | - | 0·03 | 0·03 | - | - |
| $Rb_2O$ | - | - | - | - | 0·01 | - |
| I | - | - | 0·02 | 0·02 | - | - |
| NiO | 0·01 | 0·01 | 0·01 | 0·01 | 0·01 | 0·02 |
| O/P molar ratio | - | 4·73 | - | 5·16 | - | 7·85 |

of the other elements. The homogeneity of the glasses was confirmed by compositional maps obtained with energy dispersive spectroscopy (EDS).

The water resistance of the samples was improved with changes in the compositions, achieving high water resistance. A high water resistance was reported for the samples containing an amount equal or lower than 70 wt% of $KPO_3$. There is a trend of increasing elastic modulus and hardness of the samples with amounts of fly ash and $Al_2O_3$. These properties were measured according to a continuous stiffness measurement. All these results were already reported, as well as the preparation of the samples.[10]

The current work aims to report the characteristic temperatures, optical properties, coefficient of thermal expansion (CTE), and mechanical properties of the samples. The results would be useful to compare the properties of these glasses to standard glasses used for building purposes and to suggest potential applications for them.

## 2. Methods

Making the samples: the glasses were melted in a Carbolite high temperature box furnace, model HTF 17/10, which can operate up to 1700°C. The samples were prepared in a 30 mL platinum crucible and cooled under melt quenching. The melted mixtures were poured into a stainless steel mould preheated to around 450°C and were annealed at this temperature for 3 h before cooling to room temperature inside the furnace. This annealing schedule guarantees stress-free samples, as confirmed through cross-polarised light. For the nomination of the sample, $KPO_3$ represents the amount of $KPO_3$, FA represents the amount of fly ash and $Al_2O_3$ represents the amount of $Al_2O_3$, all in wt%. For instance, the sample 55KPO₃–35FA–10Al₂O₃ contains 55% (wt) of $KPO_3$, 35% (wt) of fly ash and 10% (wt) of $Al_2O_3$.

X-ray fluorescence (XRF): was used to determine the actual composition of the samples. The investigated samples were flat, dimensions about 2·5×5 cm, and thickness of approximately 0·5 cm. The measurements were performed with a Panalytical Axios Max WD-XRF spectrometer and data evaluation was done with SuperQ5.0i/Omnian software.

X-ray powder diffraction (XRD): was used to prove the amorphous character of the samples. The samples were investigated as fine powder, homogeneised in a ceramic mortar. A thin layer of sample powder was deposited on a Si510 wafer from a powder-ethanol suspension and fixed in PMMA sample holder L510. The instrument used in the analysis was a Bruker D8 Advance diffractometer Bragg-Brentano geometry and Lynxeye position sensitive detector, using Cu $K\alpha$ radiation.

Differential scanning calorimetry and thermogravimetry analysis (DSC-TGA): was used to obtain the characteristic temperatures of the samples. Differential scanning calorimetry and thermogravimetry were performed using a Netzsch STA F3 Jupiter, using a heating rate of 10 K/min.

UV/VIS spectroscopy: was used to determine the absorption of radiation of the samples as a function of wavelength. The UV/VIS spectroscopy was performed with a spectrophotometer Perkin-Elmer Lambda 950 UV-VIS-NIR.

Nanoindentation: the measures of fracture toughness were conducted with a KLA-Tencor G200 nanoindenter using Berkovich and Knoop tips. The prepared glass samples had their surfaces polished, and then the samples were reannealed.

Microscopy: microscopic images were obtained with an environmental scanning electron microscope

(ESEM) Philips XL30 ESEM and an optical polarisation and fluorescence microscope.

Thermal expansion tests: was applied to measure the coefficient of thermal expansion of the glasses. The gauges type ZFLA-6-11 attached to the samples were connected to wires of a DAQ device of three channels. The device was also connected to a computer. The samples were placed inside a Heraeus oven type UT 12P, which is operational up to 250°C. The test ranged from 30–163°C. A bridge-compensation circuit was used outside the oven while inside it used a quarter bridge sensing gauge.

## 3. Results and discussions

### 3.1. Characteristic temperatures

Through DSC curves, it was possible to determine the characteristic temperatures of the samples. Based on these characteristic temperatures, it was also possible to determine the parameter of thermal stability of the systems, as shown in Table 3. This was achieved by comparing the $T_g$ and $T_x$ obtained for each sample. The level of thermal stability against crystallisation marks the difference between the starting crystallisation temperature and the glass transition temperature ($T_x$–$T_g$). Both temperatures were determined on the DSC curves using the tangent method. It is assumed that the glass forming tendency is proportional to the thermal stability of glass. The glass forming tendency is expressed numerically, thus enabling one to compare different kinds of glasses and the influence of various impurities.(11) Besides the $T_g$, possible inflections were observed in the DSC curves of the samples containing the lowest amounts of fly ash, remarkably the sample containing 12·5% of fly ash. We recommend repeating the thermal analysis for these samples and further investigation about possible phase separation.

The DSC-curves of the samples containing from 12·5 to 35% of fly ash (Figure 1) show that higher concentrations of fly ash and $Al_2O_3$ correlate with an increase in glass transition temperatures. In addition, thermal stability decreases with increasing concentrations of fly ash (Table 3). As the enthalpy of a transition is derived from the area under its curve, the sample with the lowest amount of $KPO_3$, shows the crystallisation peak with the highest enthalpy. This tendency to crystallise goes with the high concentrations of different oxides present in this sample, due to it containing the highest concentrations of fly ash and aluminium oxide.

*Table 3. Characteristic temperatures and thermal stability of glasses containing fly ash (FA)*

| *Sample* | $T_g$ (°C) | $T_x$ (°C) | $T_x$–$T_g$(°C) |
|---|---|---|---|
| $55KPO_3$–35FA–$10Al_2O_3$ | 456 | 510 | 54 |
| $60KPO_3$–25FA–$15Al_2O_3$ | 458 | 535 | 77 |
| $60KPO_3$–30FA–$10Al_2O_3$ | 460 | 530 | 70 |
| $60KPO_3$–20FA–$20Al_2O_3$ | 484 | 564 | 80 |
| $65KPO_3$–20FA–$15Al_2O_3$ | 451 | 569 | 118 |
| $75KPO_3$–12·5FA–$12·5Al_2O_3$ | 412 | 602 | 190 |
| $50KPO_3$–$10B_2O_3$–25FA–$15Al_2O_3$ | 475 | 514 | 39 |
| $40KPO_3$–$20B_2O_3$–25FA–$15Al_2O_3$ | 472 | 640 | 168 |

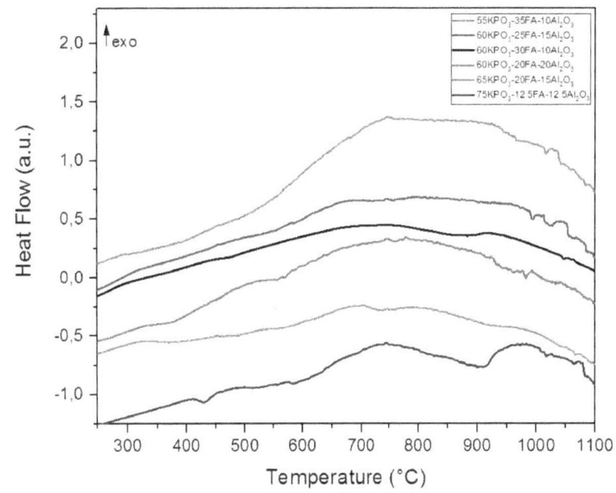


*Figure 1. DSC curves of the glasses containing fly ash, in which the amount of fly ash ranges from 12·5–35% (wt) [Colour available online]*

Two samples were produced with the addition of $B_2O_3$: $50KPO_3$–$10B_2O_3$–25fly ash–$15Al_2O_3$ and $40KPO_3$–$20B_2O_3$–25fly ash–$15Al_2O_3$. There is a marginal decrease in $T_g$ with the addition of $B_2O_3$: the glass transition temperature of the sample with 10% (wt) of $B_2O_3$ is around 475°C, while the sample containing 20% (wt) of the same oxide has a glass transition temperature of 472°C (Figure 2). A meaningful change introduced by increasing the amount of $B_2O_3$ is the crystallisation behaviour of the sample. While the sample with the lowest amount of this oxide presents two following crystallisation peaks, the first starting around 514°C and peaking around 650°C and the second one peaking around 860°C, the sample containing 20% of $B_2O_3$ presents only one peak, starting around 640°C and peaking around 810°C. Thus, the interval between the $T_x$ and

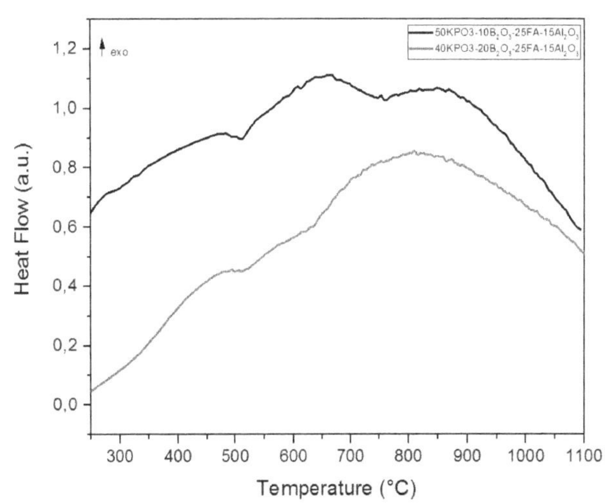


*Figure 2. DSC curves of the glasses containing fly ash and $B_2O_3$ [Colour available online]*

$T_g$ is significantly higher for the sample with the highest amount of this oxide, showing that increasing the amount of $B_2O_3$ produces samples with a higher thermal stability against devitrification.

### *3.2. Optical spectroscopy*

The samples consisting of 35% fly ash showed a green coloration and were produced at 1350°C. The samples consisting of 15% fly ash were yellow and can be produced at lower temperatures, 1100°C. Higher concentrations of fly ash made samples become darker and also increased melting temperature.

The colours of some samples containing fly ash are summarised in Table 4 and their optical spectra are shown in Figure 3. The samples contain between 1·3 and 2·5% (wt) of $Fe_2O_3$. The states of transition metals in host glass, and thus the ratio of these two iron species, not only depends on composition but also on the redox conditions of the melt, such as air atmosphere, melting temperature and melting time.[12] It is known that alkali borate and alkali silicate glass types favour the presence of transition metal ions (i.e. iron) in higher oxidation states. On the other hand, alkali phosphate glasses induce a lower oxidation state.[13]

$Fe^{2+}$ and $Fe^{3+}$ are, respectively, blue and yellow chromophores. The CT bands of $Fe^{2+}$ are located at higher energies and therefore lower wavelengths that those of $Fe^{3+}$. Ferric $Fe^{3+}$ is mainly identified by an absorption band at 380 nm. Ferrous $Fe^{2+}$ is characterised by a strong absorption band with a maximum of around 1050 nm, thus already outside the visible range. The broadness of the band may be indicative of the Jahn–Teller distortion of the octahedrally coordinated ion, an effect often found in octahedral complexes of transition metals. It is expected that the band of tetrahedrally coordinated $Fe^{2+}$ ions would be found around 2000 nm in the near infrared.[14]

The ratio between $Fe^{2+}/Fe^{3+}$ explains the colour of the glass types. While the two species can appear simultaneously in a sample, its ratio is affected by the composition of the glass and its melting conditions. The intensities of the band related to $Fe^{3+}$ are quite similar for the three samples, but the intensity of the absorption at 1050 nm is notably higher for the sample $40KPO_3–20B_2O_3–25FA–15Al_2O_3$, followed by the sample with 20% of fly ash and then 15% of fly ash. As the ratio $Fe^{2+}/Fe^{3+}$ decreases, the colour of the glass changes from dark green (20% $B_2O_3$) to yellow (15% fly ash). Green-coloured glass could therefore be associated with a relative increase in the concentration of ferrous ions, while a yellow colour is associated with a lower value of $Fe^{2+}/Fe^{3+}$, thus a relative higher amount of ferric ion.

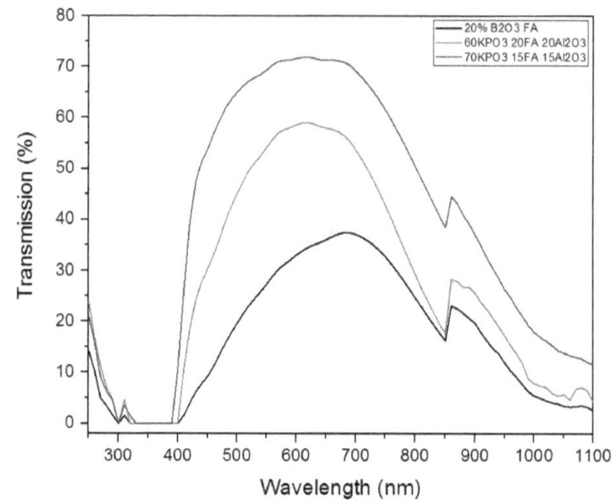


*Figure 3. Optical spectra of samples containing fly ash [Colour available online]*

The spectra exhibit a step at 850 nm, which is related to a fall from the baseline due to the replacement of the detector, as the scan is made from infrared to visible and the sensitivity of the detectors at 850 nm is different. Theoretically, with careful correction of the baseline and extracting the blank sample, it is possible to extract this event. However, in practice, there is usually some effect left. Therefore, this step is not related to the properties of the samples.

### *3.3. Thermal expansion*

Determining the thermal expansion coefficient of a glass is fundamental in order to predict the susceptibility to thermal shocks and the crack propagation after rapid cooling.

*Table 4. Colours of different glasses containing fly ash as glass beads of 3 mm [Colour available online]*

| $40KPO_3–20B_2O_3–25FA–15Al_2O_3$ | $60KPO_3–20FA–20Al_2O_3$ | $70KPO_3–15FA–15Al_2O_3$ |
|---|---|---|
|  |  |  |

*Table 5. CTE obtained for different glasses*

| Sample | $\alpha \times 10^{-6}$/°C |
|---|---|
| $40KPO_3$–$20B_2O_3$–25FA–$15Al_2O_3$ | 5·7±0·2 |
| $50KPO_3$–$10B_2O_3$–25FA–$15Al_2O_3$ | 6·2±0·7 |
| Flat glass (Kimble R-6) | 9·0±0·2 |
| $60KPO_3$–30FA–$10Al_2O_3$ | 10·4±0·3 |
| Borosilicate (B270 from Schott) | 7·6±0·9 |
| $70KPO_3$–15FA–$15Al_2O_3$ | 14·1±0·8 |

Table 5 shows the values of the coefficient of thermal expansion (CTE) measured by strain gauges for glasses containing fly ash in relation to two commercial glasses. The commercial tested glasses were a borosilicate crown glass (B270 from Schott) and a soda–lime–silica float glass (Kimble R-6).

Samples without $B_2O_3$ expand more than the other glasses. The CTE increases with the increasing amount of $KPO_3$, thus with higher amounts of $K_2O$ and $P_2O_5$. The samples with 60% of $KPO_3$ have a CTE slightly above the one of soda–lime glasses. However, the glasses containing 70% of $KPO_3$ have a CTE much higher than borosilicate or soda–lime glasses, which limit their applications in architectural engineering as they will be sensitive to temperature shocks.

An increasing CTE when alkalis are added to phosphate glasses is probably caused by the depolymerisation of anions. On the other hand, additions of $Al_2O_3$ and fly ash could increase the degree of polymerisation. Linear relations were already found between the expansion coefficient and the radius of alkali cations for a series of compositions in the systems $R_2O$–$Al_2O_3$–$P_2O_5$:[15] the expansion increases with increasing radius. This trend suggests that the increase in expansion is a result of the decrease in the strength of the cationic field and the expansion would mainly be determined by the interaction between cations and nonbridging oxygens.

The high water resistance of the developed samples containing fly ash was already reported and showed that the water resistance increases with decreasing amounts of $KPO_3$.[15] This is consistent with studies reporting an increase in expansion coefficient together with the increase of the dissolution rate of a phosphate glass, indicating that both properties are controlled primarily by the interaction of cations with nonbridging oxygens.[15] High expansion glasses with good chemical durability are desirable for applications such as bonding with plastics or metals, since the thermal expansion coefficient of the joining materials should be proximate. However, high values of CTE mean sensitivity to thermal shock which limits the applications of phosphate glasses in other areas, such as building materials. The development of new glass compositions, containing lower amounts of phosphate and higher amounts of fly ash or $B_2O_3$ demonstrated to decrease the CTE of the glasses, producing samples with a coefficient of expansion comparable to the coefficient of soda–lime and borosilicate glasses, or even lower. It expands the range of applications of typical phosphate glasses, showing potential uses in construction, such as bricks, or even in more advanced technologies which require a combination of low melting temperatures and low thermal expansion, such as 3D printing.

### 3.4. Mechanical properties

Nanoindentation has been extensively applied as a high spatial resolution micro-probe for measuring the mechanical properties of glasses at the micro-level. The mechanical properties that can be measured by this technique include: hardness, elastic modulus, and fracture toughness.[16] A crucial factor in the expansion of the use of nanoindentation is the simple sample preparation. A small glass sample requires only grinding and polishing, achieving a surface free of scratches. In addition, most indentation equipment offers experimental control of test parameters, which is necessary for the exploration of different deformation modes by changing experimental time scale, indenter tip geometry, and loading conditions.[15]

Traditional testing methods for fracture toughness determination are generally based on macroscopic crack extension experiments. However, evaluating the fracture toughness of glasses and ceramics is still problematic. Consequently, a mechanism for measuring fracture toughness of these materials has not been well determined. The nanoindentation test is an option that has been used in order to assess the fracture toughness of brittle materials; such as glasses, ceramics, films, and coatings. Where the materials exhibit brittle fracture with minimal crack tip plasticity, this crack pattern can be used to establish fracture toughness values.[18] This property can be estimated by either crack length-based method or crack energy-based method, providing useful partially quantitative information. The tests were performed under 2 N and with a distance of 200 μm between the indents. As some samples did not crack under these conditions, new tests were performed with a Knoop indenter under a load of 8·5 N.

Although somewhat problematic, fracture toughness evaluation can be quickly performed using indentation. If a sharp tip is forced into a bulk sample of a brittle material, radial cracking usually occurs after a critical load has been reached, originating from stress concentrations at the diamond facet edges, which allows an estimate of the fracture toughness based on the maximum indentation load and the crack length.[19] In indentation-based methods with sharp pyramidal indenters, the fracture toughness can be determined from measurements of the lengths of cracks emerging from the residual indentation impression.[16] An equation proposed by Evans[20] and Lawn[21] allows the calculation of $K_c$ based on indentation tests, measuring the length of cracks originating from the edges of the indent impression,

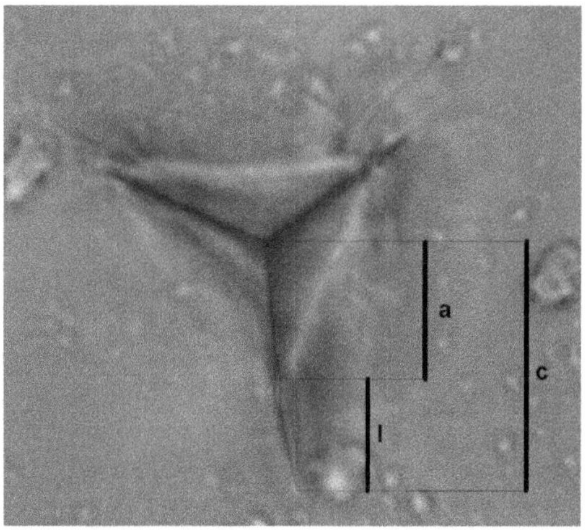


*Figure 4. Dimensions of a crack induced by a Berkovich indenter, in which c=a+l [Colour available online]*

determining that:

$$K_c = A\left(\frac{E}{H}\right)^{1/2} \frac{P}{c^{3/2}} \quad (1)$$

In which $E$=Young's modulus, $H$=hardness, $P$=load, $c$=crack length measured from the centre of the indent to the crack tip. $A$ is an empirical constant, which varies according to the indenter and the crack geometry.(19) For a Berkovich indenter, $A$=0·016.(22) The representative dimensions of the pyramidal indentation fracture testing method are shown in Figure 4.

A crack is classified as well developed if $c \gg a$. However, if the radial cracks are not well developed, the relation between $K_C$ and $P/c^{3/2}$ ceases.(23) The method presented above can be applied only to well-developed radial cracks. Some authors reported that the linear relation between $K_{IC}$ and $P/c^{3/2}$ stops when $c<2{\cdot}5a$.(24,25) Other authors claimed that the relation is valid even when $c$ is down to $1{\cdot}1a$.(26,27) The different glass compositions were analysed by optical microscopy and presented huge variations in crack lengths. At least five indents were made in each sample. In many cases, the images obtained with the use of microscopy can be related to the load–displacement curves. For instance, the sample $40KPO_3$–$20B_2O_3$–25FA–$15Al_2O_3$, in which the microscopy does not reveal the presence of cracks, and the load–displacement curve does not show any significant pop-in (Figure 5). The sample $60KPO_3$–20FA–$20Al_2O_3$ shows well-developed cracks, which can be related to the pop-ins of high energy, present in the load–displacement curve (Figure 6).

For the fracture toughness estimation based on the length of the cracks, only defined cracks, with $c/a>1{\cdot}1$, were considered. For bulk brittle materials, this method provides a calculation within 40% accuracy.(28) The most significant source of error is probably the unknown crack geometry. The models make presumptions about crack geometry, the half-penny type or Palmquist, and this assumption influences

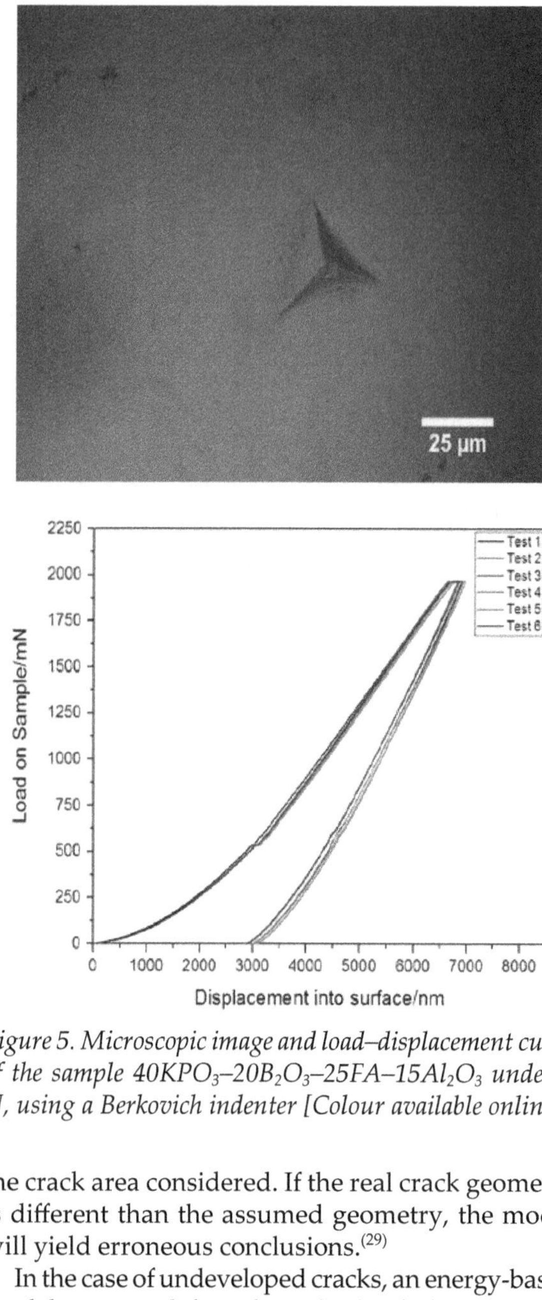


*Figure 5. Microscopic image and load–displacement curve of the sample $40KPO_3$–$20B_2O_3$–25FA–$15Al_2O_3$ under 2 N, using a Berkovich indenter [Colour available online]*

the crack area considered. If the real crack geometry is different than the assumed geometry, the model will yield erroneous conclusions.(29)

In the case of undeveloped cracks, an energy-based model was used, based on the load–displacement curve associated with the fracture. In this work, the shapes of the indentation load–displacement ($P$–$h$) curves reveal predominant elastic–plastic deformation, in which the hysteresis is caused mainly by plastic deformation mechanisms. It is opposed to the brittle behaviour, in which the indentation hysteresis is mostly generated by fracture.(17) The maximum displacement into the surface of the samples containing fly ash is significantly higher than the soda–lime sample (Figures 7 and 8). This agrees with the lower hardness values of these materials, which should allow a higher penetration depth under the same load.

Energy-based models can be used to deal with various cracking patterns. A largely applied energy-based model was proposed by Li,(30) consisting of extrapolating the loading curve where there is a fracture-induced pop-in. The $P$–$\delta$ curve is extrapo-

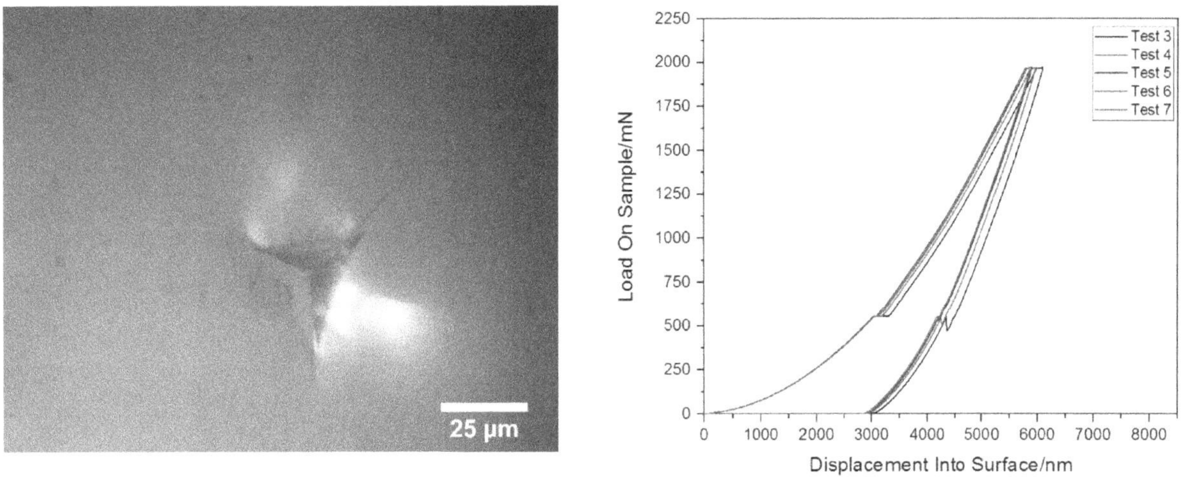


*Figure 6. Microscopic image and load–displacement curve of the sample 60KPO$_3$–20FA–20Al$_2$O$_3$ under 2 N, using a Berkovich indenter [Colour available online]*

lated from the start point (onset of the fracture) to its end point. The area under the load–displacement curve corresponds to the work performed by the indenter during the elastic–plastic deformation of the system.[30] The difference between the extrapolated curve and the measured curve is considered as the fracture dissipated energy.[23] This energy difference (*U*) was calculated for all the tests of every sample. Calculating fracture toughness by measuring fracture energy requires an estimate of the area *F* of the crack. For a Berkovich indenter, it can be adapted to:[29]

$$F = \frac{\pi c^2}{4} - \frac{a^2}{2} \tag{2}$$

Estimating the fracture toughness can be done according to the following equation, which applies to any crack shape:[30]

$$K_c = \sqrt{\frac{E}{1-v^2}\frac{U}{F}} \tag{3}$$

In which *E*=Young's modulus, *v*=Poisson ratio, *U*=crack energy from pop-in and *F*=crack area. In this work, it was assumed that *v*=0·35 for phosphate glasses and *v*=0·25 for soda–lime glass. The calculated values of fracture toughness can be found in Table 6.

Due to the absence of developed cracks in most samples, new tests were performed using a higher load and another indenter geometry, the Knoop indenter. The cracks were analysed with optical microscopy and scanning electron microscopy. Figure 9 and Table 7 show the average dimensions of the indents produced for different samples.

The dimensions of the samples varied among the different compositions. Remarkably, the soda–lime sample has a low value of *L*, while the sample containing fly ash and 20% of B$_2$O$_3$ has a remarkably low *W*. Different patterns of cracks were associated to different glass compositions.

Some of the indents in soda–lime possess extended cracks on both sides with interference fringes, indicating the presence of lateral cracks.

The samples containing 30 and 35% fly ash present

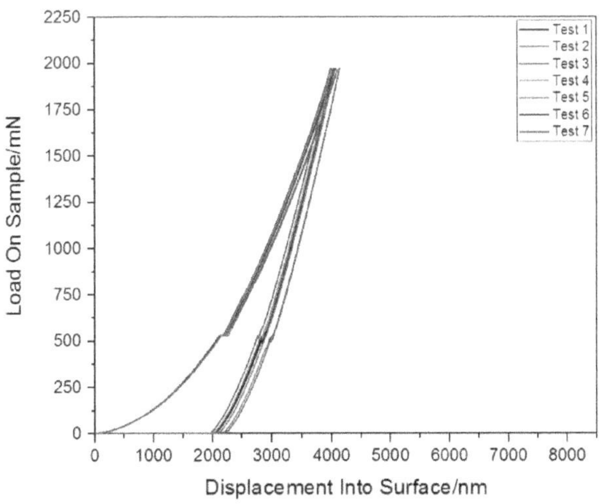


*Figure 7. Load–displacement curve of soda–lime–silica glass under 2 N, using a Berkovich indenter [Colour available online]*

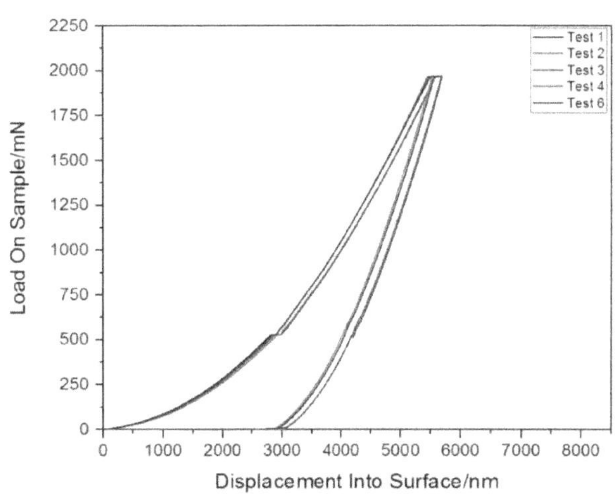


*Figure 8. Load–displacement curve of the sample 55KPO$_3$–35FA–10Al$_2$O$_3$ under 2N, using a Berkovich indenter [Colour available online]*

*Table 6. $K_c$ calculation based on the equations of Li (energy) and Lawn & Evans (crack lengths)*

| Sample | Average a (μm) | Average c (μm) | c/a | E (GPa) | H (GPa) | Kc (MPa√m) Li | Kc (MPa√m) Lawn & Evans |
|---|---|---|---|---|---|---|---|
| $40KPO_3$–$20B_2O_3$–25FA–$15Al_2O_3$ | 17·91 | 19·53 | 1·09 | 48·4 | 5·02 | 0·42 | * |
| $55KPO_3$–35FA–$10Al_2O_3$ | 18·15 | 18·15 | 1·00 | 43·5 | 4·18 | ** | * |
| $60KPO_3$–20FA–$20Al_2O_3$ | 19·13 | 30·30 | 1·58 | 41·6 | 3·94 | 0·80 | 0·62 |
| $60KPO_3$–30FA–$10Al_2O_3$ | 20·31 | 20·31 | 1·00 | 45·2 | 4·56 | ** | * |
| Soda–lime–silica (reference) | 14·33 | 24·84 | 1·73 | 74·0 | 5·50 | 0·44 | 0·94 |

*As c/a<1·1 these $K_c$ values are not reliable. **As $c=a$ there is no crack. Therefore, these $K_c$ values are not reliable and reflect a pessimistic lower boundary.

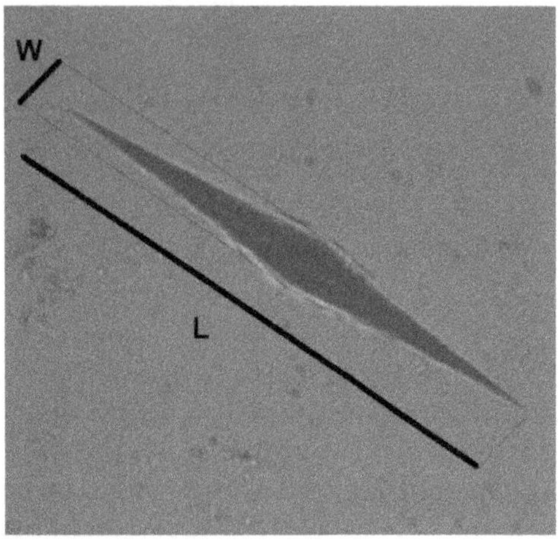


*Figure 9. Dimensions of a Knoop indent [Colour available online]*

similar fracture patterns with edge cracks, as shown in Figures 11 and 12.

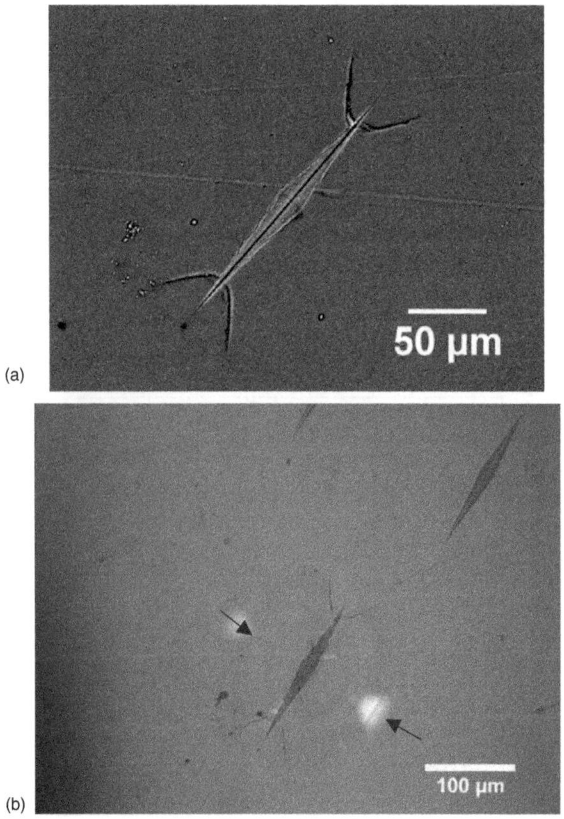


*Figure 10. Crack pattern of a silicate glass induced by a load of 8·5 N using a Knoop indenter a) indent analysed by SEM b) interference fringes (blue arrows) [Colour available online]*

*Table 7. Average dimensions of the Knoop indent in the samples*

| Sample | L (μm) | Std. Dev. (L) | W (μm) | Std. Dev. (W) |
|---|---|---|---|---|
| $40KPO_3$–$20B_2O_3$–25FA–$15Al_2O_3$ | 261·20 | 0·09 | 9·25 | 0·52 |
| $60KPO_3$–30FA–$10Al_2O_3$ | 209·20 | 3·85 | 21·00 | 1·17 |
| $55KPO_3$–35FA–$10Al_2O_3$ | 208·10 | 1·82 | 21·46 | 0·76 |
| Soda–lime–silica (reference) | 162·42 | 2·66 | 21·32 | 0·11 |

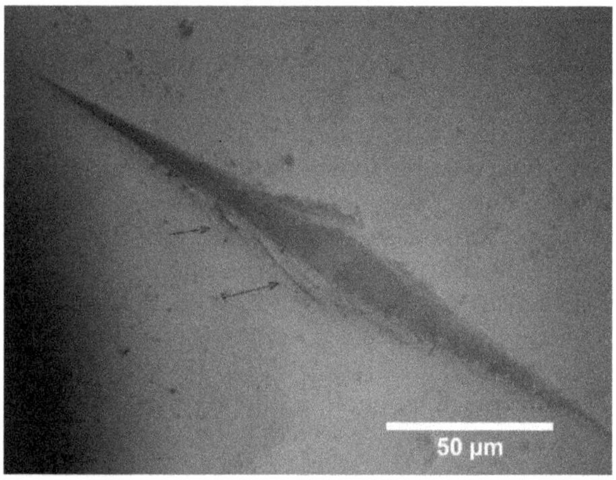


*Figure 11. Crack pattern of $60KPO_3$–30FA–$10Al_2O_3$ induced by a load of 8·5 N using a Knoop indenter. Arrows point to the edge cracks [Colour available online]*

The sample containing fly ash and 20% of $B_2O_3$ has radial cracks and an indentation pattern of an arrow, in which the dimension *W* is kept almost constant for all the extension of the sample.

The indentation pattern detailed in Figure 13 dif-

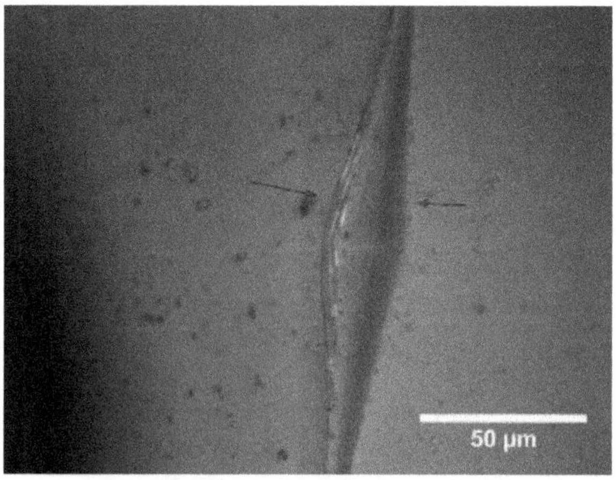


*Figure 12. Crack pattern of $55KPO_3$–35FA–$10Al_2O_3$ induced by a load of 8·5 N using a Knoop indenter. Arrows point to the edge cracks [Colour available online]*

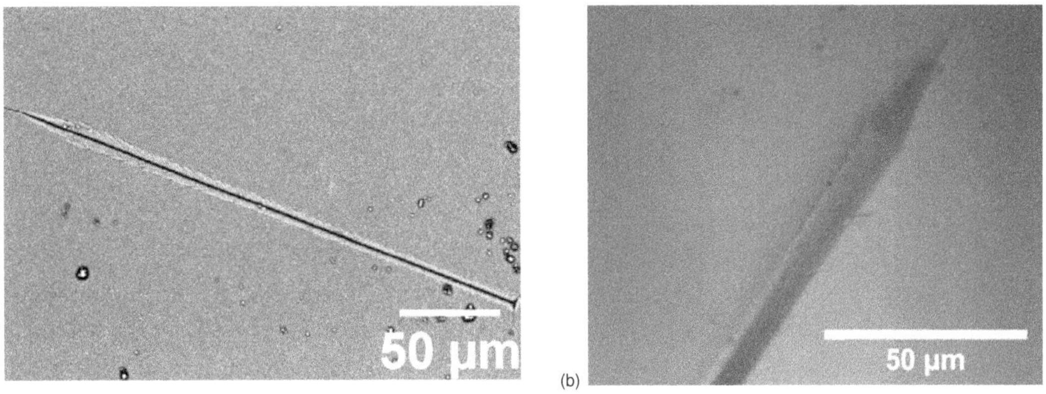

(a) (b)

*Figure 13. Crack pattern of the sample 40KPO$_3$-20B$_2$O$_3$-25FA-15Al$_2$O$_3$ induced by a load of 8·5 N using a Knoop indenter (a) indent analysed by SEM (b) close-up of the cracks analysed by optical microscopy [Colour available online]*

fers significantly from typical Knoop indents, and no reports of it were found in the literature.

Following this, new indentation tests were performed under the same conditions as the previous test, but in four different directions, as indicated in Figure 14. Three indents were performed in each direction. Different results were obtained for different directions. The DSC-curve presented in Figure 2, which has a defined glass transition temperature, combined with the x-ray diffraction pattern presented in Figure 15, which proves that it is a predominantly amorphous sample, indicates a possible anisotropic behaviour of the material.

While glasses are mostly isotropic materials, anisotropy is not incompatible with a lack of crystalline order. Anisotropy was already reported for some glasses, such as a metaphosphate glass of composition 12·5Li$_2$O–12·5Na$_2$O–12·5K$_2$O–12·5Cs$_2$O–50P$_2$O$_5$ (mol%). In this case, the enhancement of the mechanical properties was attributed to the orientation of –P–O–P– chains in the glass.[31] Alkali metaphosphate glasses, 50R$_2$O–50P$_2$O$_5$ (mol%), in which R$_2$O

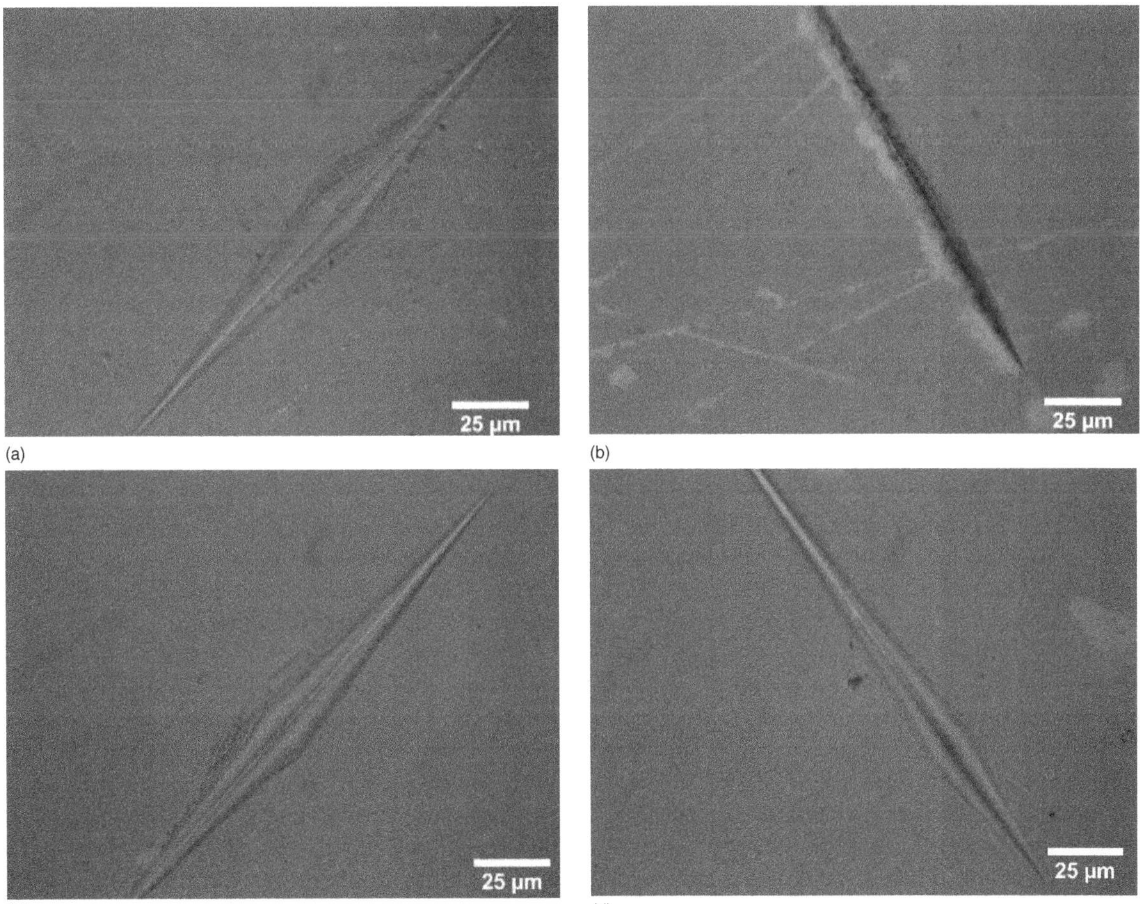

(a) (b) (c) (d)

*Figure 14. Close-up of crack patterns of the sample 40KPO$_3$–20B$_2$O$_3$–25fly ash–15Al$_2$O$_3$ induced by a load of 8·5 N using a Knoop indenter and measured at every 90° (a), (b), (c), (d) [Colour available online]*

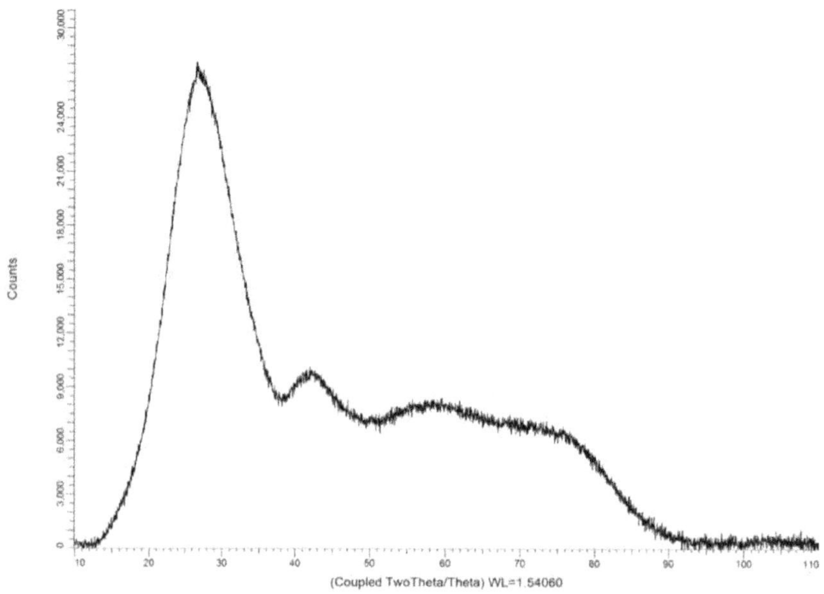


*Figure 15. X-ray diffraction pattern of the sample 40KPO$_3$–25FA–15Al$_2$O$_3$–20B$_2$O$_3$*

represents an alkali oxide, possess a structure mainly formed by one-dimensional –P–O–P– chains because their $PO_4$ tetrahedral units are connected by two oxygen atoms ($n$=2). If these chains are randomly organised, the metaphosphate glasses will be isotropic. However, $NaPO_3$ and $KPO_3$ metaphosphate crystals are anisotropic. Both have uniaxially oriented P–O–P chains, $NaPO_3$ in a spiral orientation and $KPO_3$ in a zigzag manner. When the chains in glass are uniaxially organised, such as in metaphosphate crystals, anisotropic glasses can be formed.[31] Another work from the same authors reports the occurrence of entropic elasticity in phosphate glasses with highly anisotropic structures.[9] Oxide glasses generally possess three-dimensional networks, consisting of oxygen polyhedra surrounding network-forming cations, such as $Si^{4+}$, $B^{3+}$ and $P^{5+}$. However, metasilicate and metaphosphate glasses are usually based on chain structures in which oxygen tetrahedra are connected by their two corners and the chains are composed of covalent bonds, with some degree of ionic character. Typical metaphosphate glasses have chains containing between 40 and 100 P-tetrahedra, ended by hydroxyl groups or modifiers, such as alkali ions.[9] The chains connect with each other by ionic bonding via modifiers. It is expected that oxide glasses with a rubberlike structure, containing highly polymerized chains with strong (covalent) intrachain bonds, weaker interchain interactions and moderate crosslinking may show entropic elasticity.[9] These glasses can self-organise into oriented structures by application of uniaxial stress, and then return to their original configurations when the stress is removed.

As the indentation measurements show that the glass containing 20% of $B_2O_3$ behaves differently when indented in different directions, we suggest for future works further investigation of the mechanical properties of this composition. Further investigation is recommended to check if the peculiar fracture pattern is an indicator of anisotropy or caused by other factors, for instance, an indication that the samples may not have been sufficiently flat or not properly annealed. Thus, it is recommended to repeat the nanoindentation test for this sample.

## 4. Conclusions and applications

The large and varied range of properties of the glasses developed in this work prove that the development of compositions with designed properties transforms glass into a versatile material, able to accommodate many different demands. In most of these compositions, $P_2O_5$ is the main compound, the glass former. The addition of other components, such as aluminium oxide, boron oxide, or other oxides originated from by-products, produce glasses with a higher O/P ratio. Investigations of the short- and intermediate-range structure of the glasses were not performed; however, they are encouraged, primarily in order to support a better understanding of the mechanical and optical properties.

All the produced samples have a low $T_g$ in relation to a soda-lime or a borosilicate glass. The drastic reduction of the melting temperature allows for a more energy-saving manufacturing process. Along with the incorporation of by-products, it contributes to addressing the current appeal for a more sustainable glass manufacturing process. Combined with a high thermal stability and a coefficient of thermal expansion comparable or lower than the standard glasses, these compositions could be candidates for applications in building engineering.

Although advantageous from a perspective of saving emissions and energy, the addition of by-products to the glass can be problematic as it introduces to the batch many components usually avoided by glass manufacturers, such as iron oxide. An alternative to avoid the contamination of furnaces, is to melt the glasses excluding the components present in the lowest concentrations, as it is assumed that they do not add beneficial properties to the material. For instance, eliminating all the components present in concentrations lower than 1 mol%, the $Fe_2O_3$ would be eliminated from the sample $70KPO_3–15FA–15Al_2O_3$. It brings a possibility to produce colorless glasses, and simplifying the composition to a formula containing only $SiO_2$, $Al_2O_3$, $K_2O$, CaO and $P_2O_5$. We encourage the reproduction of the samples using pure reagents instead of fly ash, and, therefore, eliminating the impurities present in the by-products, as well as a comparison of the properties of these glasses with the ones produced with the incorporation of fly ash.

The combination of low glass transition temperatures, high thermal stability and low thermal expansion, along with the production of samples based on the nominal composition, excluding impurities and using high purity reagents, eliminates variations in composition caused by the use of fly ash and could produce samples used for more advanced technologies. For instance, these glasses are candidates to be 3D printed, mainly sample $40KPO_3–25FA–15Al_2O_3–20B_2O_3$, with the lowest coefficient of thermal expansion. This promising process has distinct advantages over traditional methods of fabrication. Notably, the fact that 3D printing is capable of producing complex shapes that other manufacturing methods cannot handle, allowing the production of components such as glass connectors. 3D printing can design and print components with variable thicknesses and complex inner features, unlike glass blowing, where the inner features reflect the outer shape.[32] However, glass is inherently a very difficult material to work with. It requires a high temperature system, above 1000°C, and the viscosity changes significantly with temperature, requiring precise control of the temperature at all stages of the process.[32] The high melting temperature of glasses as fused silica combined with a high thermal expansion, that can lead to cracking, are drawbacks that have made standard glass compositions unreachable to three-dimensional printing.

High expansion glasses with good chemical durability are desirable for applications such as bonding with plastics or metals, since the thermal expansion coefficient of the joining materials should be very similar. Some of the developed samples could be used for this purpose, as they have a CTE very similar to a 446 stainless steel, which possess a CTE of about $11{\cdot}4\times10^{-6}$/°C.[33] However, samples with concentrations of $KPO_3$ higher than 70 mol% should not be designed to applications that require a high water resistance.

There is no absolute measure of chemical durability and glass types are generally graded relative to one another after subjecting them to similar experimental conditions, i.e. the nature of the experiment usually determines the relative order or durability. On the one hand, accelerated tests can be performed in a controlled environment, however, the validity of the experiment can be compromised by not reflecting real-life conditions. On the other hand, long-term tests can be conducted, however, these present a challenge to continuity and experimental logistics. Long-term tests are generally conducted under a less controlled environment and are more representative of real conditions. Other test conditions can also change: the protocols (static or dynamic) and the sample preparation can be altered, while the glass itself can be tested as a whole article or as a powder. In our work, samples were tested statically. The samples were tested under demineralised water and ambient temperature (around 20°C) in plastic vessels. These glass types demonstrated high water resistance in relation to general phosphate glass types. As an example, the reported compositions $54{\cdot}5P_2O_5–20{\cdot}5K_2O–20{\cdot}5Cs_2O–4{\cdot}5Al_2O_3$ and $50P_2O_5–16{\cdot}7Na_2O–16{\cdot}7K_2O–16{\cdot}7Cs_2O$ had a mass loss of 19·3% and 100%, respectively, when dipped in water at 18°C for 1 h,[34] while the glasses reported in this work were kept in similar conditions for 30 days without any mass loss. Finding a concentration of $P_2O_5$ low enough to produce a durable material but high enough to keep the high expansion, such as sample $60KPO_3–30FA–10Al_2O_3$, is key for applications such as glass-to-metal seals.

## Acknowledgements

Clarissa Justino de Lima is supported by a CNPq (The Brazilian National Council for Scientific and Technological Development) PhD scholarship, which is gratefully acknowledged. Hans van Limpt is acknowledged for the XRF and UV-VIS measurements. Paul Vermeulen is acknowledged for his valuable assistance with the thermal expansion measurements.

## References

1. Hoornweg, D., Bhada-Tata, P. & Kenney, C. A. Peak waste: when is it likely to occur? *J. Ind. Ecol.*, 2015, **19**, 117–28.
2. *Production and use of coal combustion products in the U.S. Market forecast through 2033*. American Road & Transportation builders association. 2015. Available at: https://www.acaa-usa.org/Portals/9/Files/PDFs/ReferenceLibrary/ARTBA-final-forecast.compressed.pdf (accessed on 23 Jan 2020).
3. Adams, T. H. *Fly ash use in concrete increases slightly as overall coal ash recycling rate declines*. American Coal Ash Association. 2020. Available at: https://www.acaa-usa.org/Portals/9/Files/PDFs/Coal-Ash-Production-and-Use.pdf (accessed on 23 Jan 2020).
4. *World Energy Outlook 2019*. ISBN 978-92-64-97300-8. 2019.
5. Rastogi, A. & Paul, V. K. A critical review of the potential for fly ash utilisation in construction-specific applications in India. *J. Environ.*

*Res. Eng. Manage.*, 2020, **76** (2), 65–75.

6. Adams, T. H. Ash at Work: Applications, Science, and Sustainability of Coal Ash. American Coal Ash Association (ACAA), 2018, Issue I.
7. Zschimmer, E. *Chemical Technology of Glass*. Society of Glass Technology, Sheffield, 2013.
8. de Lima, C. L. J., Pastena, B., Nardi, R. P. R. D., Gouvea, Jnr., J. T., Ferrari, J. L., Cassanjes, F.C. & Poirier, G. Thermal, structural and crystallisation study of niobium potassium phosphate glasses. Mater. Res., 2015, 13–16.
9. Inaba, S., Hosono, H. & Ito, S. Entropic shrinkage of an oxide glass. *Nature Mater.*, 2015, **14**, 312–17.
10. de Lima, C. J., Veer, F., Çopuroglu, O. & Nijsse, R. Innovative glass recipes containing industrial waste materials. *Challenging Glass Conf. Proc.*, 2018, **6**, 533–42.
11. Hruby, A. Evaluation of glass-forming tendency by means of DTA. *Czech. J. Phys. B*, 1972, **22**, 1187–93.
12. Babadzhanova, O. F. & Yashchishin, I. N. Low-melting glasses based on phosphate ore processing products. *Glass Ceram.*, 2000, **57** (5), 3–5.
13. ElBatal, F. H., Hamdy, Y. M. & Marzouk, S. Y. UV-visible and infrared absorption spectra of transition metals-doped lead phosphate glasses and the effect of gamma irradiation. *J. Non-Cryst. Solids*, 2009, **355**, 2439–47.
14. Möncke, D., Papageorgiou, M., Winterstein-Beckmann, A. & Zacharias, N. Roman glasses coloured by dissolved transition metal ions: redox-reactions, optical spectroscopy and ligand field theory. *J. Archaeol. Sci.*, 2014, **46**, 23–36.
15. Minamf, T. & Mackenzie, J. D. Thermal expansion and chemical durability of phosphate glasses. *J Am Ceram. Soc.*, 1977, **60**, 232–5.
16. Sebastiani, M., Johanns, K. E., Herbert, E. G. & Pharr, G. M. Measurement of fracture toughness by nanoindentation methods: recent advances and future challenges. *Curr. Op Solid State Mater. Sci.*, 2015, **19**, 324–33.
17. Oyen, M. L. & Cook, R. F. A practical guide for analysis of nanoindentation data. *J. Mech. Behav. Biomed. Mater.*, 2009, **2**, 396–407.
18. Guo, H., Jiang, C. B., Yang, B. J. & Wang, J. Q. On the fracture toughness of bulk metallic glasses under Berkovich nanoindentation. *J. Non-Cryst. Solids*, 2018, **481**, 321–8.
19. Vella, J. B., Adhihetty, I. B., Junker, K., Volinsky, A. A. Mechanical properties and fracture toughness of organo-silicate glass (OSG) low-k dielectric thin films for microelectronic applications. *Int. J. Fracture*, 2003, **120**, 487–99.
20. Evans, A. G. & Charles, E. A. Fracture toughness determination by indentation. *J. Am. Ceram. Soc.*, 1976, **59**, 1976, 371–2.
21. Lawn, B. R., Evans, A. G. & Marshall, D. B. Elastic/plastic indentation damage in ceramics: the median/radial crack system. *J. Am. Ceram. Soc.*, 1980, **63**, 574–81.
22. Chen, J. & Bull, S. J. Assessment of the toughness of thin coatings using nanoindentation under displacement control. *Thin Solid Films*, 2006, **494**, 1–7.
23. Chen, J. Indentation-based methods to assess fracture toughness for thin coatings. *J. Phys. D*, 2013, 45, 1–15.
24. Laugier, M. T. Palmqvist indentation toughness in Wc–Co composites. *J. Mater. Sci. Lett.*, 1987, **6**, 897–900.
25. Niihara, K., Morena, R. & Hasselman, D. P. H. Evaluation of $K_{Ic}$ of brittle solids by the indentation method with low crack-to-indent ratios. *J. Mater. Sci. Lett.*, 1982, **1**, 13–16.
26. Jang, J., Wen, S., Lance, M. J., Anderson, I. M. & Pharr, G. M. Cracking and phase transformation in silicon during nanoindentation. *MRS Proc.*, 2004, **795**, 313–18.
27. Scholz, T., Schneider, G. A., Munoz-Saldana, J. & Swain, M. V. Fracture toughness from submicron derived indentation cracks. *Appl. Phys. Lett.*, 2004, **84** (16), 3055–7.
28. Volinsky, A. A., Vella, J. B. & Gerberich, W. W. Fracture toughness, adhesion and mechanical properties of low-K dielectric thin films measured by nanoindentation. *Thin Solid Films*, 2003, **429**, 201–10.
29. Schiffmann, K. I. Determination of fracture toughness of bulk materials and thin films by nanoindentation: comparison of different models. *Philos. Mag.*, 2011, **91** (7–9), 1163–78.
30. Li, X., Diao, D., Bhushan, B. Fracture mechanisms of thin amorphous carbon films in nanoindentation. *Acta Mater.*, 1997, **45** (11), 4453–61.
31. Endo, J., Inaba, S. & Ito, S. Mechanical properties of anisotropic glass. *J. Am. Ceram. Soc.*, 2015, **98**, 2767–71.
32. Bradley, D. 3D printing transparent glass. *Mater. Today*, 2015, **18** (10), 531–2.
33. Brow, R. K. & Watkins, R. D. *High Expansion, Lithium Corrosion Resistant Sealing Glasses. United States Patent, 1991.*
34. *Inaba*, S. *Anisotropic glass.* United States patent application publication. 2016. Pub. No.: US 2016/0362328A1.

# International Year of Glass 2022

The International Commission on Glass (ICG), along with the Community of Glass Associations (CGA) and the Glass Art Society (GAS) promoted a United Nations International Year of Glass for 2022. The year underlines the technological, scientific, social and economic importance of glass, the transparent and enabling material underpinning many of our technologies and which can facilitate the development of more just and sustainable societies to meet the challenges of globalisation.

Glass is the main conduit of the knowledge-based society. Glass optical fibres have stimulated a change of paradigm that has led to a global communications revolution; they are the backbone of the internet and the associated globalisation of information. Glassmakers have given us touch-sensitive glass covers for our mobile phones that revolutionised the way we communicate with each other.

Glass is the transparent tool that can underpin sustainable, more developed and just societies, it is an important medium for art and its history is shared with that of humankind.

*Palace of Nations at the United Nations Geneva is the venue of the opening conference of the International Year of Glass 2022*

### *Events in the IYoG2022*

The main themed conferences for the year are:

Opening Conference in Geneva,10–11 February 2022

International Congress of ICG in Berlin, 3–8 July 2022 (celebrating the 100th Anniversary of the DGG, the German Society of Glass Technology)

The Closing Congress will take place in Japan in December 2022.

Exhibitions supporting the International Year of Glass include:

VITRUM Milano, 5–8 October 2021 with a Glass Week promoting IYOG2022.

Glasspex/Glasspro in Mumbai, 3–5 March 2022, also including a Glass Week.

The China International Glass Industrial Technical Exhibition (the China Glass Expo) in Shanghai, 11–15 April. Includes a Glass Week with Hi-tech Industrial Congress and other glass knowledge literacy events.

GLASSMAN in Monterrey, 11–12 May with satellite IYoG events.

MIR STEKLA in Moscow, 6–9 June.

Glasstech in Dusseldorf 20–23 September, will include a Glass Week.

International Glass Art/Museums, in Europe and USA:

These are taking shape across the globe and a full programme is developing.

In the UK the Henry Moore Institute in Leeds is holding a glass art exhibition from February in collaboration with the National Glass Centre, Sunderland and Corning Museum of Glass, USA.

The Society of Glass Technology will have a number of International Year of Glass supporting events including the Annual Meeting, Furnace Solutions on 8–9 June and the Physics of Non-Crystalline Materials conference (PNCS) on 10–15 July.

Preparation work for Glass and Circular Economy exhibition was underway. There would also be a Glass and Architecture exhibition focusing on sustainability.

### *Opening Conference*

The Opening Conference of the International Year of Glass will be held in Geneva on 10–11 February 2022.

The conference organising committee is arranging a series of sessions along the following topics:

- Glass History and the Arrival of the Glass Age
- Glass for Good Health and Well-being
- Affordable and Clean Energy provided by Glass
- Optical Fibres, ICT Technologies, Photonics
- Glass in Architecture, Speakers: Pritzker Award
- Glass Industry. Energy Saving and Climate Action
- Sustainable Glass in a Circular Economy
- Glass Art
- Museums and Society
- Quality Education in Glass
- Glass beyond glass

Leading voices across the glass communities will present invited talks on their areas of interest. To the audience in the United Nations venue in Geneva and by live stream across the world.

### *Special publications*

Several journals would have a special issue to celebrate the International Year of Glass.

The writing for the book to celebrate IYOG2022 had been assigned to authors. There would be 13 chapters in total with a submission deadline in October 2021. The first English version would be prepared for the opening ceremony in Geneva. There would also be a Spanish version prepared. The IYOG team would provide regional committees with the layout of the book to maintain the identical structure for translations into other languages.

### *The Grass-roots Approach*

The International Year of Glass received 1800 endorsement letters from 84 countries, with letter still coming in. From these responses 18 Regional Organisations (ROs) have been established based on geographical proximity.

It is important for ROs to have sufficient active members (6–10 depending on the size of the RO). Having members from the field of museum, art and history would be very useful.

Communication is important issue for the distributed organisation. Volunteers are needed to assist communication work. Effective dissemination of information is essential so it is important that ROs are encouraged to set up their own websites with links to the main IYOG website. The central IYOG organising team is there to report on regional activities and promote where it can but the power is very much with the regions.

**Some Planned UK Events**

Contemporary Glass Society (CGS) will hold a series of monthly talks in London plus online.

The Turner Museum of Glass at the University of Sheffield will hold monthly lunch time talks.

Exhibition: Henry Moore Institute, Leeds with National Glass Centre and Corning Museum of Glass, from February 2022.

Illumination of Catcliffe Glasshouse Cone, Rotherham will happen in March.

Exhibition: Its all in the technique (CGS), at the National Glass Centre, Sunderland on 15 January–13 March.

Exhibition: The best of the best (CGS), Newcastle, February–March.

Opening: Stourbridge Glass Museum, Stourbridge, on 24 April.

Exhibition: Craft in the Bay (CGS), Cardiff, April–May.

Exhibition: Glorious Glass, Leeds, April–July.

Exhibition: Joyful Reflections (CGS), Farnham, April–June.

Furnace Solutions (SGT), 8–9 June, venue to be confirmed.

Exhibition: The Best of the Best (CGS), Newcastle, June–July.

Exhibition: Earth/Sea/Sky, London, June–July

16th Conference on the Physics of Non-Crystalline Solids, Canterbury, 10–15 July.

Exhibition: Then, Now, The Future, Stourbridge, July–October.

UK Glass Biennale, Stourbridge, August.

International Festival of Glass, Stourbridge, 26–29 August.

Exhibition: A postcard from the Commonwealth (CGS), Stourbridge, August–September.

SGT History and Heritage, Cambridge, September.

Exhibition: The Best of the Best (CGS), Newcastle, September–October.

Exhibition: Bedazzled 25 glassmakers impress (CGS), York, September–October.

Exhibition: Razzle Dazzle FIVE X FIVE (CGS), York, September–October.

Exhibition: CGS at 25, London, September–October.

Glass Focus, British Glass Annual Conference & Awards, November.

Pyrex 100 year celebration, Sunderland.

# Using near infrared thermal imaging to optimise production and quality from oxy-gas borosilicate furnace

AMETEK Land has provided in-furnace thermal imaging to optimise the borosilicate glass production at a facility in France.

SGD Pharma chose Land's Near Infrared Borescope (NIR-B-656-Glass) in-furnace thermal imaging system to optimise the production and quality of borosilicate glass for pharmaceutical packaging from its newly rebuilt 50 tpd oxy-gas furnace at its plant in St Quentin Lamotte. The new system replaces the company's existing CCTV system.

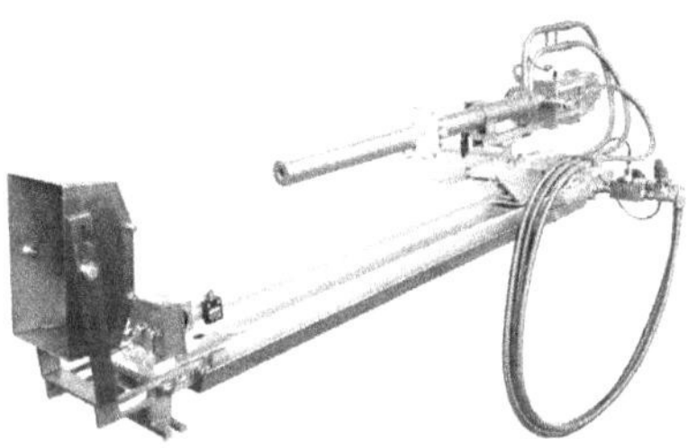

Obtaining clear images from the furnace using the CCTV system proved challenging due to the highly aggressive furnace atmosphere, which often resulted in blurry images and inconsistent, inaccurate measurements.

The NIR-B-656-Glass solution provides SGD Pharma with a true-temperature radiometric image, so live continuous temperature values can be obtained. The high quality of the purge applied on the lens delivers a clear image, providing the same views as the previous CCTV system managed at its best.

The most important benefit is the capability to obtain a furnace thermal profile continuously in the oxy-gas furnace, confirming the hot spot locations are well aligned with the furnace design and batch line. This supports the optimisation of the production and quality of the borescope glass while also helping extend the furnace's campaign life.

Francois Deblock, Glass Melting Director at SGD Pharma, said: "We are very pleased with the results we have achieved so far from AMETEK Land's Near Infrared Borescope, as it has allowed us to improve response times, identify and troubleshoot furnace operations to achieve improved yield and higher pull, as well as lower specific energy usage. We expect this will mean high cost and efficiency savings over the lifetime of the furnace."

The use of AMETEK Land's NIR-B-656-GLASS can result in increased asset life for furnaces and potentially achieve future energy optimisation and cost reductions. What can clearly be seen is that temperature measurement at critical locations in SGD Pharma's production process is essential for efficient control and optimisation of the glass melt process – and today, the technology to enable glass producers to do this is better than ever.

*NIR-B Glass inserted through the furnace wall above the throat in the furnace's centre line to provide a good view of the refractories, including crown, sidewalls, burner blocks and batch line. Dedicated infrared shields protect the NIR-B's retraction mechanism against the radiation coming from the hole camera block and the working zones around the instruments.*

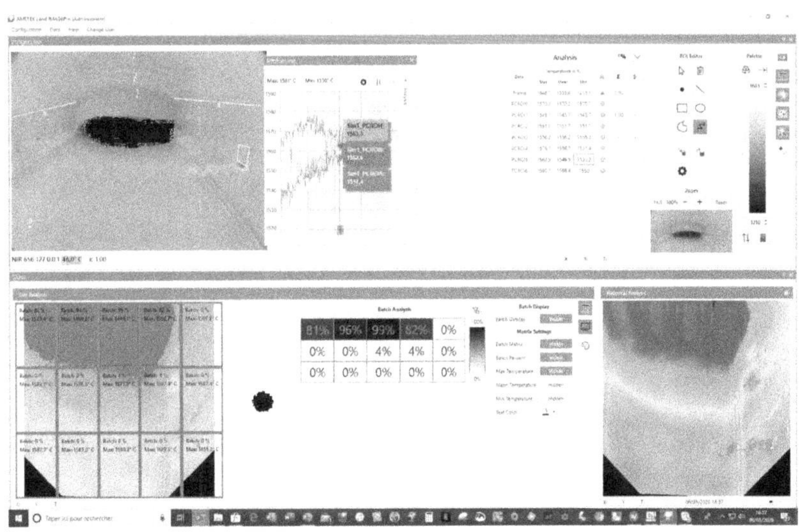

*Batch Line Location - Batch flows are initially impacted by the charging control and potentially the flames. However, the flow patterns are driven by the thermal flows/convection currents. SGD Pharma uses the IMAGEPro software that provides batch coverage data based on a grid with rows and columns for better batch tracking. Since the image is based on thermal data, SGD Pharma can add areas and apply alarms if the cold batch reaches a certain point. Whenever an alarm is triggered, the image is recorded for QA and troubleshooting purposes. Utilising a specific thermal palette and adjusting the temperature bands can identify which flames and which blocks are the most intense or hottest, generating the best suitable flames pattern and heat transfers.*

# World first as 100% hydrogen fired at Pilkington UK, St Helens

In a world first, and just one week after the Government's Hydrogen Strategy was released, trials began in the Liverpool City Region to produce float glass using hydrogen.

The natural gas fossil fuel, normally used in the manufacturing process, was completely replaced with hydrogen, indicating how industry can significantly cut carbon emissions and take a big step towards reaching net zero.

The trials took place at Pilkington UK's Greengate factory in St Helens, where the company first began making glass in 1826.

In order for the UK to decarbonise, complete transformation of almost every sector of the economy is required. Industry accounts for around 25% of all the UK's greenhouse gas emissions and reducing these emissions is vital if the country is to reach 'net zero'.

However, cleaning up energy intensive industry stands as one of the toughest challenges to meet. Emissions from industry, such as glass making, can be especially hard to reduce.

The ground-breaking 'HyNet Industrial Fuel Switching' project, led by Progressive Energy, with hydrogen being provided by BOC, will provide confidence that low carbon hydrogen from HyNet can replace natural gas. This is believed to be the first large-scale demonstration of 100% hydrogen firing in a live float glass production environment anywhere in the world.

The Pilkington UK trial is one of several projects underway in North West England to test how hydrogen can replace fossil fuels in manufacturing. A further HyNet trial is being held at Unilever in Port Sunlight later this year. Together, the demonstrations will support industry across sectors such as glass, food, drink, power and waste to convert to low carbon hydrogen, replacing their use of fossil fuels. Both trials are using hydrogen provided by BOC.

The HyNet Industrial Fuel Switching programme was awarded funding of £5.3 million from the UK government through its Energy Innovation Programme in February 2020.

HyNet will begin to decarbonise the North West from 2025. By 2030, it will be capable of removing up to 10 million tonnes of carbon from across North West England and North East Wales each year, the equivalent of taking four million cars off the road annually.

*Steve Rotheram, Metro Mayor of the Liverpool City Region, visiting the site*

*The Greengate site*

The HyNet project is also developing the UK's first low carbon hydrogen production plant at Essar Manufacturing Complex in Stanlow, which aims to be producing hydrogen that will be fuelling this switch from 2025.

Steve Rotheram, Metro Mayor of the Liverpool City Region said: "I want the Liverpool City Region to be at the forefront of the Green Industrial Revolution to take advantage of the jobs, investment and innovation that it will bring.

"Just as we led the first industrial revolution, our region is leading the green one too. HyNet has the potential to be a transformative project for our region, cutting carbon emissions across the North West by 25% and helping us take big a step towards our climate targets."

David Parkin, Director of Progressive Energy and Project Director of HyNet North West said: "Industry is vital for the economy, but is difficult to decarbonise. HyNet is focused on removing carbon from industry through a range of technologies including the capturing and locking up of carbon and the production and use of hydrogen as a low carbon fuel.

"HyNet will bring employment and economic growth to the North West region, kick-starting a low carbon hydrogen economy. We are focused on reducing emissions and protecting the North West's existing 340 000 manufacturing jobs and creating over 6000 new, permanent jobs, setting the region on course to be the world leader in clean energy innovation."

Matt Buckley, UK MD of Pilkington United Kingdom, part of the NSG Group, said: "Pilkington UK, and St Helens, are once again at the forefront of innovation in industry with this world first hydrogen trial on a float glass line. The HyNet project would be a massive step to support our decarbonisation activities. This full-scale production trial over several weeks successfully demonstrated that it is possible to use hydrogen to safely and effectively fire a float glass plant. We now look forward to the HyNet concept becoming a reality."

HyNet North West is a ground-breaking clean energy project that will unlock a low carbon future for the North West and North Wales. It will place the region at the forefront of the UK's journey to net zero and help to decarbonise multiple sectors of the economy from 2025 onwards.

The project will produce clean hydrogen to replace the fossil fuels we use today for industry, transport and homes. HyNet will also capture and store carbon dioxide ($CO_2$) emissions produced by energy intensive industries. We are in a climate emergency, and we need to act quickly to reduce our emissions. HyNet is an ambitious and viable answer to reducing $CO_2$ emissions across the North West England and North Wales region.

The project is made up of several different elements, including upgrades to existing facilities as well as the development of new infrastructure. The breadth of the project makes HyNet a leader in the creation of the UK's low carbon economy, bringing economic and environmental benefits to the local area and across the UK.

In the trial the NSG Pilkington managed to achieve a seamless transition between its current main fuel, natural gas, and hydrogen, with the expertise overcoming many challenges. The regenerators were used throughout to preheat the air for combustion. The trial proved that hydrogen was as capable as natural gas in achieving excellent melting performance and that it could be possible to operate the furnace with vastly reduced carbon emissions. It is a key step in the Group's plans to decarbonise. The switch to hydrogen to power the production from natural gas means that float glass furnaces, which account for the majority of the Group's overall carbon emissions, would be able to operate with hugely lower emissions.

With the Medium Term Vision of becoming a global glass supplier contributing to the world with high value-added glass products and services, the NSG Group aims to achieve 2% per year reduction in carbon emissions during the three-year period of the Revival Plan 24, with a view to achieve 21% reduction in 2030 compared to the 2018 levels and further to lay out its roadmap to achieve carbon neutrality by disruptive innovation.

---

Encirc has announced its firm commitment to decarbonising by the middle of this decade, using hydrogen in its furnaces to create billions of ultra-low-carbon glass bottles. The availability of hydrogen will enable the further expansion of Encirc's Elton facility, leading to the creation of at least 200 jobs, while futureproofing existing roles. Encirc employs more than 1000 people at its Elton facility will use the supply from the HyNet North West project in its furnaces to lead the decarbonisation of glass packaging.

*Adrian Curry (Encirc), Dave Dalton (British Glass), Anne-Marie Trevelyan MP and Richard Katz (Glass Futures)*

The Rt Hon Anne-Marie Trevelyan MP, the UK's Minister of State for Business, Energy and Clean Growth, visited the Cheshire-based manufacturer and bottler to learn more about the economic and environmental impact on the region if businesses like Encirc were able to be connected to hydrogen.

After touring Encirc's factory and learning more about how the glass manufacturer and bottler could take advantage of a new supply of hydrogen, Trevelyan said: "I'm delighted Encirc has announced its ambition to create the first ultra-low carbon glass by 2025 as part of the HyNet North West project.

"Working together with Government and ahead of COP26 in Glasgow, I'm delighted to see British industries are building back greener, meaning firms can play their part in creating jobs and meeting our ambitious climate commitments."

# UK glass industry net zero strategy summary

## British Glass consultation plots the route to net zero

British Glass and the glass industry have set out a strategy to achieve net zero $CO_2$ emissions by 2050. In 2015, in conjunction with the Department for Business, Energy & Industrial Strategy (BEIS), British Glass and the glass industry published a decarbonsation roadmap with the aim to reduce carbon emissions by 2050. Since then, the UK government has committed to the UN Climate Change Paris Agreement which will require the UK to reduce greenhouse gas emissions to net zero by 2050.

The glass industry has already made significant strides to reduce $CO_2$ and improve resource efficiency, with: an increase of 50% in the energy efficiency of UK glass furnaces over the past 40 years; and investment in waste heat recovery for UK plant from two multinational glass companies in the past three years.

### *Glass helps achieve net zero*

Recycling: each time one tonne of glass is recycled, approximately 580 kg of $CO_2$ is saved through the supply chain, air pollution is reduced by 20% and water pollution is cut by 50%.

The container glass sector has set itself a 90% target for container glass collected for recycling by 2030 to further the GHG reductions from the sector.

The flat glass sector is working to increase the recycled content of flat glass, which reduces the carbon footprint of glass.

Double glazing: double glazing and triple glazing has the potential to significantly increase energy efficiency in buildings, reducing energy use and $CO_2$.

Wind Turbines: continuous filament glass fibre is used in the construction of wind turbine blades which provide renewable energy.

Light weighting vehicles: continuous filament glass fibre is used as an alternative material to reduce the weight of vehicles, improving efficiency.

Solar: glass products are used for solar products which provide renewable energy.

Glass wool insulation: used for commercial and domestic buildings for improved thermal efficiency to reduce energy use and $CO_2$.

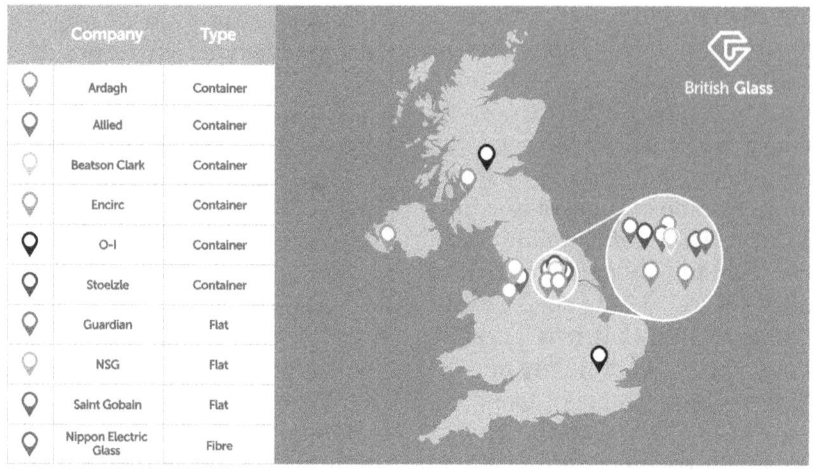


*The UK large scale glass manufacturing industry is significant and includes 10 companies with 17 sites throughout England, Scotland and Northern Ireland manufacturing container, flat and continuous filament fibre glass. The glass industry contributes almost £2 billion to the UK economy each year and directly employs around 6000 people, with an estimated 150000 more indirect jobs.*

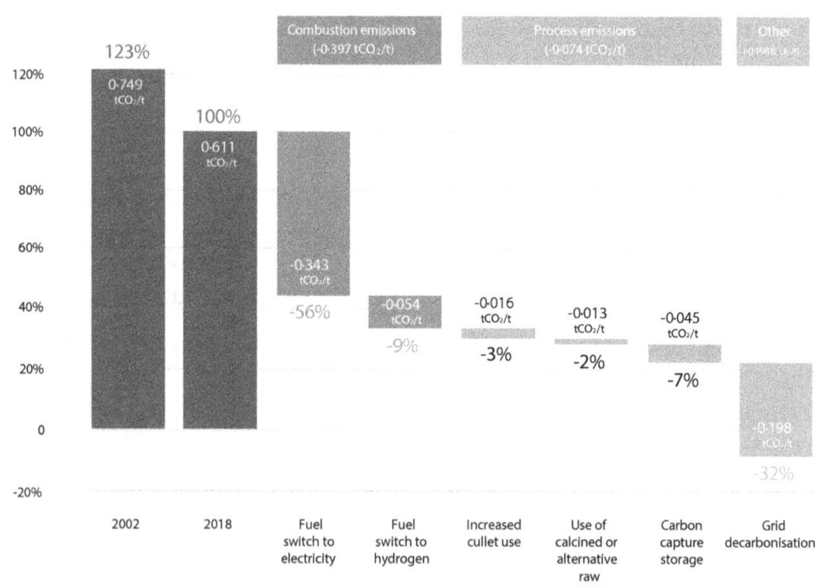


*As part of the net zero strategy, British Glass has developed a model which looks at how the glass sector can work towards net zero by 2050. This model assumes all necessary research and large-scale industrial demonstrations of new technologies take place; the grid decarbonises and becomes carbon negative by early the 2030s; and government implements the policies outlined in the strategy. This will enable the industry to transition to net zero at a cost that remains competitive.*

**Fuel Switch to electricity: saving 0·343 ton of $CO_2$ per ton of glass**

Improvements to the electricity grid, government support for upgrading site connections and policies on electricity pricing for large industrial users enable the sector to gradually transition from natural gas to electricity. Successful large scale industrial demonstration of oxy-fuel hybrid furnaces capable of running on up to 80% electricity enables this technology to be adopted in the container glass sector.

Hybrid furnaces with a lower electricity input can be utilised by the flat glass sector. Some smaller furnaces (<250 t/d) can be converted to all-electric.

**Fuel switch to hydrogen: saving 0·054 ton of $CO_2$ per ton of glass**

Innovation projects followed by large scale industrial demonstration show that hydrogen is a viable option to replace natural gas. Hydrogen is initially available inside some clusters before 2030 and a hydrogen distribution network to reach all glass sites could be built by 2050. Sites using hybrid furnaces gradually transition to hydrogen from biofuel or natural gas as it becomes available.

### *Financial viability*

Energy costs account for around a third of overall manufacturing costs. The high cost of alternative fuels which are typically over three times that of natural gas, currently prohibit manufacturers from fuel switching.

The Committee on Climate Change recommended establishing a funding mechanism to support operational, as well as capital costs associated with switching to low carbon fuels such as hydrogen, biofuels, and electricity. Electricity pricing should be reformed to reflect the much lower costs of supplying low carbon electricity in the mid-2020s and beyond. Policies to reduce UK manufacturing emissions must ensure that it does not damage UK manufacturers' competitiveness and drive manufacturing overseas.

Taxpayer funding or innovative financial schemes should be used to support deep decarbonisation in manufacturing sectors at risk of carbon leakage.

### *Infrastructure*

Most sites would need to upgrade their connection for hydrogen and electricity.

Upgrading and future-proofing electricity capacity and networks is needed to enable electrification of glass manufacturing and more generally society as it moves to electric forms of transportation. The same could be said for hydrogen infrastructure so that it allows non-cluster connections by 2035.

### *Compatibility with future technologies*

Up to 2050, new technology will be introduced, and it is essential compatibility is considered and incorporated if possible. Many glass manufacturing sites have limited space for additional saving technologies, and it is essential new processes, fuels and materials are trialled before they can be utilised on a full-scale manufacturing site.

Glass Futures in one initiative that offers the glass industry the chance to trial new ways of working across a scaled down factory space. Before supported demonstartions across a full production scale.

### *Security of supply*

There is uncertainty over the long-term availability of biofuels. Availability may be affected by increased use of biofuels in other industries such as aviation and transport. The electricity supply line across the English Channel has been affected by fire and will not be back to full capacity before 2022, such interruptions lead to spikes in wholesale energy prices. Consistent energy supply ensures a stable manufacturing environment, a major prop for sustainability.

### *Availability of raw materials and cullet*

Any new or established materials would need to be available in sufficient quantities with no interruption in supply. There may be competition from other foundation industries for supply. The composition of such materials can be variable and new technical solutions are required to address such challenges.

There is a shortage of good quality cullet for glass manufacturers to maximise the $CO_2$ saving. With the container industry aiming for a 90% recovery of glass packaging by 2030, measures need to be made to ensure this all returns to the container and not diverted to aggregate. Deposit return schemes need to recognise the established glass recycling infrastructure. Continued innovation in the recycling system to reduce losses to a minimum and increase yields of glass available for remelt.

Flat glass recovery needs wholesale revision with recognition by the construction industry. Incentives for recycling through public sector contracts and end of life regulations can set the climate. Glass is too easily lost within hardcore and circularity has to be promoted.

### *Health and safety*

New fuels and materials will impact on H&S risks and mitigation controls on production sites and will need to be considered.

### *Education*

There remains a lack of public awareness around the value of closed loop recycling compared to low value recycling such as glass being 'recycled' as aggregate. There is also a perception in the construction and demolition industries of glass being difficult to recycle.

# SGT Furnace Solutions training seminars

The usual Training Day held on the day before Furnace Solutions was cancelled due to the pandemic, instead the Society of Glass Technology have a series of five, one-off training seminars held online via Zoom on the second Wednesday of the month, starting in October. Times of presentations are 13.00 to 14.00 UK time, BST in October, GMT afterwards. The speakers and topics are presented below.

Wednesday 13th October 2021
Thermal balance and furnace screening: a periodic audit to reduce emissions and increase energy efficiency presented by Walter Battaglia of Stazione Sperimentale del Vetro (SSV).

Wednesday 10th November 2021
Forehearth Audits – is your glass conditioning equipment at its best? Presented by Alan Stephens of Fives Stein.

Wednesday 8th December 2021
Forehearth Colour Technology: Today's performance needs and new challenge presented by Pauline Darbouret of Ferro.

Wednesday 12th January 2022
NOx emission and possibilities to reduce presented by Matthias Hagen of Luft- und Thermotechnik Bayreuth (LTB).

Wednesday 9th February 2022
Challenges of Tin Bath Blocks in the "Float Process" and how to test them? Presented by Jan Theron of Lucideon.

www.sgt.org

---

# Furnace Solutions & Training Day

The Melting Technical Committee (MTC) of the Society of Glass Technology (SGT) invite you to mark your diaries for the 2022 Furnace Solutions on 9th June and Training Day on the 8th June.

Lucideon has converted their conference room to office space so the venue for the return of the person-to-person event is still to be confirmed.

Further details (including the Geoff Evans Dinner and hotel accommodation) will be announced on the website
http://www.furnacesolutions.co.uk/

Camera and check inspection solutions specialists, **Iris Inspection** machines and **Heye International** have gathered together in a cold end alliance, named WENSPECT. In working together they have overcome challenges at Carib Glassworks. Located in Port of Spain, Trinidad, Carib Glassworks has a rich history, dating back to 1948. The specialist glassworks is a member of the ANSA McAL Group of Companies, its founding fathers having laid the platform for a highly automated manufacturing facility, employing some of the best regional technical professionals.

Carib produces standard and custom-made containers in flint, green and amber for customers in the soft drinks, beer, food, alcohol and juice sectors. In 2016, the glassworks doubled production capacity to 70,000 tons/year via the commissioning of a second melting furnace and three production lines.

Carib Glassworks started working with Heye International almost 50 years ago equipping one furnace with Heye IS machines and Heye carousel machines for the cold end.

In 2015, as they were facing new quality challenges, Iris Inspection machines delivered non-contact inspection solutions for five production lines. Sharing the same passion for collaboration, Iris and Heye, now joined in WENSPECT alliance, have supported the glassmaker's business expansion initiatives, in a close working relationship.

"This is the kind of cooperative dynamic that we establish with our customer. So we have been delighted to find this level of support and service with IRIS and Heye." says Ms Tahira Khan, Project Manager at Carib Glassworks.

Having encountered wire edge and overpress defects last year, the glassmaker relied on IRIS solution, with the installation of the Wire Edge module on two production lines. Carib Glassworks has been impressed with the results achieved, so much so that at the beginning of 2020, all the impacted lines were equipped with the dedicated wire edge module.

**Pip & Nut** have transitioned their whole range of nut butters from plastic to embossed glass jars made by **Ardagh Glass**, following consistent feedback from the company's active consumer community.

Founder, Pip Murray, explains why they made the change: "As a B-Corp, we're always striving to improve on how we can make our products better for the planet. With the move to glass jars, we've improved the recyclability of our products as glass is more widely recycled in the UK than plastic."

The jars are made from a minimum of 50% recycled glass and have a fully recyclable label and lid. Working with Pip & Nut, and design agency **B&B**, the Ardagh team has developed a bespoke, embossed jar, making them desirable for reuse.

"This has been an incredible two-year creative journey. The result is a sustainable jar with a distinctive design which we are all proud of", commented Jolanda Raas, Sales Account Manager from AGP-Benelux.

The move supports Pip & Nut's commitment to sustainability and ongoing 'nut butter done better' mission.

Pip Murray adds: "We always listen closely to what our fans want, and they clearly asked us to move to glass, so we are delighted that day is finally here. As a brand we have promised our community that we'll always deliver on Better Nuts, Better Taste, Better Business and this packaging relaunch underlines our commitment to this third pillar and to our overall goal of making nut butter better."

Following the move into glass, the brand is now furthering their B-Corp promise of 'better business' by working on their commitment to setting science-based targets in 2021 to reduce their carbon emissions and align to net zero by 2030.

The Pip & Nut peanut butter is available in 300g jars and the almond butter in 170g jars.

**Ahlstrom-Munksjö** works with a wide range of industries to provide sustainable, fibre-based alternatives to traditional products and technologies. The latest of these alternatives is Glass-Gard®, a line of automotive glass interleaving papers.

Glass-Gard interleaving papers protect automotive glass applications such as windshields, backlights, sunroofs, or side windows from scratching, breakage, abrasion and hazing prior to their installation in vehicles. They prevent staining and provide cushioning for glass products during handling, storage, and transportation. The papers are engineered to reduce alkalinity and absorb moisture, protecting the product from corrosion and tarnishing of metalized components. This highly innovative material maintains glass clarity by preventing the transfer of resins and other compounds to the surface of the product.

In addition to automotive glass, Glass-Gard papers can be used to protect a variety of other products, including microscope slides, solar glass, and museum glass.

**Fluke Process Instruments** has launched the latest Datapaq Furnace Tracker System complete with the TP6 data logger, thermal barriers and intuitive software. The system is designed to be used repeatedly, in-process in the most hostile manufacturing environments, providing accurate and reliable through-process temperature surveys. Each system can help users improve process performance, decrease downtime, reduce temperature uniformity survey (TUS) reporting times and more.

The first component of the system is the Datapaq TP6 Data Logger, which can withstand the longest, most demanding operating conditions. The Datapaq TP6 features a 316-grade, stainless-steel case and a IP67 rating, making it water resistant. This datalogger is available with either 10 or 20 thermocouple inputs and can be specified for use with base or noble metal thermocouples.

The latest thermal barrier models are ideal for high-temperature heat treat processes and have been designed to provide maximum thermal protection, including applications such as vacuum and controlled atmosphere, slab reheat and solution ageing, among others. The new patented quenchable thermal barriers eliminate the need for fiber blankets. Fluke Process Instruments also offers bespoke design services to create a solution that specifically fits a customer's process.

To complete the system, Fluke Process Instruments offers the Datapaq Insight Software which transforms raw data into actionable analytics, clear user interface, context sensitive help screens and assistance options for infrequent users. Three software options are available and the complimentary Datapaq Insight for mobile is offered for free. This mobile application allows users to reset, download and analyze right on the factory floor and check thermocouple operation before a thermal profile run. The advanced TUS software also provides full temperature profiles and traceable reports for industry regulations like AMS2750 and CQi-9. www.flukeprocessinstruments.com

US hair care brand **Prose** has chosen **PONT Emballage** to supply them with **Beatson Clark**'s 60 ml amber glass jars for the launch of its new Root Source personalised hair growth supplements.

Prose, based in Paris and Brooklyn, New York, has developed a two-step process that works 'from the inside out to help balance the scalp and improve hair health and growth'.

Prose Cyrille Deschamp, VP of Operations at Prose, said: "We started working with PONT Emballage specifically for the launch of Root Source.

"We chose to use Beatson Clark's amber glass for this product for two reasons: first, it allows us to protect the plant-based ingredients from UV ray exposure, keeping the product fresh and effective; second, caring for the environment is one of our core brand values and using glass is a more sustainable packaging option.".

Charlotte Pike, Marketing Manager at UK glass manufacturer Beatson Clark, agrees that amber glass containers are perfect for hair, beauty and cosmetics products. "Amber glass offers a luxurious, premium look to any product, and because glass is inert, we can be sure there's no risk that the contents inside will be altered or damaged in any way," she said. "Amber glass is also great at protecting the product inside from harmful ultraviolet light."

*The batteries reduce how much power the Collingwood plant pulls from the grid whenever peak times are predicted*

**NSG Group** has partnered with **Convergent Energy + Power** (Convergent) to provide a battery storage system for its Pilkington plant in Collingwood, Ontario, Canada in an initiative that reduces energy consumption and cuts $CO_2$ emissions.

The battery storage installation represents NSG Group's first mutually beneficial cooperation between an energy storage developer and a glass manufacturing plant. The 5 MW/10 MWh battery energy storage system is fully operational and uses Convergent's PEAK IQ® proprietary software to reduce the facility's electricity usage during the most costly, carbon-intensive grid consumption peaks.

As the project was constructed during the 2020–21 winter season, in the height of the COVID-19 pandemic, construction teams adhered to particularly strict safety regulations. In order to achieve the projected savings of ~$450,000 per MW during the 2021 summer, Convergent coordinated with local partners to ensure pandemic-related travel restrictions did not slow down construction while navigating extensive supply chain disruptions.

"Locally, this battery storage system is part of our commitment to support our community in Collingwood and the broader efforts in Ontario to create a more efficient, sustainable grid," said Ron Young, Collingwood's Engineering Group Leader.

"We value Convergent's expertise from system design to dispatch, bringing the battery storage system online as quickly and safely as possible during the pandemic," added John Wilgar, NSG Group Head of Procurement.

"NSG Group has been an excellent partner from start to finish and demonstrated a deep commitment to all of its stakeholders," said Johannes Rittershausen, Convergent's CEO. "We are proud to build on our track record of successful, operating energy storage assets for commercial and industrial facilities here in Ontario and across North America."

**Encirc** at Ince in Cheshire offers customers a complete beverage supply chain solution for glass products. The company's 360 model allows global brands to bulk ship their wine to be filled into bottles made at the same facility and then delivered directly to retailers.

Having experienced its most successful year to date in 2020, with its filling lines being at almost full Encirc's upcoming investment in its Beverages facilities will increase its filling capacity by up to 40%. The plans will mean the introduction of a new filling line at the Cheshire site among other new projects.

Adrian Curry, Managing Director of Encirc, said: "This will be one of the most significant increases in UK filling capacity in the industry for many years. Our model is the most sustainable way to fill wine in the UK and I am pleased to say that, with this significant investment, it is a model that is here to stay."

**Glass Futures** in St Helens will be a 160,000 sq. ft Centre of Excellence, looking at how carbon can be reduced from the glass production process.

Landowner and developer **Network Space** has appointed **Civil & Remediation**, an earthworks and remediation contractor to prepare the wider site for future development and significant progress is being made. What was once a colliery before becoming a glassworks, has left a legacy of heavy metal concentrations together with gas and asbestos contamination. This phase of the project will now take around a month to complete before the main construction works start in late autumn.

The development is backed by St Helens Borough Council who committed funding to cover pre-planning and planning costs – while the scheme also received a major boost after Network Space secured a £9 million contribution from the Liverpool City Region Combined Authority.

Commenting at a recent site visit, St Helens Borough Council Leader David Baines said: "St Helens is steeped in glass manufacturing, and we want to keep it that way and at the heart of global innovation – and having Glass Futures here will put us at the centre of sustainable glass making for decades to come which also supports our climate emergency declaration."

Glass Futures is expected to be completed in 2022, bringing a number of high-skilled jobs to St Helens.

The first-place winner of the 2021 Glass Packaging Design Competition at **Michigan State University** (MSU) School of Packaging is the student team of Brendan Tulpa, Mitchell VanNote, Maddie Vasquez, Reed Vogt, and Kailey Wood. The winning team beat out 43 other student teams with their Red Dagger Hot Sauce packaging design.

"I am excited to congratulate the 2021 winners of the MSU School of Packaging student design competition," said Scott DeFife, President of the **Glass Packaging Institute** (GPI). "Getting to see the next generation put their heads together to design premier food and beverage packaging in glass makes certain that the future of packaging design is in good hands. Thank you again to

*The façade of the Shinhan Bank, Myeongdong station branch, Seoul, Korea is composed of a 41·52 mm thick insulating structure, made up of multiple layers, such as double glazing and argon gas, Low-E coating to block radiant heat, and dot frits for shading. After undergoing mock-up tests including reflection values and projection, a glass curtain of various layers was manufactured and constructed of 320 curved glass modules of 14 types. For modules and mullions, a dual clamp method that binds the parts together with those made using 3D printing was employed, and for transoms as horizontal members, which bear the weight of curved members, introduced a built-up method that assembles a CNC-processed steel plates by welding. Also, in order to best exhibit the neat and clean curved surfaces, all the structures that support the elevations were designed in a hidden bar type, and the space between glasses was filled with caulking. The curtain wall frame that supports the two-dimensional curved glass modules is made of high-strength steel so as to reduce the cross-sectional dimensions of each member as much as possible, and the steel mullions (45 mm wide and 150 mm long) are manufactured by bending. However, it was not easy to implement gentle waving glass elevations in reality. The steel formwork was manufactured to improve the accuracy of connecting the curtain wall frame to the framework, but there were cases in which they exceeded the permissable error. Although it took a lot of effort, the construction proceeded by reworking the pre-fabricated double-steel brackets and putting two structural elements together. Before initiating construction at the site in Myeongdong, we developed a two-story-high 1:1 mock-up model and conducted performance tests on the structural stability, airtightness, and watertightness against wind loads.*

---

the MSU School of Packaging and Ardagh Glass Packaging – North America (AGP – NA) for making this wonderful partnership possible for another year."

The winning student design features flint glass and is designed to look like a dagger. The blade, handguard and neck details are all embossed features of the bottle. To truly cap off the dagger look, the neck and closure are designed to look like the handle and pommel, by which consumers could wield the bottle.

AGP – NA hosted the winning team and their instructor Angela Brann for a tour of the company's Winchester, Ind., glass manufacturing facility and welcomed the students to its Indianapolis headquarters for meetings with the Ardagh executive team. The students were presented with a Lucite model of their winning design at a special dinner ceremony.

The 44 student teams, all randomly assigned, from the Fall 2020 and Spring 2021 Packaging with Glass and Metal class participated in the competition, working on the project throughout the semester as they learned about the design and manufacture of glass bottles and jars.

The top 10 submissions (5 from each semester) were judged by an industry panel including Jonathan Crowe, AGP-NA; Doug Laib, O-I Americas North; Bill Eaton, Pepsico; Dave Wengerhoff, Food Partners Consulting; and Rich Crawford.

The **O-I** *Glass 2021 Sustainability Report* examines the vision, innovation and transformation behind the challenge to be the chosen supplier of brand building packaging solutions. The report highlights the refresh of O-I's business strategy, elevating sustainability within the company.

"While we believe glass packaging is already the most sustainable packaging solution, sustainability at O-I is about more than what we make. It is also about how we make it," said Andres Lopes, Chief Executive Officer for O-I. "We enhanced matters, from our approach to governance and management of sustainability to the number of initiatives and quantitative goals. We are excited to share our elevated ambitions and what they mean for O-I and its stakeholders."

In 2020, O-I introduced structural enhancements to deepen and accelerate sustainability progress, including the appointment of its first Chief Sustainability Officer, to create an integrated approach

*AGC is co-sponsoring the Japan Pavilion at Expo 2020 Dubai to be held in Dubai, United Arab Emirates, from October 1, 2021 to March 31, 2022. The Dubai Expo is the first ever World Expo to be held in the Middle East, Africa, and South Asia regions and is expected to attract about 25 million visitors. The theme of the Japan Pavilion, "Where ideas meet" provided for the buildings which expresses the long historical ties between Japan and the Middle East. AGC supplied low-E insulating double glazing glass with high solar control and thermal insulation, also tempered laminated glass with superior strength and safety in case of breakage. In the exhibition area of the pavilion, our mirror glass and Glascene F, a transparent screen film product, are used to create a fantastic space with dynamic images and light.*

around shared goals and collaborative changes in the company.

Among the new sustainability goals for O-I, recycled content, water reduction, renewable energy and zero waste have been added to the ambitious and aspirational goals to achieve by 2030.

"This report highlights the transformational aspects of our business, providing a holistic view of how O-I and glass packaging will play a role in the sustainable future we all seek," said Randy Burns, Chief Sustainability Officer for O-I. "O-I's vision and journey for this sustainable future is not a blue-sky vision. It is here, today."

The sustainability report provides insight into the ways that O-I is reimagining and reinventing the business model for glass packaging, working toward a future where the circularity of glass meets industry-changing technology that will disrupt how glass is made and sold.
o-i.com/sustainability

**NSG** has completed the transfer its battery separator business to **ENTEK Technology Holding**, a wholly owned subsidiary of **ENTEK Technology Holdings**. NSG Group will continue to cooperate with ENTEK in managing and developing the business by holding minority stake and appoint a director to the Board of Directors as a commitment to retain and grow trust from the customers.

**Foamit Group** is expanding in the Nordic countries by acquiring **Glasopor**, a Norwegian foam glass manufacturer. The acquisition strengthens Foamit's position as the largest foam glass manufacturer in the Nordic countries. Foamit will operate in the Finnish, Swedish, Danish and Norwegian markets in the future. "The acquisition of Glasopor will give us access to new markets, more manufacturing capacity and new expertise in utilising foam glass in various locations. The availability of raw materials will also improve. Foam glass crumb is a very environmentally friendly and lightweight building material made of cleaned recycled glass, and its demand is supported by an increase in demand for circular economy products," says Kalle Härkönen, CEO of Foamit Group.

**AGP E-Glass**, a subsidiary of **AGP Group**, a Belgium-based manufacturer of glass products for the automotive, marine, and security markets, plans to establish a North American Tech Center in the Charter Township of Canton, Michigan. The roughly $15 million project is expected to create 71 jobs.

The company has been awarded a $550,000 Michigan Business Development Program performance-based grant to support this expansion in Canton. In addition, Canton Charter Township has offered a 50% property tax abatement in support of the project.

According to state officials, "with the auto industry continuing to experience increased demand in electrification, connectivity, shared mobility, autonomous driving and performance, AGP plans to establish a North American Tech Center in the Charter Township of Canton, where it will develop and exhibit the latest technologies for clients, partners, suppliers and university officials to see the products in use."

AGP provides high-tech glazing solutions to more than 20 automotive brands that are shaping the future of mobility through electrification and autonomous driving. The tech centre will continue to grow Michigan's mobility and electric vehicle sectors and will further underscore the state's position in mobility and automotive R&D, state officials said.

The problem of reusing glass bottles has led the **Cetie Users and Glass Manufacturers's Group** to draw up essential recommendations in order to guarantee optimal quality and safety of the returnable container (maximum number of trips, traceability, etc). This data sheet also includes for the moment the first chapter on handling, unloading and cleaning, and a second one on filling, capping and labelling.
*DT14.01: GUIDELINES FOR THE USE OF RETURNABLE GLASS CONTAINERS* – provides information on the proper use of returnable glass containers.

Contributors are invited to join the group and be at the heart of the implementation of these best practices.
Contact us via www.cetie.org

**SCHOTT** pioneer Marga Faulstich, invented the first lightweight lens for spectacles, "Schwerflint 64", almost 50 years ago. This invention led to a new lightweight generation of glasses and sunglasses, today's standard. Augmented Reality (AR) glasses are called "the next big thing" in consumer tech, and the company is now building on Marga's legacy by introducing

*With a new 20% lighter RI 1.9 high-index glass wafer, SCHOTT can reduce an AR wearable's overall weight by roughly 7%. SCHOTT RealView® 1.9 lightweight doesn't sacrifice image quality or durability, even with the dramatic reduction in density. Photo: SCHOTT*

a new lightweight and more immersive optical lens designed to bring AR to the masses.

With Schott RealView® 1.9 lightweight the new optical glass shaves 5 g off the weight of the waveguides currently used in AR devices, which means a lighter, more comfortable, more immersive experience for the user. Manufacturers designing consumer-grade products target a total device weight of 70 g, so saving 5 g by incorporating the new lightweight glass is a milestone in material and device development.

The weight cut is mainly due to a reduced density of the material. The result: AR devices that offer highest wearing comfort while not compromising in image quality thanks to the 1.9 high-refractive index.

"Every optical component in AR must push the boundaries of technology to keep manufacturing costs low, while also achieving a high-quality image, a large field of view, and a full immersion experience in a compact form factor without added weight or bulk," said Dr. Rüdiger Sprengard, Vice President Augmented Reality at SCHOTT. "With those goals in mind, we created an all-new glass formulation that was tailor-made for AR use while pushing the limits of what was possible for density, clarity, and robustness."

The **Gerresheimer** plant in Lohr plans to build a glass melting furnace for white glass in 2022 which will save significantly more $CO_2$. The company has submitted a funding application to the German Federal Ministry for the Environment for this innovative project.

"This pioneering technology project serves to strengthen Germany as an industrial and glass location. With this innovative technology we will set new standards in terms of sustainability and avoidance of emissions," said Andreas Kohl, who as Global Senior Vice President Operations is responsible for worldwide container glass production and its technical development in the Gerresheimer Group.

The Lohr plant produces more than one billion glass containers annually for the pharma and food industries with two melting furnaces for clear and amber glass. "By using this sustainable furnace technology in conjunction with green energy for the glass melting process, we will significantly reduce $CO_2$ emissions in production by around 25,000 tons per year," says Jörg Buchmayer, Head of Production and Technology, who is responsible for the project. "We are using hybrid technology, which makes it possible to cover half of the required melting energy in the furnace with green electricity and the other half with natural gas."

Glass can be recycled an infinite number of times. However, high temperatures are required for glass melting. This costs energy and generates corresponding $CO_2$ emissions. For more than a decade, the Gerresheimer Group has been working on using no more energy than absolutely necessary for the sustainable production of its glass containers for the pharma, cosmetics and food industries.

As part of its global sustainability strategy, Gerresheimer has set itself the goal of reducing its $CO_2$ emissions by 50% by 2030 compared with 2019. The resource-saving production of glass plays a decisive role in this. The innovative technology planned for the plant in Lohr is pioneering in this respect.

**SCHOTT** is investing a total of €70 million in the expansion of its Indian tubing site in Jambusar, Gujarat, following several million in investments over the last few years. "Against the backdrop of the growth trend in the Indian pharmaceutical business and the pandemic, we want to commit to secure the supply of pharma glass," explains Dr Patrick Markschläger, Executive Vice President of SCHOTT's Business Unit Tubing. "The increase of over 30% in the facility's overall Indian production capacity is at the same time a commitment to supporting the government's vision of India becoming a global pharmaceutical hub," adds Pawan Shukla, Managing Director Schott Glass India, during the groundbreaking event at the site.

The additional tank is scheduled to go into operation at the beginning of 2023, with the second one following a year later. The expansion in Jambusar will create new jobs for around 225 employees and is part of a more than $1 billion strategic investment program of SCHOTT through 2025, leveraging the global pharma tubing and packaging business.

The development closely follows a recent doubling of capacity at the Junyin, China borosilicate glass tubing factory.

**Owens Corning** has introduced its latest advancement: PINK Next Gen™ Fiberglas™ insulation. Featuring advanced fibre technology, this insulation solution enables up to 23% faster installation compared to existing products.

The advanced fibre technology creates a tightly woven network of soft, fine fibres to form a super-resilient blanket of insulating micro-pockets. The new product is safe for installers and residents and faster to install than competitive insulation products and spray foam insulation. Because no hazardous chemicals are required to be mixed on site, there is less potential for human error, increasing peace of mind for installers and residents.

"Contractors today have high performance expectations and extremely tight timelines," said Joe Wagner, Insulation marketing director. "This product evolution is a significant step-change in meeting the needs of contractors, as well as installers, builders and homeowners."

The product is designed to meet building codes and help professionals manage through tight labor timelines and budgets by reducing install time.

PINK Next Gen Fiberglas insulation focuses on solving problems in four meaningful areas: Safety by offering 99%-safer fire performance and no chemical fire retardants added to fibreglass, made with safe, proven ingredients, no formaldehyde added, no hazardous chemicals mixed on-site. Precision: made with advanced fibre technology for up to 23% faster installation compared to our existing products, exceptional product recovery and improved stiffness perfectly fills standard cavities and stays put, fewer clumps and chunks for fast, clean cutting and splitting. Comfort: the texture is soft as cotton, less dust for more comfortable installation and less cleanup, shed-resistant insulation stays on the batt, not on installers, and improved indoor comfort and sound reduction for homeowners. Sustainability: made with 100% wind-powered electricity, with a new industry standard for recycled content, saves 12 times the energy required to make it in the first year of use.

**Guardian** is responding to demand from architects who are pushing design boundaries with buildings that maximize views and natural light for occupants in interesting, forward-thinking facades while also reaching energy performance codes and standards.

"Super jumbo helps our fabrication customers improve yields on many common commercial sizes and produce oversize insulating glass units that aren't possible with standard jumbo coated glass," said Chris Dolan, Commercial Segment & Technical Services Director, Guardian Glass North America.

Independent Guardian Select fabricator **Press Glass** is capitalising on this advantage. Its Ridgeway, Va. factory was built to accommodate 130″×240″ glass. The fabricator has some of the largest glass sizes available in North America.

"Our customers want larger glass, whether it's commercial or residential," explains Piotr Styk, Senior Estimator/Project Manager at Press Glass. "Everyone is calling for expansive views, and the bigger the better. When we were deciding on equipment to purchase for our new facility, we researched the maximum glass size available in North America. Guardian offered the largest size at 240″, so we made sure to equip our facility with machines that could handle that size."

The Press Glass' Ridgeway facility boasts 231″×115″ triple insulating glass units that include Guardian SunGuard SNX 51/23 coated glass and SunGuard SuperNeutral 68 coated glass, while the under-construction Terminal 5 at Chicago O'Hare International Airport, uses SunGuard SNX 62/27 coating on UltraClear low-iron glass in a large variety of sizes.

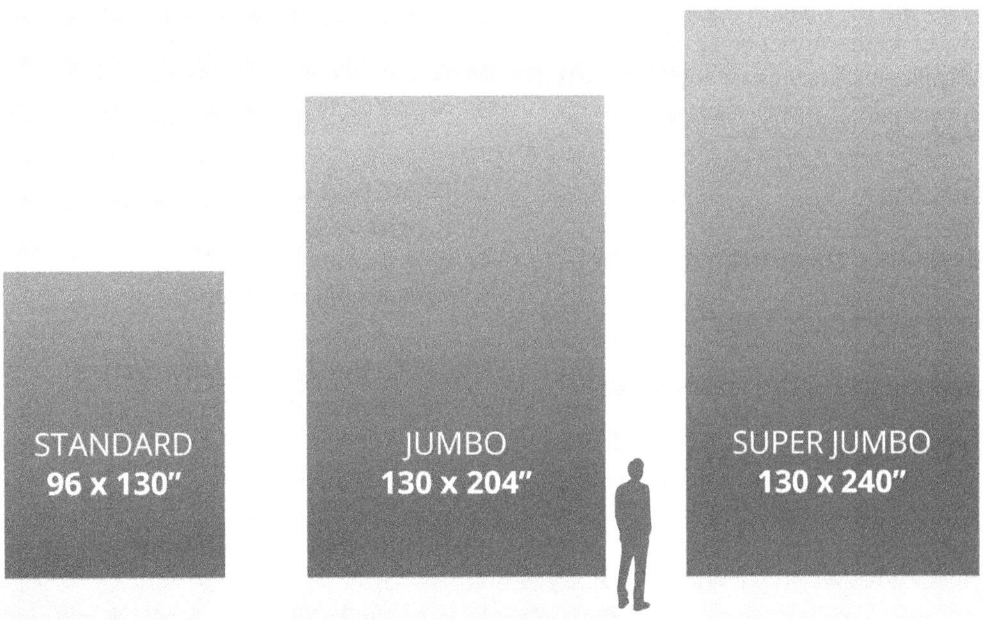


*The progression of glass sizes to Super Jumbo*

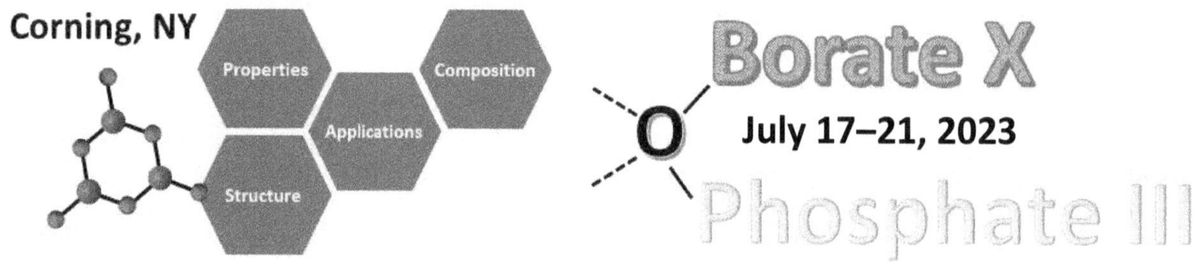


## Revised Conference Dates

The 10th International Conference on
**BORATE GLASSES, CRYSTALS AND MELTS**
**July 17–19, 2023**

This marks the 10th in the series of conferences on borate glasses, crystals and melts, and will return to New York, USA close to where the first Borate Conference was held (Alfred, 1977). The most recent meeting took place in Oxford, UK (2017). The Borate Conference is a forum for the discussion of current scientific research on borate and B-containing materials. The conference will be dedicated to Dr. E.I. (Stratos) Kamitsos from the National Hellenic Research Foundation (Athens), to honor his lifetime of work in the field of borate glasses.

The 3rd International Conference on
**PHOSPHATE GLASSES, CRYSTALS AND MELTS**
**July 19–21, 2023**

This conference continues the tradition established with the first two conferences held in Pardubice, Czech Republic (2014) and Oxford, UK (2017). It will cover all aspects of phosphates and P-containing materials, including P-O, P-S, P-Se and P-O-N networks. Topics of interest will cover basic science, applications and technologies. The Phosphate Materials conference will be dedicated to Professor Ladislav Koudelka, University of Pardubice, Czech Republic, recognizing his career in phosphate and borophosphate glasses.

### Conference Themes

This joint meeting, to be held at the Corning Radisson Hotel, is comprised of two conferences that will run sequentially. On the third day of the meeting there will be a combined day dedicated to borophosphate glasses followed by a conference dinner to mark the close the Borate Conference and the start of the Phosphate Conference. Topics will encompass anything related to borate and phosphate materials, including:

- **Energy**
- **Healthcare**
- **Optical and data storage materials**
- **Property-structure relationships**
- **Fundamentals of the glassy state**
- **Novel glasses, crystals and melts**
- **Technology and Industry**

### Conference Chair

Dr. Randall Youngman
Corning Incorporated
youngmanre@corning.com

### Local Organizing Committee

R. Youngman (chair), J. Wu (Corning), L. Ma (Corning), L. Huang (RPI), A. Goel (Rutgers), D. Möncke (Alfred), A. Clare (Alfred)

***www.boratephosphate2020.org***

# CONFERENCE DIARY

**27 September–1 October**
CelSian training course - general glass technology training, Eindhoven, The Netherlands.
*CelSian Glass BV Email infodesk@celsian.nl Web www.celsian.nl/training-and-rd/*

**4–6 October**
Sixth International Conference on Competitive Materials and Technology Processes (ic-cmtp6), Miskolc-Lillafured, Hungary.
*Email cmtp.conf@gmail.com Web www.ic-rmm1.eu/ic-cmtp/*

**5–8 October**
Vitrum 2021, Milan, Italy.
*Fiera Milano, Via Carlo Ilarione Petitti, 16, 20149 Milano MI, Italy. Email vitrum@vitrum-milano.it Web www.vitrum-milano.com*

**6–8 October**
Intersolar 2021, Munich, Germany.
*Web www.intersolar.de/en/home*

**12–15 October**
FESPA Global Print Expo, Amsterdam, The Netherlands.
*Web www.fespaglobalprintexpo.com/*

**12 October**
CelSian training course: heat transfer in glass melting furnaces, Eindhoven, The Netherlands.
*CelSian Glass BV Email infodesk@celsian.nl Web www.celsian.nl/training-and-rd/*

**13 October**
Furnace Solutions training seminar: Thermal balance and furnace screening: a periodic audit to reduce emissions and increase energy efficiency presented by Walter Battaglia of Stazione Sperimentale del Vetro (SSV).
*Web www.sgt.org*

**13–14 October**
CelSian training course: burners, combustion and NOx emissions, Sheffield, UK.
*CelSian Glass BV Email infodesk@celsian.nl Web www.celsian.nl/training-and-rd/*

**16–17 October**
Icon Bath 2021 Conference: Ceramics, Glass and Stained Glass, Bath UK.
*The International Institute for Conservation of Historic and Artistic Works, 3 Birdcage Walk, Westminster, London, W1H 9JJ, UK. Email IconBath2021@gmail.com Web www.iiconservation.org*

**17–21 October**
ACerS 123rd Annual Meeting with Materials Science & Technology 2021, Pittsburgh, USA.
*The American Ceramic Society, 550 Polaris Parkway, Suite 510, Westerville, OH 43082, USA. Email customerservice@ceramics.org Web www.ceramics.org*

**20–22 October**
Glass Performance Days, Tampere, Finland.
*Brown Onduso, GPD, Vehmaistenkatu 5, 33730 Tampere, Finland Email brown.onduso@gpd.fi Web gpd.fi/*

**26–27 October**
CelSian training course: refractory selection, maintenance and related defects, Eindhoven, The Netherlands.
*CelSian Glass BV Email infodesk@celsian.nl Web www.celsian.nl/training-and-rd/*

**27–30 October**
Seventh International Conference on Multifunctinal, Hybrid and Nanomaterials, Genoa, Italy.
*Web www.elsevier.com/events/conferences/international-conference-on-multifunctional-hybrid-and-nanomaterials*

**28–29 October**
CelSian training course - flat glass melting and forming, Eindhoven, The Netherlands.
*CelSian Glass BV Email infodesk@celsian.nl Web www.celsian.nl/training-and-rd/*

**30 October–5 November**
Sixth Wuhan ICG Winter School, Wuhan, China.
*Prof. Kai Xu: SMART, Wuhan University of Technology, 1040 Heping Ave, Wuchang Qu, Wuhan Shi, Hubei Sheng, China. Email kaixu@whut.edu.cn Web www.icglass.org*

**30 October–5 November**
82nd Conference on Glass Problems, Columbus, USA.
*Donna Banks, GMIC, 550 Polaris Parkway, Suite 510, Westerville, OH 43082, USA. Email dbanks@gmic.org Web glassproblemsconference.org*

**5–6 November**
Glass in architecture from the pre- to the post-industrial era: Production, use and conservation, Vitromusée Romont, Switzerland.
*Vitrocentre Romont Au Château P.O. Box 150 CH-1680 Romont, Switzerland. Email sarah.tabbal@vitrocentre.ch Web vitrocentre.ch*

**10 November**
Furnace Solutions training seminar: Forehearth Audits – is your glass conditioning equipment at its best? Presented by Alan Stephens of Fives Stein.
*Web www.sgt.org*

**~~16–18 November~~ postponed to 26–28 October 2022**
Glasstech Asia 2021, Bangkok, Thailand.
*Web www.glasstechasia.com.sg/*

**17–18 November**
Glass and Modern Technologies - XXI, Moscow, Russia.
*Organizing Committee Email forum@steklosouz.ru Web www.steklosouz.ru*

**18 November**
Glass Focus Awards 2021, UK.
*British Glass, 9 Churchill Way, Chapeltown, Sheffield, South Yorkshire S35 2PY, UK. Web www.britglass.org.uk/glass-focus-2021*

**23–25 November**
Celsian training: glass melting and process improvement for Indian glass industry, Mumbai, India.
*CelSian Glass BV Email infodesk@celsian.nl Web www.celsian.nl/training-and-rd/*

**5–8 December**
International Commission on Glass (ICG) Annual Meeting, Incheon, South Korea.
*Secretariat, Room 403, Meolijae Building, 76 Bangbae-ro, Seochu-gu, Seoul, Republic of Korea. Email secretariat@icg2021.org Web www.icg2021.org/*

**8 December 2021**
Furnace Solutions training seminar: Forehearth Colour Technology: Today's performance needs and new challenge presented by Pauline Darbouret of Ferro.
*Web www.sgt.org*

**12–17 December (new dates)**
14th Pacific Rim Conference on Ceramic and Glass Technology (PACRIM 14) including Glass & Optical Materials Division 2021 Annual Meeting (GOMD 2021), Vancouver, BC, Canada.
*Andrea Ross, The American Ceramic Society, 550 Polaris Parkway, Suite 510, Westerville, OH 43082, USA. Email aross@ceramics.org Web www.ceramics.org*

**2022 UN International Year of Glass**
**Postponed to 2022 no date announced**
13th International Symposium on Crystallization in Glasses and Liquid, Paris, France.
*Laurent Cormier, CNRS, Paris, France. Email crystallization@sciencesconf.org Web crystallization.sciencesconf.org/*
ACerS Pan American Ceramics Conference, PACC-FMAs, Panama City, Panama.
*Andrea Ross, The American Ceramic Society, 550 Polaris Parkway, Suite 510, Westerville, OH 43082, USA. Email aross@ceramics.org Web www.ceramics.org*

**12 January**
Furnace Solutions training seminar: NOx emission and possibilities to reduce presented by Matthias Hagen of Luft- und Thermotechnik Bayreuth (LTB).
*Web www.sgt.org*

**9 February**
Furnace Solutions training seminar: Challenges of Tin Bath Blocks in the "Float Process" and how to test them? Presented by Jan Theron of Lucideon.
*Web www.sgt.org*

**10–12 February**
Opening Ceremony of the International Year of Glass 2022, Geneva, Switzerland.
*Web www.IYoG2022.org*

**18–19 February**
Glassman Asia, Seoul, Korea.
*Ken Clark, Quartz Business Media, Quartz House, 20 Clarendon Road, Redhill, Surrey RH1 1QX, UK Email ken-clark@quartzltd.com Web www.glassmanevents.com*

**3–5 March**
glasspex INDIA 2022, Mumbai, India.
*Vivek Bohra, Senior Project Manage Email bohrav@md-india.com Web www.glasspex.com*

**6–10 March**
Optical Fiber Communications Conference and Exhibition (OFC), San Diego, California USA.
*The Optical Society, USA. Email custserv@osa.org Web www.ofcconference.org*

**15–18 March**
Unified International Technical Conference on Refractories (UNITECR 2022) 17th Biennial Worldwide Congress on Refractories, Chicago, USA.
*The American Ceramic Society, 550 Polaris Parkway, Suite 510, Westerville, OH 43082, USA. Email aross@ceramics.org Web www.ceramics.org*

**2–3 April**
Deco '22, The Society of Glass and Ceramic Decorating Products annual conference, Columbus, OH, USA.
*Myra Warne, Society of Glass and Ceramic Decorating Products, PO Box 2489, Zanesville, OH 43702, USA. Email info@sgcd.org Web www.sgcd.org*

**3–6 April**
TMS World Congress on Artificial Intelligence in Materials and Manufacturing (AIM 2022), Pittsburgh, Pennsylvania, USA.
*TMS Meeting Services, 5700 Corporate Drive Suite 750, Pittsburgh, PA 15237, USA. Web www.tms.org/AIM2022*

**5–6 April**
GlassPrint 2022, Düsseldorf, Germany.
*Email sales@glassworldwide.co.uk Web www.glassprint.org*

**13–16 April**
China Glass 2022, Shanghai, China.
*Web www.chinaglass-expo.com*

**18–20 April**
Int. Conf. on Glass Science and Technology from the Golden Era of the Pharaohs to a Sustainable Future, Egypt.

**19–22 April**
13th European Conference on Industrial Furnaces and Boilers, Algarve, Portugal.
*Raquel Reis, Cenertec, Rua Gago Coutinho, 185-187, 4435-034 Rio Tinto, Portugal. Email raquel.reis@cenertec.pt Web infub.pt*

**11–12 May**
Glassman Latin America, Monterrey, Mexico.
*Ken Clark, Quartz Business Media, Quartz House, 20 Clarendon Road, Redhill, Surrey RH1 1QX, UK Email ken-clark@quartzltd.com Web www.glassmanevents.com*

**18–20 May**
REFRA PRAGUE 21, 21st Conference on modern refractory materials and key achievements in high temperature technologies, Prague, Czech Republic.
*Ing. Jan K. Luxa, Silikátová společnost České republiky, Novotného lávka 5, 110 01 Praha 1, Czech Republic. Email hluxa@seznam.cz Web www.silikaty.cz*

**18–21 May**
GAS 2022, Glass Art Society conference, Tacoma, WA, USA.
*Glass Art Society, 2208 NW Market Street, Suite 200, Seattle, Washington 98107, USA. Email info@glassart.org Web www.glassart.org*

**22–26 May**
2022 Glass and Optical Materials Division Annual Meeting (GOMD2022), Baltimore, USA.
*Andrea Ross, The American Ceramic Society, 550 Polaris Parkway, Suite 510, Westerville, OH 43082, USA. Email aross@ceramics.org Web www.ceramics.org*

**6–9 June**
Mir Stekla 2022, Moscow, Russia.
*Ekaterina Razumeitseva, Project Head, EXPOCENTRE, Moscow, Krasnopresnenskaya nab., 14, 123100, Russia. Email re@expocentr.ru Web www.mirstekla-expo.ru/en/*

**8–9 June**
Furnace Solutions and Training Day, venue TBC, UK.
*Christine Brown, Society of Glass Technology, 9 Churchill Way, Chapeltown, Sheffield S35 2PY, UK. Email christine@sgt.org Web www.sgt.org*

**16–17 June**
XXXIV ATIV Int. Conf., Where Glass Science, Art and Technology Meet Together, Parma, Italy.
*ATIV Secretariat, MV Congressi spa Via Marchesi 26 d 43126 Parma, Italy. ativ@ativ-online.it Web www.ativ.eu*

**21–24 June**
Techtextil 2022, Frankfurt, Germany.
*Email techtextil@messefrankfurt.com Web techtextil.messe-frankfurt.com*

**23–24 June**
Challenging Glass Conference 8: International Conference on the Architectural and Structural Application of Glass, Ghent, Belgium.
*Email cgc8@challengingglass.com Web www.challenging-glass.com*

**3–8 July**
26th International Congress on Glass, Berlin, Germany. Hosted by the DGG in their centenary year.
*Annette Doms, Deutsche Glastechnische Gesellschaft, Siemenstrasse 45, D-63071 Offenbach, Germany. Email dgg@hvg-dgg.de Web www.hvg-dgg.de*

**4–8 July**
Ninth International Conference on Optical, Optoelectronic and Photonic Materials and Applications (ICOOPMA) and EuroDIM: 14th Europhysical Conference on Defects in Insulating Materials, Ghent, Belgium.
*Web icoopma-eurodim.ugent.be*

**10–15 July**
Physics of Non-Crystalline Solids, now also including Society of Glass Technology Annual Meeting Canterbury, UK.
*Christine Brown, Society of Glass Technology, 9 Churchill Way, Chapeltown, Sheffield S35 2PY, UK. Email christine@sgt.org Web www.sgt.org*

**10–14 July**
Ninth Int. Congress on Ceramics (ICC9), Krakow, Poland.
*European Ceramic Society AISBL, Av. Gouverneur Cornez 4, B-7000 Mons, Belgium. Email ecers@bcrc.be Web ecers.org/*

**24–29 July**
Sol-Gel 2021, International Sol-Gel Society Conference, Lyon, France.
*Université Claude Bernard Lyon 1, Cellule Congrès - SOLGEL 2021, 43, bd du 11 Novembre 1918, 69622 Villeurbanne, France. Email solgel2021@univ-lyon1.fr Web www.solgel2021.fr*

**27–30 August**
International Festival of Glass, Stourbridge, UK.
*Ruskin Glass Centre, Wollaston Road, Amblecote, Stourbridge DY8 4HF, UK. Email ifg@rmlt.org.uk Web www.ifg.org.uk*

**20–23 September**
Glastec 2022, Düsseldorf, Germany.
*Anke Seltmann, glastec, D-40474 Düsseldorf, Am Staad (Stockumer Höfe), Germany. Email seltmanna@messe-duesseldorf.de Web www.glasstec-online.com/*

**9–13 October**
ACerS 124th Annual Meeting with Materials Science & Technology 2022, Pittsburgh, USA.
*The American Ceramic Society, 550 Polaris Parkway, Suite 510, Westerville, OH 43082, USA. Email customerservice@ceramics.org Web www.ceramics.org*

**11–13 October**
GlassBuild America 2022, Las Vegas, USA.
*National Glass Association, 1945 Old Gallows Road, Suite 750, Vienna, VA 22182, USA. Web www.glassbuildamerica.com/*

**19–21 October**
XXXIV ATIV Int. Conf., Where Glass Science, Art and Technology Meet Together, Parma, Italy.
*ATIV Secretariat, MV Congressi spa Via Marchesi 26 d 43126 Parma, Italy. ativ@ativ-online.it Web www.ativ.eu*

**26–28 October 2022**
Glasstech Asia 2022, Singapore.
*Web www.glasstechasia.com.sg/*

**2023**

**5–9 March**
Optical Fiber Communications Conference and Exhibition (OFC), San Diego, California USA.
*The Optical Society, USA. Email custserv@osa.org Web www.ofcconference.org*

**4–9 June**
2023 Glass and Optical Materials Division Annual Meeting (GOMD 2023), New Orleans, USA.
*The American Ceramic Society, 550 Polaris Parkway, Suite 510, Westerville, OH 43082, USA. Email customerservice@ceramics.org Web www.ceramics.org*

**10–14 July**
Nonlinear Optics, Honolulu, Hawaii, USA.
*The Optical Society, 2010 Massachusetts Ave, NW, Washington, DC 20036 USA. Email custserv@osa.org Web www.osa.org*

**17–21 July**
Tenth International Conference on Borate Glasses, Crystals and Melts and Third International Conference on Phosphate Materials, Corning, USA.
*Dr Randall Youngman, Corning Incorporated Email youngmanre@corning.com Web boratephosphate2020.org/*

**~~August~~ (moving to 2024)**
NCM15, Structure of Non-Crystalline Materials 15, combined with the SGT Annual Meeting, UK.
*Christine Brown, Society of Glass Technology, 9 Churchill Way, Chapeltown, Sheffield S35 2PY, UK. Email christine@sgt.org Web www.sgt.org*

**2024**

**June**
Tenth International Conference on Optical, Optoelectronic and Photonic Materials and Applications (ICOOPMA), Pardubice, Czech Republic.
*Center of Materials and Nanotechnology and Department of General and Inorganic Chemistry, Faculty of Chemical Technology, University of Pardubice, Cs. Legii sq. 565, 532 10 Pardubice, Czech Republic. Email info@icoopma.com Web www.icoopma.com*

**2–6 July**
XVIII ECerS Conference, Lyon, France.
*European Ceramic Society AISBL, Av. Gouverneur Cornez 4, B-7000 Mons, Belgium. Email ecers@bcrc.be Web ecers.org/*

**14–19 July**
Tenth Int. Congress on Ceramics (ICC10), Montreal, Canada.
*The American Ceramic Society, 550 Polaris Parkway, Suite 510, Westerville, OH 43082, USA. Email customerservice@ceramics.org Web www.ceramics.org*

**20–23 August**
Materials Challenges in Alternative & Renewable Energy 2023 combined with the Sixth Annual Energy Harvesting Society Meeting (EHS 2023), Bellevue, Washington, USA.
*The American Ceramic Society, 550 Polaris Parkway, Suite 510, Westerville, OH 43082, USA. Email customerservice@ceramics.org Web www.ceramics.org*

ERNST-ABBE-FONDS
IM STIFTERVERBAND

# CALL FOR APPLICATIONS

## OTTO SCHOTT RESEARCH AWARD

In 2022 the Ernst-Abbe-Fonds will be presenting 17$^{th}$ Otto Schott Research Award. Since 1991 the prize honours biannually excellent researchers for their scientific and technological achievements in basic and applied research in the fields of glass, glass-ceramic and related advanced materials.

The Otto Schott Research Award is awarded for individual achievements, but also the work done by smaller groups. It is endowed with a total of 25,000 euros.

Scientists from universities, non-university research institutes and industrial research institutions worldwide are eligible. Candidates can be nominated by an internationally renowned scientist or engineer. Moreover, self-applications are allowed, but must be accompanied by a letter of support.

Central criteria for the awarding of the prize are the candidate´s scientific activity in recent years, the quality and originality of his/her original scientific publications, and the extent how he/she has influenced the research field.

The board of trustees of the Ernst-Abbe-Fonds is responsible for selecting the laureates. It will base its decision on the recommendations of a prize jury which is composed of internationally renowned material scientists. The board´s decision is final.

Each application/nomination should include:

» the candidate´s five top ranked and/or most influential publications in the field of glass sciences or related fields;
» one or more letters of support from outstanding scientists acknowledging the candidate´s scientific achievements;
» metrics (h-index) of scientific activity over the last 5 and 10 years;
» a list of other activities that highlight the candidate´s scientific impact (e.g. prizes, peer review, editorships, committee functions);
» the candidate´s curriculum vitae.

Applications should be submitted only via this website until
**3rd October 2021.**

If you have any questions please contact

Klaus Kuli
Ernst-Abbe-Fonds, Deutsches Stiftungszentrum
Postfach 16 44 60 . 45224 Essen
T ++49 201 8401-160 . klaus.kuli@stifterverband.de

The Ernst-Abbe-Fonds is a foundation that is dedicated to scientific research in the fields of fine mechanics, optics, electronics, glass, and ceramic glass. It was founded in 1989 by the Carl Zeiss Foundation in honor of its 100th anniversary. The Carl Zeiss Foundation is the only shareholder of SCHOTT AG, Mainz, Germany, and Carl Zeiss AG, Oberkochen, Germany. For further information on the Ernst-Abbe-Fonds please refer to www.abbe-fonds.de.

STIFTERVERBAND
Bildung. Wissenschaft. Innovation.

# BOOK REVIEW

**The First Thousand Years of Glass-Making in the Ancient Near East: Compositional Analyses of Late Bronze Age and Iron Age Glasses** by Wendy Reade

Paperback; 205×290 mm; 274 pages; 204 figures, 72 tables. 731 2021. Available both in printed and e-versions. Printed ISBN 9781789697032. Epublication ISBN 9781789697049. Printed Price £45.00. EPublication Price £16.00.

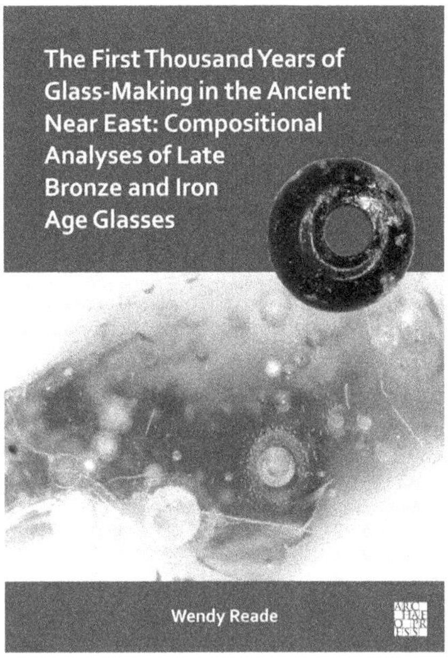


This book, which is based on the author's doctoral thesis completed in 2009, contributes very nicely to the expanding field of compositional studies of Near Eastern archaeological glasses dating to the Late Bronze Age and the Iron Age. Variations in raw glass recipes for the selected area and periods are examined on contemporary and diachronic scales through the detection of major, minor and trace elements by SEM-EDS (scanning electron microscopy with electron dispersive x-ray analysis), EPMA (electron probe micro-analysis) and ICP-MS (solution inductively coupled plasma mass spectrometry). Specialists who want to delve deeper in the topic of mid-second and first millennia BC glass chemistry, production and technology are the prime target audience.

The book is structured in 11 chapters and six appendices. Chapter 1 outlines the aims of the research, providing the reader with an overview of the history and technology of early glass in Egypt and the Near East, alongside reviews of typological and chemical studies of Late Bronze Age and Iron age glasses. Ample space is devoted to a detailed exploration of the raw ingredients employed in glass-making. Chapter 2 thoroughly describes the analytical methods and protocols applied for the collection and subsequent organisation of data. Chapters 3–7 are dedicated to the compositional analyses of glasses from Nuzi, Iraq (Late Bronze Age), Pella, Jordan (Late Bronze Age and Iron Age), and Nimrud, ancient Mesopotamia (Iron Age). The results obtained for each period are then interpreted through a wider geographical lens in Chapters 8 and 9, by respectively comparing the Late Bronze Age samples with glass retrieved in Egypt and Syria, and the Iron Age glasses from Pella and Nimrud with specimens from France, Iran and Egypt. The picture emerging from the conclusions (Chapter 10) is one of relative compositional uniformity for Late Bronze Age plant-ash glasses, in contrast with the experimentation and innovation that characterises the Iron Age, primarily through the use of natron as a flux. An epilogue (Chapter 11) with a review of the relevant literature published between 2009 and 2019, is very welcome, as it complements and builds upon the topics explored in the volume. The appendices contain the results of the compositional analyses provided in raw values and a model for the compositional investigation of ancient glasses. The latter is not only very useful to illustrate the rationale behind the collection of data, but it could also help archaeologists who approach glass chemistry for the first time to build an analytical framework.

This volume comes with a wealth of well-presented data in the form of plots and tables, photographs and micrographs of selected samples, adding to our understanding of the subject and offering material for comparative studies. Clearly written and well argued, this book is also suitable for researchers who may be relatively new to either scientific or archaeological approaches to glass.

Ele Montanari, PhD Student, School of History, Classics, and Archaeology, Newcastle University

# 1. A BRIEF HISTORY OF GLASS AND GLASSMAKING

Glass has a very long history and throughout the ages its unique combination of properties has made it one of the most useful classes of material known. This chapter covers only a brief overview of the history of these unique materials. More comprehensive information on the history, relating to specific technologies, is provided in the chapters that follow.

It is known that the natural glass obsidian was used as spear and arrow heads and as sharp cutting tools by Stone Age people as long as 9000 years ago. Volcanic glasses, known as tektites, may also have been used for the same purpose. An example of an obsidian arrow head is shown in Figure 1.1. According to the Roman historian Pliny the Elder, writing in the first century AD, the discovery of synthetic glass is thought to have occurred accidently in Syria around 7000 years ago with Phoenician merchants (Dodsworth, 1982). Having landed on a sandy shore, these merchants possibly rested cooking pots on natron blocks (sodium carbonate then used as a cleaning medium), which they were transporting at the time, by their fire, having not been able to find any suitable rocks to rest their pots on. The heat of the fire melted the blocks which then reacted with the sand to form a viscous liquid; on cooling, a glossy solid remained which, unknown to the merchants, would have been an impure form of sodium silicate glass. This is, of course, just a story, but undoubtedly a congenial one, although it is highly debatable whether or not the temperature of the fire would have been high enough to accomplish the formation of glass. It may also have been created accidentally in the form of a glassy slag as a by-product of early metal working operations, with this speculation being the more likely. Another story is that synthetic glass was first formed, again accidently, as a glaze on pots when sand inadvertently entered a kiln and stuck to the wet clay before the pots were fired. Much later, salt used in this way led to the production of salt-glazed stoneware pottery. It is known with some certainty, however, that synthetic glassy materials in the form of decorative glazes appeared around 6000 years ago in Mesopotamia and later in Egypt, glazed stones having been found by archaeologists and dated to this period.

*This is a chapter taken from The Science and Technology of Inorganic Glasses and Glass-Ceramics by Ian Donald, published by the Society of Glass Technology: https://sgt.org/store/viewproduct.aspx?id=15456681*

*Figure 1.1. Obsidian arrow head (www.brucker.com)*

The earliest known primary glass factory was discovered at Qantir-Piramesses on the Nile Delta in Egypt in 2005. There was evidence to suggest that large quantities of glass were made there out of raw materials as early as 1250 BC. Many crucible fragments were found at this site with the remains of glass still adhering to them. The crucibles were broken to remove glass ingots that would have subsequently been transported to individual workshops to be made into practical and artistic objects. It was originally assumed that sand was used as a raw material in glass production, but evidence from the site has shown that the Egyptians generally used quartz pebbles ground down to a fine powder. Plant ash which contains various proportions of potassium and sodium hydroxide and carbonate together with minor quantities of chloride and sulphate salts with magnesium, calcium, copper, iron, zinc and lead metal species and phosphorus and silica, depending on the origin of the plants, was believed to have been used with the quartz powder (Babayemi *et al*, 2010; Tite *et al*, 2006; Rehren, 2008). The drawing of softened glass into fine fibres was also practiced in ancient Egypt. Many Egyptian vessels were made by winding coarse glass fibres onto clay mandrels, subsequently heating the assembly to soften the glass so that the fibres fused together. On cooling, the clay core would be removed to produce a viable glass vessel (Loewenstein, 1983). The Egyptians initially used a method known as core moulding around 1500 BC for making small vessels to hold ointments and perfumes (Dodsworth, 1982). In this technique, a former (core) was made from mud or clay in the shape of the vessel required. This was then attached to the end of a metal rod and dipped into molten glass. The rod was subsequently allowed to cool and

Figure 1.2. Roman glass
(Vassil, JC.Musée Saint-Remi à Reims, Marne, France; Wikipedia Open Source)

the core taken out leaving a hollow glass article. This method of producing hollow ware was eventually superseded by the art of glass blowing (with the use of a blowpipe), thought to have originated in Syria around 50 BC. Using this technique glass would be blown into simple moulds, most likely made of clay, to produce a variety of shaped thin-walled articles.

The interest in glass and glass making rapidly expanded under the auspices of the Roman Empire, spreading from Egypt and Syria to other areas of the Roman Empire including Germany, France and eventually Britain. The new art of glass blowing brought a new dimension to glass production, lifting glass from a luxury product to an everyday item. Many different types of glass article were produced in quantity including household items such as bottles, jars, vases, jugs and cups, and also more quality items such as decorated glass vessels, ornaments, jewellery and cameos, as illustrated in Figure 1.2. Pliny the Elder noted that in Roman times sands from the Belus River in Israel and the Volturno River in Italy were used in glass production, with certain areas of both Spain and France also offering the availability of suitable raw materials, although there is no archaeological evidence for this. It is likely that glass making would have been concentrated in regions where suitable sands were available, as it would have been costly to transport large quantities of sands to other areas. A recent study involving melting mixtures of various sands from a number of sources has concluded that sands suitable for glass making are quite rare and only six areas can be identified where they would have been found in these times (Brems *et al*, 2012). It was noted that the best sands occur in the Basilicata and Puglia regions of southeast Italy and after fluxing with natron these were found to produce glasses with compositions close to that of known Roman natron glass. Other suitable regions were confirmed to include areas of southeast Spain and southeast France.

Unfortunately, following the collapse of the Roman Empire in the fifth century, glass making in Europe went into decline, although it continued for some time in the Middle East (Dodsworth, 1982). With the later decline of Islam in the Middle East, Venice became the centre of quality glass making from around 1400 to 1700. It is not known precisely when Venetian glass making actually began, but as early as 1291 glass making had been moved from the central islands of Venice to the island of Murano situated in the lagoon of Venice, ostensibly because of the risk to Venice itself of fire due to foundry operations. Glass making operations were very strictly controlled, with raw materials supplied only through the government, and Venetian glass workers forbidden to take their skills abroad. A statute was drawn up in the thirteenth century outlining the regulations to be followed by the glass workers, covering all phases of glass making including the production and selling of artifacts, worker and owner relations, and the collection of taxes. During this period, some remarkable glass articles were made by these skilled workers. Clear glasses were originally produced including 'cristallino' glass, so named due to its

Figure 1.3. Decorated Venetian glass c. 1330.
(The Aldrevandini beaker with enamel decoration, The British Museum)

*Figure 1.4. Filigree glass c. 1600 (Mayflower Antiques: PO Box 6828, Bewdley, Worcestershire, DY12 9AU)*

*Figure 1.5. Waldglas vessel (Ulrich Mayring, Wikipedia, 2007. GNU Free documentation Licence)*

very clear nature similar to rock crystal. Coloured glasses were later produced, some with religious, mythological and heraldic scenes, and items became increasingly elaborate, as highlighted in Figures 1.3 and 1.4 which show an example of a decorated glass and filigree glass, respectively (the filigree method uses twisted canes of glass on a clear glass base). By the eighteenth century an opaque white glass known as 'lattimo' was produced to imitate porcelain.

Despite the penalties involved for taking their expertise abroad, some Venetian glass workers did move elsewhere and by the end of the sixteenth century glass factories had become well established in Austria, Germany, Bohemia, Belgium, England, Spain and the Netherlands producing Venetian-style glass. Earlier European glass was sometimes known as 'waldglas' or 'forest glass', produced in Northern Europe from around 1000–1700, and employing wood in the firing process as a source of potash. This glass was often of poor quality when compared with Syrian or Roman glass, containing less silica. A good example of a higher quality waldglas vessel from the seventeenth century is shown in Figure 1.5.

Although glass making had been introduced into Britain in Roman times, British expertise lagged behind the rest of Europe until the Venetian, Jacopo Verzelini, arriving in London in 1572, brought knowledge of the Venetian technology to Britain. (Dodsworth, 1982). These early glass makers used wood as fuel and this led James I in 1615 to ban wood as fuel due to the unacceptable rate of deforestation taking place. Glass makers subsequently used coal as fuel and moved to areas where coal was plentiful, including London, Bristol, Stourbridge, Newcastle and SE Scotland. Although some reports suggest that glassmaking in Britain declined by the middle of the seventeenth century there is evidence to show that this was not the case and that the industry in fact flourished with influence from the Italian glassmaking centre at Altare (Brain & Brain, 2014).

A major event in British glassmaking occurred in 1674 with George Ravenscroft's patenting of lead glass. This was more robust and less expensive than the fragile Venetian glass and pushed Britain to the forefront of glass technology. At first this new glass was susceptible to 'crizzling' whereby a series of very fine cracks develops on the surface of the glass leading to a cloudy surface appearance and a loss in transparency; however, further development produced a glass that was less susceptible to crizzling when compared with many of the earlier European and Venetian glasses (Brill, 1975). The phenomenon of crizzling is treated in more detail in Chapter 13. Lead glass is also known as 'crystal glass' reflecting the glassmakers aim of producing a colourless glass to mimic rock crystal. Today, lead or lead crystal glass is made all over the world, and by law must contain at least 30 mass% lead oxide. Some beautiful glass articles were made in Britain from the seventeenth century on, with some examples of English drinking

*Figure 1.6. Examples of English drinking glasses; (a) L. M. Bickerton, 1984; (b) a glass belonging to the author*

*Figure 1.7. Medieval Glass melting facility (E. G. West, 1978)*

*Figure 1.9. Catcliffe glass cone (© Richard Cross, 2009. Licensed for reuse under the Creative Commons Licence)*

glasses shown in Figure 1.6. Crystal glass also flourished in Ireland, France and central Europe during this period (Von Kerssenbrock-Krosigk, 2012).

An illustration of an early glass manufacturing facility from the medieval period was published in Agricola's *De Re Metallica* of 1556, as shown in Figure 1.7. This provides considerable detail as to the layout of the time, and highlights the apparent total lack of health and safety considerations. Broken glass litters the floor where bare footed glass 'technicians' wearing no personal protective equipment work, and there is a drinking parlour next to the furnace, where workers can quench their thirst, presumably with an alcoholic beverage! By the early nineteenth century glass factories in Britain consisted of conical brick structures which acted as a giant chimney for the furnace, whilst the glass makers worked inside the cone between the furnace and the inner walls of the cone. The glassmakers would be arranged in groups of four workers at a number of positions around the furnace. Each worker in a group had a specific task, depending on the item to be made. Although not now used, some glass cones do still exist, with four remaining in the UK. One of these, situated at Wordsley near Stourbridge, is shown in Figure 1.8 and is now a museum. Another example, shown in Figure 1.9, is situated at Catcliffe near Rotherham and Sheffield. This was built about 1740 and was a working facility until the early twentieth century. It served as a prison during the First World War and as a canteen during

*Figure 1.8. Early nineteenth century glassmaking factory – Stuart Crystal's Red House Glass Cone, Wordsley (© Phil Loach, 1976)*

*Figure 1.10. Illustration of a nineteenth century glass workshop (www.lostindustry.org)*

*Figure 1.11. Illustration of the Crystal Palace, London, as it would have been seen c. 1851 (www.telegraph.co.uk)*

the General Strike of 1926 (www.treetonweb.co.uk). The conical glassmaking structures were superseded after around 1830 by large sheds, often with several furnaces connected to a common chimney, as illustrated in Figure 1.10. Glass making expanded during the Victorian era, culminating in construction of the Crystal Palace, built in Hyde Park, London in 1851, to house the Great Exhibition. This building, shown as an illustration in Figure 1.11, utilized 299,655 panes of glass produced using the blown cylinder method (described in Chapter 3). Further information on the history of British glass can be found in a number of publications (e.g. Dodsworth, 1982; Honey, 1987; Brooks, 1981; Bickerton, 1984).

The first glassmaking operation began in North America around 1608 on Jamestown Island. This was established by Dutch, German and Polish immigrants who used local oyster shells as a source of calcium oxide. They made bottles, window glass and drinking vessels, some of their products being shipped back to England. In 1640 Evert Duycking established the first glass factory in New York (then known as New Amsterdam), manufacturing mainly window glass (www.historical-american-glass.com). It was not until the eighteenth century, however, that major factories were established in New York and elsewhere. The first known Canadian glass factory was established in 1839 at Mallorytown in Upper Canada, but closed down in 1840 (www.thecanadianencyclopedia.com). A number of additional glass factories subsequently were established in the mid-to-late 1880s manufacturing mainly window glass and bottles together with plates, dishes, mugs and vases. Serious glassmaking in Japan did not start until 1873 when a glassworks to produce window glass was established with British assistance at Shinagawa near Tokyo (Haden, 2013).

Early innovative products in glass include lenses which go back to ancient Egyptian times, where they were most likely used for burning and lighting fires. Alhazen, who is regarded by many as the father of optics, was born in Basra in around 965, and in his seven volume treatise on optics, noted that the convex shape of a lens produced an enlarging effect. It was not until towards the end of the thirteenth century, however, that lenses first appeared in spectacles. The use of lenses in microscopes and telescopes followed much later in the seventeenth century, first with the construction of what may be regarded as the first microscope by a Dutch spectacle maker, Zacharias Jannsen, in around 1600; however, this was not a practical instrument, the images generated suffering from serious aberration and distortion, and it was not until later in the seventeenth century that practical microscopes appeared. The practical science of microscopy began with Robert Hooke's publication of *Micrographia* and *Microscopium* in 1665 and 1669, respectively, and the construction of powerful microscopes by Anton van Leeuwenhoek in the years that followed. The invention of the 'spyglass' by a Dutch spectacle maker, Hans Lippershey, in 1608 was Galileo Galilei's inspiration in constructing a practical telescope in 1609 (the word 'telescope' being coined first by Demisiani in 1611), with which he noticed that the moon did not have a smooth surface, as was widely believed, but was in fact rough and covered by 'cavities'. Later, in 1610, with a more powerful telescope, he discovered three of the largest moons of Jupiter, noting that they seemed to be revolving around the planet, a fact which eventually changed our perception of the universe forever and destroyed the myth that the earth was at the centre of the universe (Boltz, 1983; Johnson, 1983).

Other inspirational glass products include stained glass which has been used extensively since the tenth century and possibly even as early as the sixth century, notably in church windows. It is a generic name for decorative windows made from coloured glass pieces fitted together by grooved strips of lead to form patterns and biblical scenes, the whole window being supported in a rigid frame. As noted by Whitehouse (2006), 'stained' glass is an inaccurate term because the glass is often coloured throughout by use of metal oxides, rather than by 'staining' which colours only the surface of the glass.

*Figure 1.12. Stained glass window, Troyes Cathedral, late 13th. century (Vassil, Wikipedia Open Source)*

Iron and manganese oxides were often impurities in early glass and caused the glass to be coloured from green to purple depending on the melting conditions. Specific additives were subsequently discovered to impart colours to glass; for example, red or blue glass could be made by adding copper or cobalt oxides to the glass batch. Stained glass was also made by painting pictures onto the surface of pieces of glass and firing in a furnace to produce a variety of colours depending on the painting medium, which would consist of various metal oxides, glass powder and a binder. Figure 1.12 shows an example of a thirteenth century stained glass window from Troyes cathedral in France. Further more comprehensive details on stained glass can be found elsewhere (e.g., Kunicki-Goldfinger, 2014; Raguin, 2008; Marks, 1993; Hutter, 1964; Wikipedia). The composition of stained glass windows has been determined in a number of instances; for example, fragments of eighteenth century windows from Girona Cathedral, Reus Priory Church and Vitoria Cathedral in Spain have been analyzed and shown to be ostensibly an alkali calcium silicate glass in the case of Girona Cathedral and soda–lime–silica glasses in the case of Reus and Vitoria, coloured by CuO, CoO, NiO, MnO and $Fe_2O_3$ (Palomar *et al*, 2011). Seventeenth century glass from Christ Church Cathedral, Oxford, is also primarily of the lime–silica family (Scott *et al*, 2012).

Other forms of coloured glass have been made from early times, with the nineteenth century being a particularly productive period. Glass containing uranium, also known as 'vaseline glass' due to its greenish yellow colouration (an earlier description was 'canary glass'), was produced from the 1830s up until the 1940s, although there is some evidence to suggest that glass containing uranium was also produced in Roman times (Fajfar *et al*, 2013). Some uranium glass is markedly greener in colour than yellow due to the presence of iron in addition to uranium, and such glass is usually referred to as 'depression glass' rather than vaseline glass. Vaseline glass contains typically around 0.06 to 0.63 mass% $UO_2$ (although glasses with higher concentrations are known) and is therefore mildly radioactive. It has been shown that as long as reasonable precautions are taken uranium glass does not pose a significant health hazard, as radiation levels are usually very low (Lopes *et al*, 2008). Vaseline glass generally fluoresces quite spectacularly under ultraviolet light, with an

*Figure 1.13. A uranium-containing drinking glass under uv light (from the author's collection)*

*Table 1.1. Compositions of a selection of some early glasses*

| Designation | Composition (mass%) | | | | | | | | | | | | | | Reference |
|---|---|---|---|---|---|---|---|---|---|---|---|---|---|---|---|
| | $SiO_2$ | $Na_2O$ | $K_2O$ | MgO | CaO | $TiO_2$ | MnO | $Fe_2O_3$/FeO | $Al_2O_3$ | PbO | $P_2O_5$ | $Cl^-$ | $SO_2$ | Others | |
| Architectural metallurgical slag | 34.95 | 0.48 | 1.44 | 2.10 | 31.09 | 0.57 | 0.77 | 11.2 | 13.98 | - | 3.50 | - | - | - | Vernay-Carron *et al*, 2008 |
| Architectural metallurgical slag | 61.28 | 0.33 | 0.51 | 5.72 | 24.06 | 0.02 | 3.27 | 1.46 | 3.42 | - | - | - | - | - | Vernay-Carron *et al*, 2008 |
| Roman glass | 70.70 | 19.52 | 0.38 | 0.38 | 4.94 | - | - | 0.30 | 1.78 | - | - | 1.00 | 0.26 | $Sb_2O_3$ 0.83 | Vernay-Carron *et al*, 2008 |
| Roman natron glass | 69.60 | 16.60 | 0.80 | 0.50 | 2.40 | 0.10 | 0.08 | 1.30 | 2.40 | - | 0.04 | - | - | - | Brems *et al*, 2012 |
| Roman mosaic glass | 69–77 | 9.4–19.1 | 0.5–2.0 | - | 5.6–11.7 | - | - | - | 2.1–4.3 | - | - | - | - | - | Gedzeviciuté *et al*, 2009 |
| Athenian Agora | 69.49–70.65 | 18.63–19.3 | 0.40–0.51 | 0.62–0.66 | 6.11–6.34 | - | - | 0.44–1.39 | 1.81–1.94 | - | - | - | - | $Sb_2O_3$ 0.53 | Paterakis, 2003; Weinberg & Stern, 2009 |
| Girona Cathedral | 66.68 | 2.20 | 22.60 | 0.15 | 5.29 | - | 0.02 | 0.12 | 1.57 | - | - | 0.51 | 0.83 | BaO 0.02 | Palomar *et al*, 2011 |
| Reus Priory | 66.72 | 17.41 | 0.81 | 0.22 | 9.79 | - | 0.15 | 1.40 | 1.66 | - | - | - | 0.21 | BaO 0.04<br>$Cr_2O_3$ 0.76 | Palomar *et al*, 2011 |
| Vitoria Cathedral | 69.53 | 12.51 | - | 0.14 | 15.46 | - | 0.70 | 0.19 | 0.53 | 0.01 | - | 0.14 | 0.56 | BaO 0.08 | Palomar *et al*, 2011 |
| Christ Church Cathedral, Oxford | 59.56 | - | - | - | 20.57 | 0.18 | 0.76 | 0.91 | 3.05 | - | - | - | - | - | Scott, *et al*, 2012 |
| Venetian glass c. 1709 | 76.00 | 0.13 | 18.70 | 0.04 | 1.30 | - | 0.01 | 0.05 | 0.10 | 2.22 | - | - | - | $As_2O_5$ 1.66 | Brill, 1975 |
| French glass c. 1600–1650 | 70.70 | 17.70 | 3.42 | 0.10 | 1.53 | - | 0.38 | 0.10 | 0.50 | 0.01 | - | - | - | - | Brill, 1975 |
| French glass c. 1750 | 76.00 | 0.89 | 18.70 | 0.10 | 0.22 | - | 0.45 | 0.05 | 0.50 | < 0.01 | - | - | - | - | Brill, 1975 |
| American c. 1790–1810 | 69.00 | 2.62 | 24.20 | 0.20 | 3.19 | - | < 0.01 | 0.05 | 0.55 | < 0.01 | - | - | - | - | Brill, 1975 |
| Silesian c.1710–1730 | 73.50 | 0.39 | 16.60 | 0.03 | 1.61 | - | 0.04 | < 0.01 | 1.03 | 5.88 | - | - | - | - | Brill, 1975 |
| Anglo-Saxon | 67.97 | 18.28 | 0.68 | 0.58 | 7.34 | 0.11 | 0.83 | 0.67 | 2.11 | < 0.2 | 0.13 | 0.84 | 0.46 | - | Freestone *et al* |
| Anglo-Saxon | 64.59 | 18.36 | 0.90 | 1.17 | 8.49 | 0.18 | 1.50 | 1.00 | 2.80 | < 0.2 | - | 0.58 | 0.44 | - | Freestone *et al* |
| Anglo-Saxon | 66.18 | 18.07 | 1.58 | 2.02 | 7.74 | 0.16 | - | 0.83 | 2.44 | < 0.2 | 0.33 | 0.85 | 0.37 | - | Freestone *et al* |
| 18th century English window glass | 63.9 | 7.8 | 4.4 | 5.6 | 10.7 | 0.13 | 0.09 | 1.02 | 2.1 | - | 1.4 | 0.8 | - | SrO 0.37<br>$ZrO_2$ 0.05 | Dungworth, 2006 |
| Early Islamic glass | 72.40 | 20.61 | 0.64 | 1.14 | 6.69 | - | - | - | 1.44 | - | - | - | - | - | Kato *et al*, 2009 |
| Archaeological model glass | 69.2 | 18.1 | 1.0 | 0.5 | 6.6 | - | 0.6 | - | 4.0 | - | - | - | - | - | McLoughlin *et al*, 2006 |
| Archaeological model glass | 51.1 | 0.7 | 14.2 | 4.4 | 19.6 | - | - | 1.2 | 3.3 | - | 5.6 | - | - | - | McLoughlin *et al*, 2006 |

*Figure 1.14. Salt-glazed stoneware bottles (Theroadislong, 2012, Wikipedia, Licensed for reuse under the Creative Commons Licence)*

example shown in Figure 1.13.

Salt-glazed pottery, also known as stoneware, is another innovation. This was produced in Germany as early as 1280. By around 1350 many small industries with kilns flourished along the Rhine and were producing stoneware. The method was thriving in England by the seventeenth century and in the United States by the eighteenth century. It came later to Australia in around 1850. The process involves introducing salt into the kiln (usually by simply throwing it in via an access port) when the temperature was at its highest, around 1400°C (the boiling point of sodium chloride is 1413°C). This resulted in vaporization of the salt and reaction with silica in the clay vessel to form a sodium silicate based glass, or glaze, on the surface of the vessel. Depending on the constituents present in the clay, the final fired article could range in colour from various shades of brown, to blue or purple (due to the presence in the clay of iron, cobalt or manganese, respectively). Alternative methods of adding the salt include adding a salt solution to the kiln at the appropriate temperature and soaking the article in brine prior to firing. The advantage of the glaze is to provide a non-porous surface that is resistant to most acids and other chemicals. An example of salt-glazed stoneware bottles is shown in Figure 1.14. The compositions of a selection of early glasses are provided in Table 1.1.

Additional information on various aspects of the history of glass can be found in many publications. A selection follows, to which the interested reader is directed (e.g. Wikipedia; Brooks, 1981; Wills, 1981; Dodsworth, 1982; Mehlman, 1985; Newton, 1985; Honey, 1987; Kingslake, 1989; Watts, 1990; Lloyd and Klein, 2000; Macfarlane and Martin, 2002; Raguin and Higgins, 2008; Jones, 2011; Rasmussen, 2012; Chopinet, 2012; Cable 2013).

## References


**Agricola, G.** *De Re Metallica*, translated from the first Latin edition of 1556 by H. C. Hoover and L. H. Hoover, Dover Publications, Inc. (New York), 1950.

**Babayemi, J. O.**, Dauda, K. T., Nwude, D. O. and Kayode, A. A. A. Evaluation of the composition and chemistry of ash and potash from various plant materials – a review, *J. Appl. Sci.*, **10**, (2010), 1820–1824.

**Bickerton, L. M.** *English drinking glasses 1675-1825*, Shire Album No. 116, Shire Publications Ltd. (Aylesbury), 1984, 32 pp.

**Boltz, C. L.** 'And yet it moves', *New Scientist*, 7 April 1983, 25–29.

**Brain, A. J.** and Brain, C. Crystal glass-making in London 1642-1672: 'These twenty years last past much improved themselves', *Glass Technol.: Eur. J. Glass Sci. Technol. A*, **55**, (2014), 153–165.

**Brems, D.**, Degryse, P., Ganio, M. and Boyen, S. The production of Roman glass with western Mediterranean sand raw materials: preliminary results, *Glass Technol.: Eur. J. Glass Sci. Technol. A.*, **53**, (2012), 129–138.

**Brill, R. H.** Crizzling – A problem in glass conservation, Conservation in Archaeology and the Applied Arts, Stockholm Congress, 1975, International Institute for Conservation of Historic and Artistic Works, pp. 121–134.

**Brooks, J.** *The Arthur Negus guide to British glass*, Hamlyn Publishing Group Ltd. (London), 1981, 176 pp.

**Cable, M.** The classic texts of glass technology, *Glass Technol.: Eur. J. Glass Sci. Technol., A*, **54**, (2013), 57–65.

**Cable, M.** The world's first successful regenerative furnace, *Glass Technol.: Eur. J. Glass Sci. Technol. A*, **54**, (2013), 93–99.

**Chopinet, M.-H.** Developments of Siemens regenerative and tank furnaces in Saint-Gobain in the XIX[th] century, *Glass Technol.: Eur. J. Glass Sci. Technol. A*, **53**, (2012), 177–188.

**Dodsworth, R.** *Glass and Glassmaking*, Shire Publications Ltd., Shire Album No. 83, 1982.

**Dungworth, D.** Composition of early eighteenth century window glass from Silkstone, Yorkshire, English Heritage Research Department Report 18/2006, 2006.

**Edison, T. A.** Electric-Lamp, U.S. Patent No. 223,898, Jan. 27, 1880.

**Emsley, J.** Glass: past elegant, future thin, *New Scientist*, 8 December 1983, 728–732.

**Fajfar, H.**, Smit, Z. and Kos, M. PIXE-PIGE analysis of coloured historic glass, *Glass Technol.: Eur. J. Glass Sci. Technol. A*, **54**, (2013), 218–225.

**Freestone, I. C.**, Hughes, M. J. and Stapleton, C. P. The composition and production of Anglo-Saxon glass, Catalogue of Anglo-Saxon Glass in the British Museum, https://www.britishmuseum.org.

**Gedzeviciuté, V.**, Welter, N., Schussler, U. and Weiss, C. Chemical composition and colouring agents of Roman mosaic and millefiori glass, studied by electron microprobe analysis and Raman microspectroscopy, *Arch. Anthrop. Sci.*, **1**, (2009), n15–29.

**Haden, S. E.** They went "to larn 'em": British glassmakers help to establish Japan's first western-style glassworks, 1874-1883, *Glass Technol.: Eur. J. Glass Sci. Technol. A*, **54**, (2013), 25–30.

**Henderson, J.** *Ancient glass an interdisciplinary exploration*, Cambridge University Press, 2013.

**Honey, W. B.** *English glass*, Braken Books (London), 1987, 48 pp.

**Hutter, H.** *Medieval Stained Glass*, Crown Publishers, 1964.

**Johnson, N.** The art of seeing the very small, *New Scientist*, 19 May 1983, 468–475.

**Jones, D. M.** *Archaeological Evidence for Glassworking*, English Heritage Publishing (Swindon), 2011, pp. 37.

**Kato, N.**, Nakai, I. and Shindo, Y. Change in chemical composition of early Islamic glass excavated in Raya, Sinai Peninsula, Egypt: on-site analyses using a portable X-ray fluorescence spectrometer, *J. Arch. Sci.*, **36**, (2009), 1698–1707.

**Kingslake, R.** *A History of the Photographic Lens*, Academic Press (San Diego), 1989.

**Kunicki-Goldfinger, J. J.**, Freestone, I. C., McDonald, I., Hobot, J. A., Gilderdale-Scott, H. and Ayers, T. Technology, production and chronology of red window glass in the medieval period – rediscovery of a lost technology, *J. Arch. Sci.*, **41**, (2014), 89–105.

**Lloyd, W.** and Klein, D. *The History of Glass*, Little, Brown and Co., 2000, 288 pp.

**Loewenstein, K. L.** *The manufacturing technology of continuous glass fibres, Glass Science and Technology 6*, Elsevier (Amsterdam), Second Edition, 1983, 354 pp.

**Lopes, F.**, Ruivo, A., Muralha, V. S. F., Lima, A., Duarte, P., Paiva, I., Trindade, R. and de Matos, A. P. Uranium glass in museum collections, *J. Cultural Heritage*, **9**, (2008), 64–68.

**Macfarlane, A.** and Martin, G. *Glass: A World History*, University of Chicago Press, 2002, 288 pp.

**Marks, R.** *Stained glass in England during the middle ages*, Taylor & Francis, 1993, 362 pp.

**McGregor, J.** *What is it and how do you recognize it?* Prosit, Stein Collectors International, March 2005.

**McLoughlin, S. D.**, Hyatt, N. C., Hand, R, J. and Lee, W. E. Corrosion of archaeological model glasses after 32 years of burial at Ballidon, *Mater. Res. Soc. Proc.*, **932**, (2006), 162.

**Mehlman, F.** *The illustrated guide to glass*, Peerage Books (London), 1985, 256 pp.

**Newton, R. G.** Eighth W. E. S. Turner Memorial Lecture, *Glass Technol.*, **26**, (1985), 93–103.

**Palomar, T.**, Agua, F., García-Heras, and Villegas, M. A. Chemical degradation and chromophores of 18th century window glass, *Glass Technol.: Eur. J. Glass Sci. Technol. A*, **52**, (2011), 145–153.

**Paterakis, A. B.** The conservation of ancient glass, *Mater. Res. Soc. Proc.*, **757**, (2003), Paper II2.3.

**Raguin, V. C.** and Higgins, M. C. *The History of Stained Glass*, Thames & Hudson Ltd., 2008, 288 pp.

**Rasmussen, M.**, Greenman, M. and Brown, J. Glass art and glass science: a mutually beneficial exchange, 64th Conference on Glass Problems, Ceramic Engineering and Science Proc., vol. 25, edited by W. M. Kriven, the American Ceramic Society (Westerville, OH), 2004, pp. 267–279.

**Rasmussen, S. C.** *How Glass Changed the World: The history of glass from antiquity to the 13th century*, Springer, 2012, 85 pp.

**Rehren, Th.** A review of factors affecting the composition of early Egyptian glasses and faience: alkali and alkali earth oxides, *J. Arch. Sci.*, **35**, (2008), 1345–1354.

**Savage, G.** and Newman, H. *An illustrated dictionary of ceramics*, Thames and Hudson Ltd. (London), 1974.

**Schmid, R. E.** Dated 1250BC, world's oldest glass factory found in Nile delta, *The Independent*, 18 June 2005.

**Scott, R. B.**, Shortland, A. J., Degryse, P., Power, M., Domoney, K., Boyen, S. and Braekmans, D. *In situ* analysis of ancient glass: 17th century painted glass from Christ Church Cathedral, Oxford and Roman glass vessels, *Glass Technol.: Eur. J. Glass Sci. Technol. A*, **53**, (2012), 65–73.

**Taite, H.** Ed., *Five thousand years of glass*, The British Museum Press (London), 1999.

**Tite, M. S.**, Shortland, A., Maniatis, Y., Kavoussanaki, D. and Harris, S. A. The composition of the soda-rich and mixed alkali plant ashes used in the production of glass, *J. Arch. Sci.*, **33**, (2006), 1284–1292.

**Vernay-Carron, A.**, Gin. S. and Libourel, G. A fractured roman glass block altered for 1800 years in seawater: Analogy with nuclear waste glass in a deep geological repository, *Geochim. Cosmochim. Acta*, **72**, (2008), 5372–5385.

**Von Kerssenbrock-Krosigk, D.** Glass of the alchemists: a retrospect on an exhibition at the Corning Museum of glass 2008, *Glass Technol.: Eur. J. Glass Sci. Technol. A*, **53**, (2012), 74–76.

**Warmus, W.** A brief history of glass, http://contempglass.org/2012-celebration/info/a-brief-history-of-glass.

**Watts, D. C.** Why George Ravenscroft introduced lead oxide into crystal glass, *Glass Technol.*, **31**, (1990), 208–212.

**Weinberg, G. D.** and Stern, E. M. Vessel Glass (Excavations in the Athenian Agora), American School of Classical Studies, Vol. XXXIV, 2009, 250 pp.

**West, E. G.** Textbooks and teachers – thoughts on education of metallurgists, The Metallurgist and Materials Technologist, August 1978, 427–431.

**Whitehouse, D.** *Glass: A pocket dictionary of terms commonly used to describe glass and glassmaking*, The Corning Museum of Glass (Corning, New York), 2006.

**Wikipedia,** History of glass, http://www.britglass.org.uk/history-of-glass.

**Wikipedia,** Alhazen, http://en.wikipedia.org/wiki/Alhazen.

**Wikipedia,** http://en.wikipedia.org/wiki/Galileo_Galilei.

**Wills, G.** *The country life collector's pocket book of glass*, Country Life Books (Richmond upon Thames), 1981, 317 pp.

## International Year of Glass 2022 Fundraising campaign launched

In May 2021, the United Nations formally agreed to endorse 2022 as the International Year of Glass, following a campaign supported by almost two thousand letters from 85 countries.

So, 2022 is an opportunity to celebrate all things 'glassy' and recent social isolation has left a thirst for group activities. Our flagship event is the Opening Ceremony in the Palace of Nations, Geneva from 9–11th February. This will showcase an array of internationally recognised speakers primed to place Glass at the heart of the UN 2030 Humanitarian Goals. The audience will include important political figures and noteworthy members of the glass community, although in numbers possibly restricted by COVID. The proceedings will be live streamed for those unable to attend and available later as a recording. A book highlighting the same aspirations will be released simultaneously.

We are committed to financing the Opening Ceremony without recourse to Fees. So, the support of our endorsers: institutions, companies, artists and educators is vital. A fundraising campaign has been launched through our website, https://donate.iyog2022.org/; individual donors can participate using PayPal or Stripe (credit or debit card). After completing your data, you will receive the corresponding invoice by email.

We are now focusing our campaign on companies, associations and institutions able to support IYOG with larger gifts. Such donations will be invaluable and will decide whether we can proceed. There are different ways to give support, by sponsoring specific aspects of the event. Alternatively, you can become a 'Friend of IYOG'. All gifts will be formally acknowledged and much appreciated.

Many other events are also planned: trade fairs, conferences, magazine articles, exhibitions in art galleries and museums, talks, factory open days, newspaper articles, etc.

For more information, the donations tool and news items see our web site (www.IYOG2022.org). Shortly links to key contacts and a searchable database of planned activities will be added. Do take every opportunity you can to make your colleagues and friends aware of what is happening, speak to your local press, radio and TV service and use social media to pass on the news. To join a mailing list please register your interest on the web www.iyog2022.org

Administration has rested largely on the International Commission of Glass with considerable assistance from the Community of Glass Associations (COGA) and the International Committee for Museums and Collections of Glass (ICOM-Glass). Additional assistance is always welcome.

Yours sincerely

Prof Alicia Durán

Diamond Sponsorship - €50,000 (limited to three) includes:

- An opportunity for a senior representative from the sponsor organisation is invited to address the attendees at either the IYOG opening or closing sessions.
- Acknowledgement of Sponsorship throughout the event.
- Company name and logo on all event literature.
- Tag lines mentioning sponsorship through social media activities of the IYOG 2022 ceremony.
- Corporate banners will be displayed in a prominent position.
- A full-page advertisement in the attendees' programme.
- A promotional paragraph of 150 words and contact details in the attendee's programme.
- An opportunity to include two inserts into attendees' bags: either corporate literature (A4 size or smaller) OR promotional merchandise.
- Use of the IYOG2022 major sponsor logo during and after 2022.
- A Quartz ICG corporate membership for 2022

Avventurina Sponsorship - €25,000 (multiple sponsorships are available) includes:

- Acknowledgement of sponsorship during the Opening and Closing Sessions.
- Company name and logo appearing on all event literature: The IYOG event website (with a link to the company website), email teasers (with a links), event features in various glass magazines, event programmes, event banners and signage displayed throughout the event space.

• Corporate banner to be displayed in a prominent position.
• A promotional paragraph of 100 words and contact details showcasing your company in the event programme.
• An opportunity to include one (1) insert into attendee bags: Either corporate literature (A4 size or smaller) OR promotional merchandise.
• Use of the IYOG2022 Official sponsor logo during and after 2022
• A Quartz ICG corporate membership for 2022

Cristallo Sponsorship - €5000 (multiple sponsorships are available) includes:
• Acknowledgement of sponsorship during the Opening and Closing Sessions, and at appropriate times throughout the event.
• Company name and logo will appear on all event literature: The event website, email teasers, event features in various glass sector magazines, event programmes, banners and signage displayed throughout the event space.
• Corporate banner will be displayed in a prominent position.
• A promotional paragraph of 50 words and contact details showcasing the sponsor in the attendee's programme.
• An opportunity to include one (1) insert into attendee's bags: EITHER corporate literature (A4 size or smaller) OR promotional merchandise.
• Use of the IYOG2022 Official sponsor logo during and after 2022
• Glass ICG corporate membership for 2022
• Acknowledgement of sponsorship during the Opening and Closing Sessions, and at appropriate times throughout the event.

Lattimo Sponsorship - €2500 (multiple sponsorships are available) includes:
• Acknowledgement of sponsorship during the Opening and Closing Sessions, and at appropriate times throughout the event.
• Company name and logo to appear on all event literature: The IYOG event website, email teasers, event features in various glass magazines, event programme, event banners and signage displayed throughout the event space.
• Corporate banner to be displayed in a prominent position.
• A full-page advertisement in the attendee's programme.
• One (1) A4 insert into the attendee's bag.
• Use of the IYOG2022 Official sponsor logo during and after 2022.

## Other sponsorship opportunities

There are several other sponsorship opportunities available: Pre-Gala Cocktails and the Gala Dinner on 10 February; Welcome Reception on 9 February; the six Keynote Speakers; Lunches; Charging Stations; Photography; Lanyards; Pens and Notebooks.

### Some Planned UK Events

Contemporary Glass Society (CGS) will hold a series of monthly talks in London plus online.

Exhibition: Henry Moore Institute, Leeds with National Glass Centre and Corning Museum of Glass, from February 2022.

Illumination of Catcliffe Glasshouse Cone, Rotherham will happen in March.

Exhibition: Its all in the technique (CGS), at the National Glass Centre, Sunderland on 15 January–13 March.

Exhibition: The best of the best (CGS), Newcastle, February–March.

Exhibition: Craft in the Bay (CGS), Cardiff, April–May.

Exhibition: Glorious Glass, Leeds, April–July.

Exhibition: Joyful Reflections (CGS), Farnham, April–June.

Furnace Solutions (SGT), 8–9 June, venue to be confirmed.

Exhibition: The Best of the Best (CGS), Newcastle, June–July.

Exhibition: Earth/Sea/Sky, London, June–July

16th Conference on the Physics of Non-Crystalline Solids, Canterbury, 10–15 July.

Exhibition: Then, Now, The Future, Stourbridge, July–October.

UK Glass Biennale, Stourbridge, August.

International Festival of Glass, Stourbridge, 26–29 August.

Exhibition: A postcard from the Commonwealth (CGS), Stourbridge, August–September.

SGT History and Heritage, Cambridge, September.

Exhibition: The Best of the Best (CGS), Newcastle, September–October.

Exhibition: Bedazzled 25 glassmakers impress (CGS), York, September–October.

Exhibition: Razzle Dazzle FIVE X FIVE (CGS), York, September–October.

Exhibition: CGS at 25, London, September–October.

Glass Focus, British Glass Annual Conference & Awards, November.

Pyrex 100 year celebration, Sunderland.

# Rebuilding the UK telecommunications industry

Professor David Richardson has been at the forefront of photonics research for the past 30 years. He has led various large collaborative optical communications research programmes in the UK and abroad, as well as being involved in pioneering research in hollow-core optical fibres, telecommunications, high power fibre lasers and optical fibre amplifiers. He has also co-founded two photonics spin out companies. As Deputy Director of the Optoelectronics Research Centre (ORC) at the University of Southampton and a Future Photonics Hub Co-Investigator, he shares his insights on the current state of our telecommunications networks and what needs to be done to ensure they keep up with society's demand.

The telecommunications industry contributes 4–5% of the UK's GDP and plays a vital role in the economic and social well-being of the country. Consumers have an abundance of choice at their fingertips when it comes to mobile phone networks, broadband providers and cable TV providers.

Up until the late 1990s/early 2000s, the UK had a very healthy telecommunications industry. It was doing world-leading research and was home to large UK-owned manufacturing companies spanning the full range of what was needed to build high performance telecommunications networks. In the late 1990s/early 2000s the telecoms market really took off, the early phase of the so-called dot-com bubble. There was large investment and many new companies came to market. To remain competitive and to cut down on costs, many of these companies started to move their manufacturing activities offshore, mostly to Asia. In addition, some of the big telecoms companies, both carriers and equipment manufacturers, reduced their research and development in the photonics and optics space and started to look internationally to where they were going to buy their technology. Ultimately, this created an oversupply, not just in the UK but around the world and the bubble collapsed. Many companies went out of business, some almost overnight, and much consolidation amongst those companies that survived ensued.

Fast forward to 2021, we're now in a position where there are few major UK-owned telecoms companies, and we are left with relatively modest telecoms-centric manufacturing capability. Moreover, as a country we have fallen behind many others in terms of the performance of our communications networks whilst the importance and demands on our digital infrastructure becomes ever more critical. However, the demand for telecommunications is only going to keep rising. For example, every time we undergo a generational change in mobile communications, from 4G to 5G presently, and in the future to 6G, the underlying optical network capacity requirements increase massively. Every year the traffic demand increases by around 40–50%. Industry needs to sustainably support that growth without a significant increase in costs to the consumer, presenting huge technological, economic and logistical challenges.

*Professor David Richardson, Optoelectronics Research Centre (ORC), University of Southampton*

## We are meeting the challenge

There are a number of activities being done to address this challenge and we are at the forefront. Fibre optic cables form the pipeline through which nearly all communications get sent and received and silicon photonics provides a means to increase the performance and to reduce the cost of equipment.

Current fibres have a glass diameter of around 125 microns and contain a single core in the centre, about 10 microns in diameter, where the signal, encoded in laser light, flows. This represents a single signal pathway. Research teams are exploring ways to transmit signals over multiple optical pathways through that single core simply by making it bigger. We are also trying to increase the number of cores in the fibre cross section to provide additional signal pathways. Both approaches, and their combination, enable much more efficient use of the fibre cross-section and potentially allow the amount of data transmitted in each fibre to be increased by factors of up to 100.

It is to be appreciated that when a signal propagates down a fibre, energy is unavoidably lost, so every 50–100 km, the signal needs a boost. In current networks this is provided by the erbium doped fibre amplifier, a device invented at Southampton in the mid-1980s. We are now working on optical amplifiers that can simultaneously amplify signals on multiple pathways to suit the new multipath transmission fibres discussed above. So far, we have demonstrated the amplification of 42 individual spatial channels in a single device, removing the need for 42 amplifiers of conventional fibre design.

This is exciting research that has a lot of potential but on its own it will not have a big enough impact. We need a combination of investment and an integrated innovation network to ensure research is translated into real-world applications. Programmes like the Future Photonics Hub enable an efficient and direct link to industry partners and play a crucial role in moving a lab-based discovery to a commercial platform. The Hub provides expertise in the preparatory work required when working with industry partners and provides a well-defined translational interface between companies and our core government-funded research programmes.

## A national strategy

The pandemic has shown us just how important and critical our digital infrastructure is. The UK has a huge diversity of world class expertise in telecoms. There are good reasons to regenerate the telecoms industry in the UK, both from a business opportunity and a UK supply chain/secure sovereign capability perspective. However, to do this will require both a national strategy and substantial investment. Greater coherence and co-ordination across government, as well as the industrial and academic sectors will be needed to make the most of the commercial opportunity for the UK and to ensure a leading edge national infrastructure long into the future.

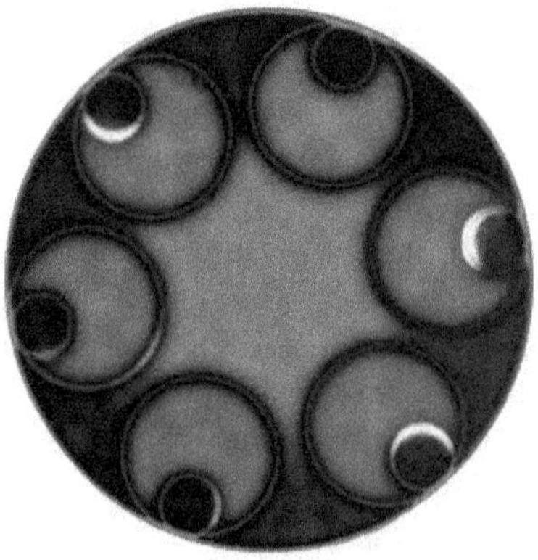

*A cross-section of Lumenisity's hollowcore NANF fibre*

A lot of discussion is currently underway at the national level as to what the UK telecoms strategy might look like across several government departments, industry and the academic community, all of whom have important roles to play. It remains to be seen where this ultimately leads. In the meanwhile, Southampton along with other UK academic institutions, continue to develop new technologies to meet the challenges the telecommunications industry and UK more broadly faces. They are developing these new technologies through new spinout companies, such as Lumenisity Ltd based in Southampton, who are developing leading-edge hollow-core optical fibre cables to improve the responsiveness of communication networks.

## An open market

Importantly, there has been a material shift in the way that telecom systems are employed and installed, which has had a major effect on the marketplace. Through open standards, the internationally agreed performance standards, companies are able to buy and deploy devices and components for the network instead of relying on large system integration companies with bespoke solutions. They know that they will work when they integrate them together. This is both driving the cost down and removing the barrier to entry for competition. Smaller companies now have better opportunities to enter the market and to address particular components, or needs, within the network. This competition should ultimately prove a good thing for UK industry and will hopefully allow vendors and operators to look at homegrown as well as overseas suppliers.

## The future

There's world-leading capabilities in the UK research space, particularly in universities, that is available to help regenerate the UK communications capability. This includes technology, research and world class facilities. Training and knowledge exchange are essential to develop skilled people who are then able to go out and join companies, or indeed to form their own, as is having people from industry come and work in an academic environment. The transferring of skills and expertise between people creates a healthy ecosystem that should enable UK industry to thrive in the future.

# SGT Furnace Solutions training seminars

The usual Training Day held on the day before Furnace Solutions was cancelled due to the pandemic, instead the Society of Glass Technology have a series of five, one-off training seminars held online via Zoom on the second Wednesday of the month, starting in October. Times of presentations are 13.00 to 14.00 UK time, BST in October, GMT afterwards. The speakers and topics are presented below.

Wednesday 13th October 2021
Thermal balance and furnace screening: a periodic audit to reduce emissions and increase energy efficiency presented by Walter Battaglia of Stazione Sperimentale del Vetro (SSV).

Wednesday 10th November 2021
Forehearth Audits – is your glass conditioning equipment at its best? Presented by Alan Stephens of Fives Stein.

Wednesday 8th December 2021
Forehearth Colour Technology: Today's performance needs and new challenge presented by Pauline Darbouret of Ferro.

Wednesday 12th January 2022
NOx emission and possibilities to reduce presented by Matthias Hagen of Luft- und Thermotechnik Bayreuth (LTB).

Wednesday 9th February 2022
Challenges of Tin Bath Blocks in the "Float Process" and how to test them? Presented by Jan Theron of Lucideon.

www.sgt.org
Recordings can be seen on the SGT YouTube channel:
https://www.youtube.com/channel/UCe2tywIcr11dor_0cdrT5Vg

---

# Furnace Solutions & Training Day

The Melting Technical Committee (MTC) of the Society of Glass Technology (SGT) invite you to mark your diaries for the 2022 Furnace Solutions on 9th June and Training Day on the 8th June.

Lucideon has converted their conference room to office space so the venue for the return of the person-to-person event is still to be confirmed.

Further details (including the Geoff Evans Dinner and hotel accommodation) will be announced on the website
http://www.furnacesolutions.co.uk/

**British Glass** CEO Dave Dalton provides an update on the ongoing energy and gas crisis and its effect on the UK glass industry:
"We are bitterly disappointed at the lack of action from both the Secretary of State and wider government to contain the escalating energy costs that are deeply affecting glass manufacturers across the UK.

"It has been three weeks since we set out proposals, alongside the Energy Intensive User Group (EUIG), that would help lessen the burden that is currently being placed on glass manufacturers, but we are yet to see any solutions or even propositions from government, while manufacturers that are vital to the UK economy continue to suffer extortionate costs that put their future at risk.

"Further inaction and a lack of support from government to address the energy crisis only deepens the impact on businesses.

"We need to ensure that manufacturers do not have to cease production and guarantee that the growing disparity between UK industry and international competitors does not widen even further, but we cannot do this without some form of plan from government, which currently does not exist."

**Beatson Clark** has committed £1 million in expanding the on-site recycling plant in Rotherham, South Yorkshire, installing new, more efficient equipment. The company partnered with **Redwave** to complete a turnkey plant expansion which began operation in October. The project has created 12 full time jobs which were previously contracted. The recycling plant used to generate on average 50% of this cullet; following these investments it now generates over 75% of the cullet.

As part of the investment, Beatson Clark added optical sorters, vibratory feeder conveyors, new screening machines and a JCB telehandler. Additional processing stages have been introduced using vibrating screens and dedicated optical sorters, improving quality and reducing waste.

The new equipment and the additional inspection stages have reduced the false rejects of glass which were removed with CSP (ceramics, stones and pottery) by over 50%. This makes the process more sustainable and means less waste glass going to aggregate for reuse rather than in the furnaces.

'Addition of new optical sorters makes the process more precise and reduces the amount of good glass which gets rejected and sent to waste,' said Patrick Pagdin, recycling plant manager at Beatson Clark. 'We are now seeing an increased output of both white flint cullet and green and amber cullet thanks to easier automated sorting and improvements in the colour sorting process.'

Cullet comes to the Beatson Clark recycling plant from the hospitality sector and from local authority kerbside collections of domestic glass. The post consumer recycled content rates for containers manufactured by Beatson Clark are 30% for white flint and 45% for amber glass.

**The Worshipful Company of Glass Sellers** are appealing to the British glass industry to join them in their efforts to support schools in their local area as part of the International Year of Glass.

Through the work of their charity arm, the Glass Sellers have been funding primary school STEM projects in areas such as physics, chemistry, biology, maths and ICT and are now asking industry the chance to join them in their charitable efforts.

Companies are being asked to nominate a single school or group of schools in their surrounding areas to put together a STEM project/have an ongoing STEM project with the possibility of co-funding the activities with the Glass Sellers.

The projects can range from new science equipment to helping school children study space programmes, as is the case for a group of 17 schools based in Cornwall, and everything in between.

Anyone wanting to nominate schools to work with should contact Chairman of the Charity Trustees Leigh Baildham (leighbaildham@me.com) or Barbara Beadman (brbeadman@yahoo.co.uk) for further details on the funding.

The **Scotch Whisky Association** (SWA) has unveiled a limited edition blended Scotch Whisky to be enjoyed by world leaders, delegates and visitors to the COP26 conference.

The whisky, which has a run of just 900 bottles, represents the Scotch Whisky industry's commitment to sustainability from grain to glass, and will be presented to negotiators gathering in Glasgow in November.

To mark the 26th annual Conference of the Parties, the COP26 limited edition Scotch Whisky has been blended and bottled in the host city of Glasgow, and is comprised of whiskies produced at 26 distilleries across Scotland's five whisky regions.

The bottle made from 97% recycled glass was created by **Allied Glass** in their Knottingley factory where reductions in carbon intensity and a switch to 100% renewable electricity will save 33,000 tonnes of $CO_2$ during 2021. The fully recyclable stopper was created from natural cork, recycled wood and recycled cork particles. The label paper contains 55% post-consumer recycled fibres, including leftover barley from the production process, and the carton is made from 100% recycled fibres.

A COP26 Spokesperson said: "World leaders coming to Glasgow for COP26 will receive a warm UK welcome which will include the chance to sample a limited edition whisky blend produced sustainably in Scotland, especially for COP.

"The path towards a more sustainable future will only be achieved through collaboration across societies and sectors. The efforts of the Scotch whisky industry, which is world-renowned, to reach net zero is a great example of the change we need to see."

UK telecoms provider **BT** has achieved a new milestone in the development of quantum-secure communications by conducting the world's first trial of quantum key distribution (QKD), a method of ultra-secure communications, over hollow core fibre cable, from **Southampton University** spin out, **Lumenisity**.

BT kicked off trials of a new type of optical fibre, nested anti-resonant nodeless fibre (NANF) hollow core fibre, at the BT Labs in Ipswich to test the potential benefits of deploying this technology for a variety of scenarios, including secure communications. The trials used cable developed and manufactured by Lumenisity to address the need for high-speed transactions and bandwidth increases in advanced communications systems, which has also been used for applications such as Data Centre Interconnects (DCIs), Edge and 5G xHaul.

In the latest trial, BT researchers successfully operated a state-of-the-art QKD system using commercial equipment over a 6 km long Lumenisity CoreSmart cable with a hollow, air-filled centre, revealing potential benefits such as reduced latency and no appreciable crosstalk, the effect of a transmitted signal interfering with the transmission of another signal.

In most optical fibre communications, high-speed signals are sent over a solid piece of glass using different wavelengths of light to deliver high capacity transmissions. In QKD systems, quantum light is transmitted on a single photon channel, traditionally necessitating use of a separate fibre, due to 'crosstalk' an effect that causes the light from high-speed data channels to spread their wavelengths, interfering with a quantum signal carried over the same fibre, as the change in frequency can cause channels of light to leak into other channels. The effect is similar to having a whispered conversation next to an orchestra – it can be hard to hear the other person's voice over the music.

Hollow core fibre doesn't have internal material, it is filled with air, so there is less light scattering and less crosstalk between channels, even at a single photon level. This clearer separation makes it easier to deliver both a high-speed encrypted data stream, and the faint quantum signal that carries the encryption key, over the same fibre.

Professor Marek Liška has received an exceptional award for "Lifelong support of science" from the **Slovak Academy of Science**.

Prof. Liška focused his research on the properties of glasses and melts, their dependence on composition and temperature and on the application of mathematical-statistical methods in the study of these dependences, objective evaluation of glass colour, optimisation of melting units, vacuum development in solar collector tube and optimisation of chemical composition of glasses.

He is the co-author of seven monographs, the author or co-author of 250 publications, of which 171 publications have been published in international professional journals. He is also the author or co-author of 346 papers at professional scientific conferences, of which 201 were lectures and has 534 citations and 147 cited publications. He applies his scientific-research knowledge in pedagogical practice, especially in the education of doctoral students as future researchers.

Marek Liška is a regional editor of both SGT journals and contributes to the smooth passage of peer reviewed papers through the system.

**Apogee Enterprises** has entered into an agreement to sell its architectural glass facility located in Statesboro, Georgia to **Cardinal Glass Industries**. The company had previously announced its intent to close the location as a part of a broader restructuring plan. Cardinal Glass plans to repurpose the facility to support its residential glass business. Apogee expects to record a pre-tax gain on the sale of between $18 million to $20 million.

"This sale agreement is a terrific outcome for Apogee, our employees in Statesboro, and the surrounding community," said Ty R. Silberhorn, Apogee's Chief Executive Officer. "This transaction will further improve Apogee's already strong financial position and gives us more financial flexibility as we move forward with implementing our new enterprise strategy. I'd like to recognize the efforts of our Architectural Glass team for bringing this to a successful conclusion."

Kyle Petersen, President of Cardinal LG, added, "Cardinal Glass has a long history of working together with Apogee and its Viracon business. We are excited about the opportunity to purchase the Statesboro facility and expand Cardinal's residential glass capability in the Southeastern United States. We are especially impressed with the quality of the team in Statesboro and look forward to building a lasting relationship with those employees and the Statesboro community."

**ROCKWOOL** is expanding production capacity at the plant in Vyborg, Russia. The new production line uses the company's low-carbon electric melting technology, cutting the facility's carbon emissions in half.

Established in 2006, the facility in Vyborg, northwest of Saint Petersburg, supplies North West and Central Russia as well as Finland.

"We are expanding production capacity to meet customer demand for energy efficient, sustainable, and non-combustible building materials. Implementing the electric melting technology combined with renewable energy will reduce the factory's carbon emissions by 50% and also improve its circularity. This will help satisfy the growing market demand for sustainably produced, low-embodied carbon construction products while also helping to meet our own decarbonisation goals", says Rockwool Group CEO, Jens Birgersson.

In the coming years, Rockwool expects to invest up to around €200 million to expand capacity at Vyborg and upgrade equipment and processes at its other Russian facilities as well. The expansion will create more than 70 additional permanent jobs.

"Rockwool has operated in the Leningrad Region for 15 years, and is a reliable partner and well-recognised brand within the building materials market. I am glad that the company continues to develop in our region, investing in new production and creating new jobs.", said Governor of the Leningrad Region, Alexander Drozdenko.

*Left to right: ROCKWOOL Group CEO, Jens Birgersson, Alexander Drozdenko, Governor of the Leningrad Region and Carsten Søndergaard, The Danish Ambassador to Russia*

State Secretary Terhi Lehtonen, Ministry of the Environment of Finland, highlighted how the project connects with Finnish national priorities, saying, "Energy efficiency of buildings and use of sustainable building materials play a key role in reaching Finland's targets for carbon neutrality by 2035. New national legislation is currently being circulated for comments. As Finland is aiming at carbon neutrality and carbon negativity thereafter, it becomes essential to manage the emissions that are related to the full life cycle of buildings. This will cover the manufacturing of products, construction work, use of the building, renovations, and, ultimately, its deconstruction and recycling".

Global packaging supplier **TricorBraun** has entered into an agreement to acquire premium glass packaging supplier **Vetroelite** from **LBO France**. The acquisition strengthens TricorBraun's European presence and continues the company's global expansion.

Founded in 1994, Vetroelite provides premium glass packaging solutions for the fast-growing spirits, gourmet food, home fragrances, wine, and cosmetics industries. Vetroelite has a worldwide presence, operating from multiple locations across Europe and North America, and serving customers in 90 countries.

"We are very pleased to expand our European presence with the acquisition of Vetroelite," said Court Carruthers, president and CEO, TricorBraun. "The Vetroelite team has established a strong reputation for high-quality, innovative packaging solutions, fueled by deep expertise and dedicated customer service. We are proud to welcome Vetroelite to the TricorBraun family, and look forward to supporting Vetroelite's continued growth."

All Vetroelite team members, including CEO Daniele Feletto and the company's leadership team, will remain with TricorBraun and continue to work out of the company's existing locations. Effective with the closing of the transaction, Vetroelite will operate as Vetroelite, a TricorBraun company.

Since its founding, TricorBraun has partnered with management teams to successfully acquire and further accelerate growth for nearly 30 packaging companies. Vetroelite will mark TricorBraun's fourth acquisition in 2021.

**Vitro** has announced an investment of approximately $70 million for the construction of a new Container furnace at the Vitro plant in Toluca, State of Mexico. It will meet the increased demand for glass containers in the various segments in which the company participates.

"This new furnace will have the best technology for the manufacture of glass with high quality, design and a high level of sophistication. With this new furnace we will be able to offer greater availability of value-added products," said Adrián Sada Cueva, Vitro's CEO, also reiterating that this investment confirms Vitro's commitment to offer the best level of service to its customers and markets worldwide.

In 2007, Vitro began operations in one of the largest glass packaging plants in the world to meet, among others, the requirements of the packaging markets for cosmetics, fragrances, toiletries and pharmaceuticals with high quality and efficiency standards.

"The new furnace will provide additional capacity to support our customers and their growing requirements for glass containers, providing a higher level service in the regions and segments where we participate. The furnace will have state-of-the-art technology to manufacture high-value-added glass containers with sustainable processes," said Schlomo Frymerman, Director of Glass Containers.

Construction of the new furnace will begin this year and start its operation in the first half of 2023.

**Prince International**, a portfolio company of **American Securities**, is exploring a sale of its porcelain enamel and glass coatings business in the US and Europe. Prince is exploring a sale of the business in connection with its efforts to obtain regulatory clearance required to complete the acquisition of Ferro. The Ferro transaction is subject to customary closing conditions and is anticipated to close in the first quarter of 2022.

The second-largest producer of porcelain enamel globally and a major player in glass coatings. With over 110 years of expertise in frit production and glass technology, the glass and coatings business serves niche, differentiated end uses such as home appliances, cookware, and automotive. It operates three state-of-the-art facilities across North America and Europe and caters to a diverse portfolio of global blue-chip customers.

Saint-Gobain, through its building products subsidiary CertainTeed, has announced a $400 million investment plan to expand its production capacity at four key manufacturing locations in the United States. The combined investments will reduce waste by more than 50% and reduce $CO_2$ emissions by more than 20% compared to existing manufacturing technology.

"This is an investment in our company's future, and also in

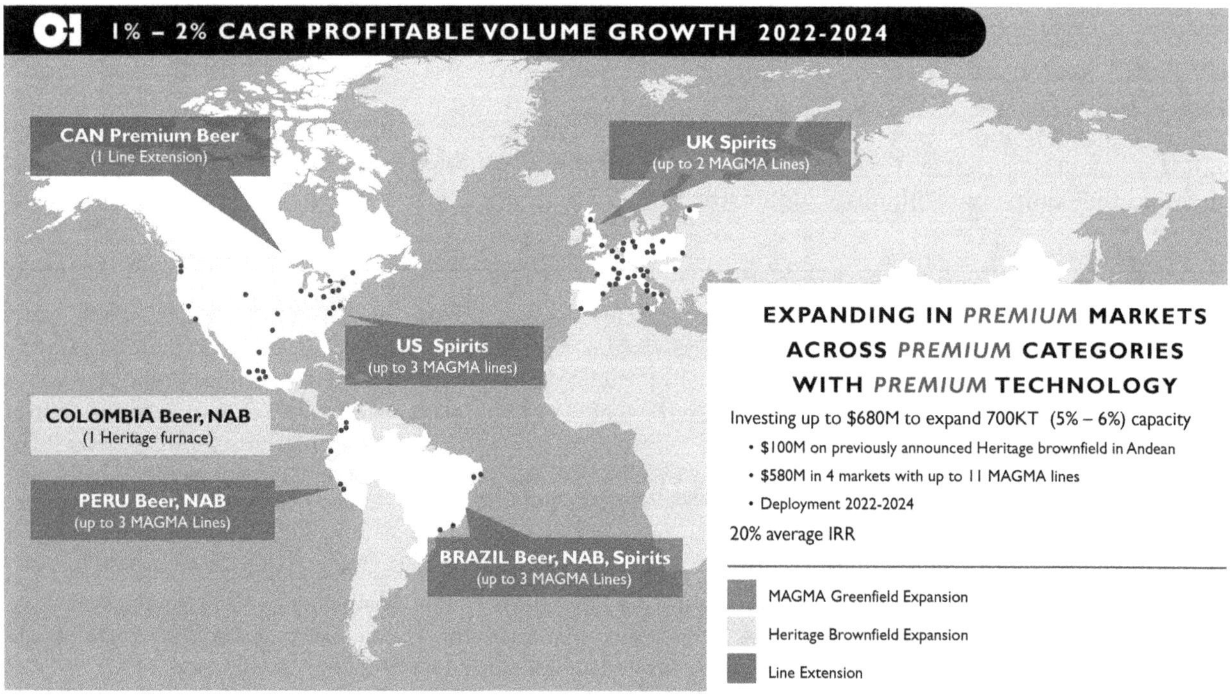


*As part of the next investment cycle O-I are backing heavily the next three generations of their MAGMA container production machines with deployment around the world*

*For half a century, engineers at* ***SCHOTT*** *have been designing the cooktops of the future. Here glass-ceramic is a key component. Since CERAN® was introduced 50 years ago, black glass-ceramic has been established in the market as the standard material for induction and electric cooktops in modern kitchens worldwide. For its anniversary year, the company is presenting the next generation: CERAN Luminoir™. The new glass-ceramic is more translucent but still deep black thanks to a new formula. The classic elegant black appearance provides virtually endless possibilities in product and lighting design and unique advantages for leading household appliance manufacturers. For users, Luminoir offers cooking surfaces that are intuitive to operate and dynamic, providing visual support when connecting with other smart kitchen devices. Light is an essential tool for transporting and exchanging information. Lighting solutions ensure that controls, cooking zones, or even digital features, are visible when switched on. To make this happen, light, emitted by LEDs below the cooktop panel, has to penetrate the glass-ceramic as much as possible. Luminoir's newly developed composition makes the glass-ceramic more translucent allowing white and blue light, in particular to shine through brighter and clearer. That's why it opens up new design possibilities for household appliance designers, at the same time reducing the production complexity of smart cooktops, ultimately saving consumers money.*

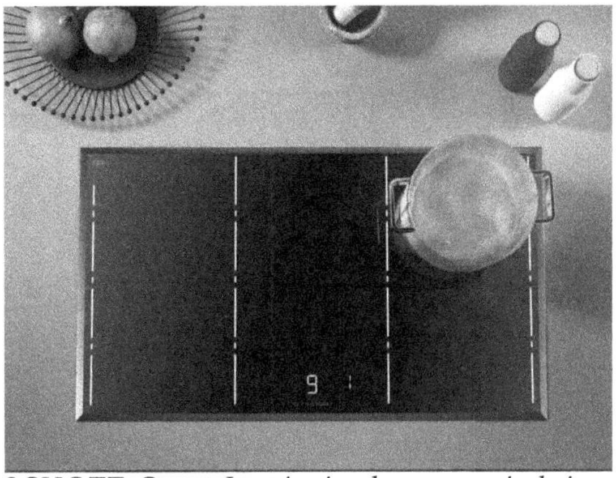

*SCHOTT Ceran Luminoir glass-ceramic brings bright, clear and true-colour lighting solutions to cooktops*

*"The new material formula enables our customers to save on production outlays and costs. This is possible because luminous and multicolored light features can now be produced without additional filter layers. This not only saves on material and production steps, but also makes innovative lighting solutions in cooktops interesting for broader assortment of customers," explains Dr Matthias Bockmeyer, Head of Development in the Cooking division at SCHOTT.*

---

our future ability to ensure security of supply for our customers," said Mark Rayfield, CEO of Saint-Gobain North America and CertainTeed. "We will continue to look for opportunities to strategically grow and expand our business, all while striving to become the world's leader in light and sustainable construction."

Approximately $100 million will be used to double production capacity at one of the roofing shingle plants in the Southeastern United States. The same region will receive $250 million to add 900 million square feet of gypsum production capacity and storage.

At the company's insulation plant in Chowchilla, California, Saint-Gobain will invest $32 million in a new oven and new furnace, which will allow the plant to expand its production by 13% at that location and will begin coming online in 2022.

At the company's plant in Athens, Georgia, Saint-Gobain will invest $30 million to start up an additional production line that will increase the plant's capacity of loose-fill insulation. This expanded production is expected to begin in the fourth quarter of 2023 and will lead to the creation of up to 20 new permanent full-time jobs.

**Forglass** has re-imagined and re-engineered the glass melting process to help its clients manage production, save money and improve the environmental impact. With smaller order volumes for specific assortments of container glass and ever shorter order lead times, glassworks all over the world are struggling with the current technology, which does not allow them sufficient flexibility to efficiently produce what their customers order.

The Forglass Multi-glass Furnace can simultaneously produce smaller volumes of two or even several glasses with energy consumption comparable to melting only one glass in a standard, single-tank furnace.

With lower energy consumption per ton of glass and reduced emissions the Multi-glass Furnace is the ideal option for both greenfield investments and modernisations in restricted spaces. For many glassworks located in urban areas, this opens up new opportunities for expansion or restructuring of their production capacities. Conversion of a standard furnace can be carried out in any glassworks by expanding the furnace (adding a second tank) only in its width, not length. It is generally the only space available in the production hall and, what's crucial, the expansion of a U-fired furnace does not involve radical rebuilding of the regenerator, which makes it inexpensive and fast.

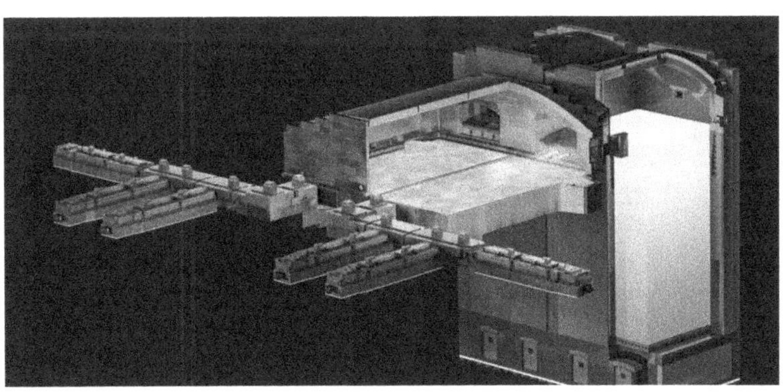

# Morley Glass triples post-consumer glass recycling capacity

Morley Glass will be recycling more post-consumer glass and supporting more local good causes than ever following the installation of new state of the art automated machinery to process waste glass collected from Uni-Blinds® installers.

The new plant commissioned at the company's HQ and factory near Leeds replaces the pioneering 'Crush' machinery, which has been retired after two reliable years of service. The new machine is designed to be smarter and more productive through automation, extracting a larger amount of annealed glass out of the frame.

This eliminates some of the more time consuming and labour-intensive stages of the glass recycling process to enable larger volumes to be handled. Through a process of separation and crushing, the glass is turned into cullet which can then go straight into the manufacture of new glass, whilst spacer bars and any other materials are channelled off into a separate waste stream.

This first stage recycling process is a joint initiative between Morley Glass and Saint-Gobain Glass whereby recycled glass cullet is collected in half tonne bags for use as a raw material in the production of new glass. Since the initiative started in 2019, more than 670 tonnes of glass cullet have been produced which has saved more than half a million tonnes of 'virgin' sand from Saint-Gobain Glass's manufacturing process.

But the benefits of recycling post-consumer glass in this way go far beyond saving sand reserves. Using waste glass in place of virgin raw materials also reduces the amount of energy needed in production, which means that 199,000 tonnes of $CO_2$ has been saved, equivalent of 332 'energy years' for an average UK household.

Morley Glass goes even further, however, to make sure the value of the recycled glass cullet can be utilised to fund environment initiatives across West Yorkshire through the GreenVision Fund. This enables local organisations involved in environmental improvement and green micro initiatives to benefit from individual grants of £500 to support the work they do.

Ian Short, MD of Morley Glass said: "We're thrilled that we can now handle three times the amount of recycled glass than we could previously, helping the industry take another important step forward in terms of sustainability. But equally important will be our increased ability to support more green initiatives than ever as the size of our GreenVision Fund grows with recycling volumes – it's a win-win."

By providing a post-consumer glass recycling service, Morley Glass helps its customers reduce the amount of waste they need to dispose of and improve the whole-life sustainability of Uni-Blinds integral blinds. It also assists Saint-Gobain Glass with their sustainability goals as they seek to increase the amount of recycled content in new float glass, which is already at the forefront of the industry.

# Fountain – the first highly durable anti-reflective coated glass offered by AGCULTURE

AGCULTURE™, a new business activity of **AGC Glass Europe**, is launching its first durable glass product, Fountain, which has been specially designed to improve crop yields for horticultural growers. Fountain is basically a low-iron float glass, one side of which is chemically treated to create the required level of light diffusion. This is sandwiched between two durable anti-reflective (AR) coatings. The side of the glass with the microstructures is positioned on the inside of the greenhouse, while the flat side is positioned on the outside. Despite this simple design, Fountain differs in several key ways compared to existing solutions on the market.

The double-sided AR coating combined with low-iron float glass provides a high level, around 96·5%, of photosynthetically active radiation (PAR) transmission, regardless of the level of diffusion. However, the sun's position changes throughout the day and direct PAR transmission (perpendicular to glass surface) alone cannot sufficiently guide growers as to which type of glass to use in order to maximise the amount of light in their greenhouses. Hence, according to NEN2675 + C1:2018, a more meaningful parameter is to consider hemispherical transmission, which is the integral value of light transmission for all angles of incidence. As a rule of thumb defined by Wageningen University & Research (WUR) for tomatoes, a 1% increase in hemispherical transmission equals a 0·8% increase in crop yield. Depending on diffusion level, the hemispherical transmission delivered by Fountain glass ranges from 85·5% to 80·6%.

The low-iron glass used in Fountain offers minimum absorption in the UV range and, when equipped with double AR, delivers 89·5% UV transmission. This high level of UV transmission is very important to the pollination process since part of the bumble bee's eye sensitivity lies in the middle of the UV range (340 nm) and high transmission in this range helps bees to find their way, thus enhancing the pollination process. Moreover, with the emergence of lettuce growing in, for example, the United States, UV light plays a significant role in the colour formation of such crops as Lollo Rossa.

The outer leaves of crop canopies typically block light from reaching deep inside the canopy, meaning inner leaves receive less light and can contribute much less to the photosynthesis process, resulting in lower yields. Moreover, intense direct light causes leaves to bend down in order to protect themselves, reducing the effective leaf area receiving light, which in turn reduces the photosynthesis rate and productivity. The microstructure of the inner side of Fountain glass can provide a wide range of hortiscatter from 15% to 63%. This high level of scatter is basically a sign of homogeneous distribution of light, which can penetrate deep inside the crop canopy and increase the photosynthesis rate and productivity by enabling the inner leaves to contribute. Furthermore, having homogeneous and uniform light inside the greenhouse causes the leaves to experience less stress, so they tend to orient themselves more horizontally. This effect not only enables light to penetrate more deeply into the canopy but also increases the effective area of each leaf, contributing to the photosynthesis process which in turn results in higher productivity. WUR has identified an empirical rule for tomatoes: a 1% increase in hortiscatter can represent a 0·3% increase in yield. Considering the wide range of hortiscatter provided by Fountain, growers can significantly improve their output by using this glass.

More info on Fountain advanced greenhouse glass on www.agculture.eu

Corning, NY

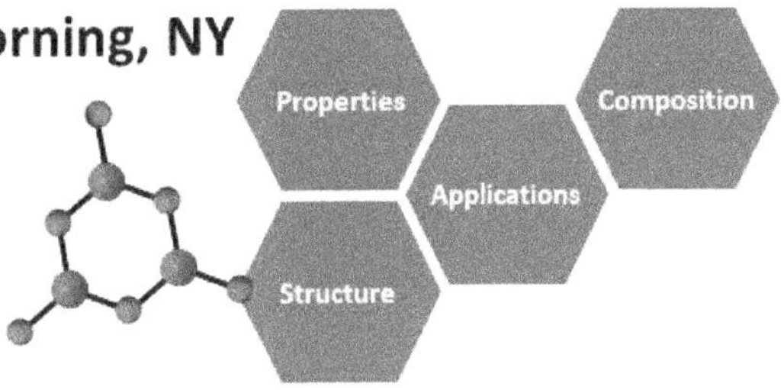


# Borate X

## July 17–21, 2023

# Phosphate III

## Revised Conference Dates

The 10th International Conference on
**BORATE GLASSES, CRYSTALS AND MELTS**
**July 17–19, 2023**

This marks the 10th in the series of conferences on borate glasses, crystals and melts, and will return to New York, USA close to where the first Borate Conference was held (Alfred, 1977). The most recent meeting took place in Oxford, UK (2017). The Borate Conference is a forum for the discussion of current scientific research on borate and B-containing materials. The conference will be dedicated to Dr. E.I. (Stratos) Kamitsos from the National Hellenic Research Foundation (Athens), to honor his lifetime of work in the field of borate glasses.

The 3rd International Conference on
**PHOSPHATE GLASSES, CRYSTALS AND MELTS**
**July 19–21, 2023**

This conference continues the tradition established with the first two conferences held in Pardubice, Czech Republic (2014) and Oxford, UK (2017). It will cover all aspects of phosphates and P-containing materials, including P-O, P-S, P-Se and P-O-N networks. Topics of interest will cover basic science, applications and technologies. The Phosphate Materials conference will be dedicated to Professor Ladislav Koudelka, University of Pardubice, Czech Republic, recognizing his career in phosphate and borophosphate glasses.

## Conference Themes

This joint meeting, to be held at the Corning Radisson Hotel, is comprised of two conferences that will run sequentially. On the third day of the meeting there will be a combined day dedicated to borophosphate glasses followed by a conference dinner to mark the close the Borate Conference and the start of the Phosphate Conference. Topics will encompass anything related to borate and phosphate materials, including:

- **Energy**
- **Healthcare**
- **Optical and data storage materials**
- **Property-structure relationships**
- **Fundamentals of the glassy state**
- **Novel glasses, crystals and melts**
- **Technology and Industry**

## Conference Chair

Dr. Randall Youngman
Corning Incorporated
youngmanre@corning.com

## Local Organizing Committee

R. Youngman (chair), J. Wu (Corning), L. Ma (Corning), L. Huang (RPI), A. Goel (Rutgers), D. Möncke (Alfred), A. Clare (Alfred)

***www.boratephosphate2020.org***

# CONFERENCE DIARY

**26–27 October**
CelSian training course: refractory selection, maintenance and related defects, Eindhoven, The Netherlands.
*CelSian Glass BV Email infodesk@celsian.nl Web www.celsian.nl/training-and-rd/*

**27–30 October**
Seventh International Conference on Multifunctinal, Hybrid and Nanomaterials, Genoa, Italy.
*Web www.elsevier.com/events/conferences/international-conference-on-multifunctional-hybrid-and-nanomaterials*

**28–29 October**
CelSian training course - flat glass melting and forming, Eindhoven, The Netherlands.
*CelSian Glass BV Email infodesk@celsian.nl Web www.celsian.nl/training-and-rd/*

**30 October–5 November**
Sixth Wuhan ICG Winter School, Wuhan, China.
*Prof. Kai Xu: SMART, Wuhan University of Technology, 1040 Heping Ave, Wuchang Qu, Wuhan Shi, Hubei Sheng, China. Email kaixu@whut.edu.cn Web www.icglass.org*

**30 October–5 November**
82nd Conference on Glass Problems, Columbus, USA.
*Donna Banks, GMIC, 550 Polaris Parkway, Suite 510, Westerville, OH 43082, USA. Email dbanks@gmic.org Web glassproblemsconference.org*

**5–6 November**
Glass in architecture from the pre- to the post-industrial era: Production, use and conservation, Vitromusée Romont, Switzerland.
*Vitrocentre Romont Au Château P.O. Box 150 CH-1680 Romont, Switzerland. Email sarah.tabbal@vitrocentre.ch Web vitrocentre.ch*

**10 November**
Furnace Solutions training seminar: Forehearth Audits – is your glass conditioning equipment at its best? Presented by Alan Stephens of Fives Stein.
*Web www.sgt.org*

**~~16–18 November~~ postponed to 26–28 October 2022**
Glasstech Asia 2021, Bangkok, Thailand.
*Web www.glasstechasia.com.sg/*

**~~17–18 November~~ Postponed to 2022**
Glass and Modern Technologies - XXI, Moscow, Russia.
*Organizing Committee Email forum@steklosouz.ru Web www.steklosouz.ru*

**18 November**
Glass Focus Awards 2021, UK.
*British Glass, 9 Churchill Way, Chapeltown, Sheffield, South Yorkshire S35 2PY, UK. Web www.britglass.org.uk/glass-focus-2021*

**23–25 November**
Celsian training: glass melting and process improvement for Indian glass industry, Mumbai, India.
*CelSian Glass BV Email infodesk@celsian.nl Web www.celsian.nl/training-and-rd/*

**2 December**
11th Workshop Laserbearbeitung von Glaswerkstoffen, Hannover, Germany.
*Katharina Rettschlag, Gruppenleiterin Glas, Abteilung Produktions- und Systemtechnik, Laser Zentrum Hannover e.V., Hollerithallee 8, 30419 Hannover, Germanay. Email k.rettschlag@lzh.de Web www.lzh.de/de/glasworkshop2021*

**~~5–8 December~~ Cancelled**
International Commission on Glass (ICG) Annual Meeting, Incheon, South Korea.
*Secretariat, Room 403, Meolijae Building, 76 Bangbae-ro, Seochu-gu, Seoul, Republic of Korea. Email secretariat@icg2021.org Web www.icg2021.org/*

**8 December 2021**
Furnace Solutions training seminar: Forehearth Colour Technology: Today's performance needs and new challenge presented by Pauline Darbouret of Ferro.
*Web www.sgt.org*

**12–17 December (new dates)**
14th Pacific Rim Conference on Ceramic and Glass Technology (PACRIM 14) including Glass & Optical Materials Division 2021 Annual Meeting (GOMD 2021), Vancouver, BC, Canada.
*Andrea Ross, The American Ceramic Society, 550 Polaris Parkway, Suite 510, Westerville, OH 43082, USA. Email aross@ceramics.org Web www.ceramics.org*

**2022 UN International Year of Glass**
**Postponed to 2022 no date announced**
13th International Symposium on Crystallization in Glasses and Liquid, Paris, France.
*Laurent Cormier, CNRS, Paris, France. Email crystallization@sciencesconf.org Web crystallization.sciencesconf.org/*
ACerS Pan American Ceramics Conference, PACC-FMAs, Panama City, Panama.
*Andrea Ross, The American Ceramic Society, 550 Polaris Parkway, Suite 510, Westerville, OH 43082, USA. Email aross@ceramics.org Web www.ceramics.org*

**5–7 January**
Recent advances on the glass problem, Lausanne, Switzerland.
*Elisabeth Agoritsas, EPFL CECAM, Avenue de Forel 2, BCH 3103, 1015 Lausanne, Switzerland. Email helpdesk@cecam.org Web www.cecam.org/workshop-details/48*

**12 January**
Furnace Solutions training seminar: NOx emission and possibilities to reduce presented by Matthias Hagen of Luft- und Thermotechnik Bayreuth (LTB).
*Web www.sgt.org*

**22–27 January**
Photonics West, San Francisco, USA.
*SPIE, PO Box 10, Bellingham, WA 98227-0010, USA. Email MeetingInfo@SPIE.org Web spie.org*

**9 February**
Furnace Solutions training seminar: Challenges of Tin Bath Blocks in the "Float Process" and how to test them? Presented by Jan Theron of Lucideon.
*Web www.sgt.org*

**10–12 February**
Opening Ceremony of the International Year of Glass 2022, Geneva, Switzerland.
*Web www.IYoG2022.org*

**16–18 February 2022**
Glass Performance Days, Tampere, Finland.
*Brown Onduso, GPD, Vehmaistenkatu 5, 33730 Tampere, Finland Email brown.onduso@gpd.fi Web gpd.fi/*

**17–18 February**
Glassman Asia, Seoul, Korea.
*Ken Clark, Quartz Business Media, Quartz House, 20 Clarendon Road, Redhill, Surrey RH1 1QX, UK Email ken-clark@quartzltd.com Web www.glassmanevents.com*

**24 February**
Simulating glasses (webinar)
*Sara Bonella, EPFL CECAM, Avenue de Forel 2, BCH 3103, 1015 Lausanne, Switzerland. Email helpdesk@cecam.org Web www.cecam.org/workshop-details/1105*

**3–5 March**
glasspex INDIA 2022 plus GlassPro India, Mumbai, India.
*Vivek Bohra, Senior Project Manage Email bohrav@md-india.com Web www.glasspex.com*

**6–10 March**
Optical Fiber Communications Conference and Exhibition (OFC), San Diego, California USA.
*The Optical Society, USA. Email custserv@osa.org Web www.ofcconference.org*

**8–10 March**
JEC World, Paris, France.
*Web www.jec-world.events/*

**9–10 March**
Sustainable Industrial Manufacturing (SIM), Brussels, Belgium.
*Nicola Davison, Quartz Business Media, Quartz House, 20 Clarendon Rd, Redhill, Surrey, RH1 1QX, UK. Email nicoladavison@quartzltd.com Web sustainableindustrial-manufacturing.com*

**15–18 March**
Unified International Technical Conference on Refractories (UNITECR 2022) 17th Biennial Worldwide Congress on Refractories, Chicago, USA.
*The American Ceramic Society, 550 Polaris Parkway, Suite 510, Westerville, OH 43082, USA. Email aross@ceramics.org Web www.ceramics.org*

**18–20 March**
2022 International Flameworking Conference, Salem, NJ, USA.
*Salem Community College, 460 Hollywood Avenue, Carneys Point, NJ 08069, USA. Email info@salemcc.edu*

**2–3 April**
Deco '22, The Society of Glass and Ceramic Decorating Products annual conference, Columbus, OH, USA.
*Myra Warne, Society of Glass and Ceramic Decorating Products, PO Box 2489, Zanesville, OH 43702, USA. Email info@sgcd.org Web www.sgcd.org*

**3–6 April**
TMS World Congress on Artificial Intelligence in Materials and Manufacturing (AIM 2022), Pittsburgh, Pennsylvania, USA.
*TMS Meeting Services, 5700 Corporate Drive Suite 750, Pittsburgh, PA 15237, USA. Web www.tms.org/AIM2022*

**3–7 April**
Photonics Europe 2022, Strasbourg, France.
*SPIE Europe, 2 Alexandra Gate, Ffordd Pengam, Cardiff, CF24 2SA Wales. Email info@spieeurope.org Web spie.org*

**5–6 April**
GlassPrint 2022, Düsseldorf, Germany.
*Email sales@glassworldwide.co.uk Web www.glassprint.org*

**13–16 April**
China Glass 2022, Shanghai, China.
*Web www.chinaglass-expo.com*

**18–20 April**
Int. Conf. on Glass Science and Technology from the Golden Era of the Pharaohs to a Sustainable Future, Egypt.

**19–22 April**
13th European Conference on Industrial Furnaces and Boilers, Algarve, Portugal.
*Raquel Reis, Cenertec, Rua Gago Coutinho, 185-187, 4435-034 Rio Tinto, Portugal. Email raquel.reis@cenertec.pt Web infub.pt*

**11–12 May**
Glassman Latin America, Monterrey, Mexico.
*Ken Clark, Quartz Business Media, Quartz House, 20 Clarendon Road, Redhill, Surrey RH1 1QX, UK Email ken-clark@quartzltd.com Web www.glassmanevents.com*

**18–20 May**
REFRA PRAGUE 21, 21st Conference on modern refractory materials and key achievements in high temperature technologies, Prague, Czech Republic.
*Ing. Jan K. Luxa, Silikátová společnost České republiky, Novotného lávka 5, 110 01 Praha 1, Czech Republic. Email hluxa@seznam.cz Web www.silikaty.cz*

**18–21 May**
GAS 2022, Glass Art Society conference, Tacoma, WA, USA.
*Glass Art Society, 2208 NW Market Street, Suite 200, Seattle, Washington 98107, USA. Email info@glassart.org Web www.glassart.org*

**22–26 May**
2022 Glass and Optical Materials Division Annual Meeting (GOMD2022), Baltimore, USA.
*Andrea Ross, The American Ceramic Society, 550 Polaris Parkway, Suite 510, Westerville, OH 43082, USA. Email aross@ceramics.org Web www.ceramics.org*

**6–9 June**
Mir Stekla 2022, Moscow, Russia.
*Ekaterina Razumeitseva, Project Head, EXPOCENTRE, Moscow, Krasnopresnenskaya nab., 14, 123100, Russia. Email re@expocentr.ru Web www.mirstekla-expo.ru/en/*

**8–9 June**
Furnace Solutions and Training Day, venue TBC, UK.
*Christine Brown, Society of Glass Technology, 9 Churchill Way, Chapeltown, Sheffield S35 2PY, UK. Email christine@sgt.org Web www.sgt.org*

**16–17 June**
XXXIV ATIV Int. Conf., Where Glass Science, Art and Technology Meet Together, Parma, Italy.
*ATIV Secretariat, MV Congressi spa Via Marchesi 26 d 43126 Parma, Italy. ativ@ativ-online.it Web www.ativ.eu*

**21–24 June**
Techtextil 2022, Frankfurt, Germany.
*Email techtextil@messefrankfurt.com Web techtextil.messe-frankfurt.com*

**23–24 June**
Challenging Glass Conference 8: International Conference on the Architectural and Structural Application of Glass, Ghent, Belgium.
*Email cgc8@challengingglass.com Web www.challenging-glass.com*

**29 June–3 July**
Glassman South America, Sao Paulo, Brazil.
*Ken Clark, Quartz Business Media, Quartz House, 20 Clarendon Road, Redhill, Surrey RH1 1QX, UK Email ken-clark@quartzltd.com Web www.glassmanevents.com*

**3–8 July**
26th International Congress on Glass, Berlin, Germany. Hosted by the DGG in their centenary year.
*Annette Doms, Deutsche Glastechnische Gesellschaft, Siemenstrasse 45, D-63071 Offenbach, Germany. Email dgg@hvg-dgg.de Web www.hvg-dgg.de*

**4–8 July**
Ninth International Conference on Optical, Optoelectronic and Photonic Materials and Applications (ICOOPMA) and EuroDIM: 14th Europhysical Conference on Defects in Insulating Materials, Ghent, Belgium.
*Web icoopma-eurodim.ugent.be*

**6–8 July**
World Glasstech-MEXICO / Glass Technology Expo Mexico, Guadalajara, Mexico.
*Glasstech Mexico, Av. Santa Fe 170, 01210 Lomas de Santa Fe, Mexico. Email info@glasstechmexico.com Web www.glasstechmexico.com*

**10–15 July**
Physics of Non-Crystalline Solids, now also including Society of Glass Technology Annual Meeting Canterbury, UK.
*Christine Brown, Society of Glass Technology, 9 Churchill Way, Chapeltown, Sheffield S35 2PY, UK. Email christine@sgt.org Web www.sgt.org*

**10–14 July**
Ninth Int. Congress on Ceramics (ICC9), Krakow, Poland.
*European Ceramic Society AISBL, Av. Gouverneur Cornez 4, B-7000 Mons, Belgium. Email ecers@bcrc.be Web ecers.org/*

**18–21 July**
65th ASGS Symposium, Sarasota, Florida, USA.
*Email philip.legge@gmail.com*

**20–28 August**
ICOM Prague 2022, Prague, Czech Republic.
*Secretariat, ICOM Czech Republic, Moravian Museum, Zelný trh 6, 659 37 Brno, Czech Republic. Email lehman-nova@icom-czechia.cz Web prague2022.icom.museum*

**24–29 July**
Sol-Gel 2021, International Sol-Gel Society Conference, Lyon, France.
*Université Claude Bernard Lyon 1, Cellule Congrès - SOLGEL 2021, 43, bd du 11 Novembre 1918, 69622 Villeurbanne, France. Email solgel2021@univ-lyon1.fr Web www.solgel2021.fr*

**27–30 August**
International Festival of Glass, Stourbridge, UK.
*Ruskin Glass Centre, Wollaston Road, Amblecote, Stourbridge DY8 4HF, UK. Email ifg@rmlt.org.uk Web www.ifg.org.uk*

**15 September**
Going Global: neue Herausforderungen an die Provenienzforschung, Graduate Workshop, Romont, Switzerland.
*Vitrocentre Romont Au Château PO Box 150, CH-1680 Romont, Switzerland. Email sarah.tabbal@vitrocentre.ch Web www.vitrocentre.ch*

**20–23 September**
Glastec 2022, Düsseldorf, Germany.
*Anke Seltmann, glastec, D-40474 Düsseldorf, Am Staad (Stockumer Höfe), Germany. Email seltmanna@messe-duesseldorf.de Web www.glasstec-online.com/*

**2–7 October**
Tenth International Conference on Multiscale Materials Modeling: including a symposium on: Multiscale Modeling of Glasses and Structurally Disordered Materials, Baltimore, Maryland, USA.
*MMM 10 Conference, 3400 North Charles Street, Department of Mechanical Engineering, Johns Hopkins University, Baltimore, MD 21286, USA. Email mmm10@jhu.edu Web mmm10.jhu.edu*

**7 October**
Celebrating the Birth of English and Irish Crystal Drinking Glass, 1640–1702, London, UK.
*Colin Brain, Association for the History of Glass. Email cbrain@interalpha.co.uk Web www.historyofglass.org.uk*

**9–13 October**
ACerS 124th Annual Meeting with Materials Science & Technology 2022, Pittsburgh, USA.
*The American Ceramic Society, 550 Polaris Parkway, Suite 510, Westerville, OH 43082, USA. Email customerservice@ceramics.org Web www.ceramics.org*

**11–13 October**
GlassBuild America 2022, Las Vegas, USA.
*National Glass Association, 1945 Old Gallows Road, Suite 750, Vienna, VA 22182, USA. Web www.glassbuildamerica.com/*

**17–22 October**
TCM-TOEO 2022 Conference: 8th International Symposium on Conductive Materials & 12th International Symposium on Transparent Oxide and Related Materials for Electronics and Optics, Heraklion, Crete.
*George Kiriakidis, Nik. Plastira 100, 70013, Heraklion, Crete, Greece. Email info@mitos.com.gr Web tcm2021.org/*

**18–20 October**
GlassBuild America, Las Vegas, Nevada, USA.
*National Glass Association NGA, 8200 Greensboro Drive, Suite 302, VA 22102 McLean, USA. Email attend@glass.org Web www.glass.org*

**19–21 October**
XXXIV ATIV Int. Conf., Where Glass Science, Art and Technology Meet Together, Parma, Italy.
*ATIV Secretariat, MV Congressi spa Via Marchesi 26 d 43126 Parma, Italy. ativ@ativ-online.it Web www.ativ.eu*

**26–28 October**
Glasstech Asia 2022, Singapore.
*Web www.glasstechasia.com.sg/*

**5 November**
International Conference on Glass in Museums: Research, Context, Communication, Frauenfeld, Switzerland.
*Email sarah.keller@vitrocentre.ch Web historisches-museum.tg.ch and vitrocentre.ch*

**9–11 November**
Recent Advances in Glass and Ceramics Conservation 2022. Sixth ICOM Interim Meeting of the Glass and Ceramics Working Group, Lisbon, Portugal and online.
*Unidade Investigação VICARTE – Vidro e Cerâmica para as Artes, NOVA School of Science and Technology, Hangar III – Campus da Caparica, 2829-516 Caparica, Portugal.*

*Email vicarte.secretariado@fct.unl.pt Web eventos.fct.unl.pt/icomcc_gc_2022*

**2023**

**3–5 February**

Future Past, Ausglass 2023, Melbourne, Australia.
*Ausglass, GPO Box 101, Canberra ACT 2601, Australia. Email secretary@ausglass.org Web ausglass.org*

**5–9 March**

Optical Fiber Communications Conference and Exhibition (OFC), San Diego, California USA.
*The Optical Society, USA. Email custserv@osa.org Web www.ofcconference.org*

**4–9 June**

2023 Glass and Optical Materials Division Annual Meeting (GOMD 2023), New Orleans, USA.
*The American Ceramic Society, 550 Polaris Parkway, Suite 510, Westerville, OH 43082, USA. Email customerservice@ceramics.org Web www.ceramics.org*

**2–7 July**

18th European Ceramics Society Conference, Lyon, France.

**10–14 July**

Nonlinear Optics, Honolulu, Hawaii, USA.
*The Optical Society, 2010 Massachusetts Ave, NW, Washington, DC 20036 USA. Email custserv@osa.org Web www.osa.org*

**17–21 July**

Tenth International Conference on Borate Glasses, Crystals and Melts and Third International Conference on Phosphate Materials, Corning, USA.
*Dr Randall Youngman, Corning Incorporated Email youngmanre@corning.com Web boratephosphate2020.org/*

**~~August~~ (moving to 2024)**

NCM15, Structure of Non-Crystalline Materials 15, combined with the SGT Annual Meeting, UK.
*Christine Brown, Society of Glass Technology, 9 Churchill Way, Chapeltown, Sheffield S35 2PY, UK. Email christine@sgt.org Web www.sgt.org*

**31 October–2 November**

GlassBuild America, Atlanta, Georgia, USA.
*National Glass Association NGA, 8200 Greensboro Drive, Suite 302, VA 22102 McLean, USA. Email attend@glass.org Web www.glass.org*

**2024**

**June**

Tenth International Conference on Optical, Optoelectronic and Photonic Materials and Applications (ICOOPMA), Pardubice, Czech Republic.
*Center of Materials and Nanotechnology and Department of General and Inorganic Chemistry, Faculty of Chemical Technology, University of Pardubice, Cs. Legii sq. 565, 532 10 Pardubice, Czech Republic. Email info@icoopma.com Web www.icoopma.com*

**2–6 July**

XVIII ECerS Conference, Lyon, France.
*European Ceramic Society AISBL, Av. Gouverneur Cornez 4, B-7000 Mons, Belgium. Email ecers@bcrc.be Web ecers.org/*

**14–19 July**

Tenth Int. Congress on Ceramics (ICC10), Montreal, Canada.
*The American Ceramic Society, 550 Polaris Parkway, Suite 510, Westerville, OH 43082, USA. Email customerservice@ceramics.org Web www.ceramics.org*

**20–23 August**

Materials Challenges in Alternative & Renewable Energy 2023 combined with the Sixth Annual Energy Harvesting Society Meeting (EHS 2023), Bellevue, Washington, USA.
*The American Ceramic Society, 550 Polaris Parkway, Suite 510, Westerville, OH 43082, USA. Email customerservice@ceramics.org Web www.ceramics.org*

*Glass Technol.: Eur. J. Glass Sci. Technol. A*, 2021, **62** (5), 160–167
*Int. Congr. Glass*, Volume 1. Invited Papers, Edinburgh, Scotland, 1–6 July 2001, 259–265

# Effect of natural convection and internal radiation on glass conditioning

*M. R. Hyre & K. Paul*
*Emhart Glass, 123 Day Hill Road, Windsor, CT 06095, USA*

A better understanding of the factors influencing the flow and heat transfer in forehearths can assist in improving the performance of existing installations and is important in any consideration of new and improved designs. Properly conditioning molten glass as it flows from the furnace to the spout requires an integrated forehearth/working end system specialised to provide predictable tonnage pulls and homogeneous glass temperature. One of the most important aspects of the conditioning system performance concerns the delivery of glass at a near uniform temperature. This temperature is dictated by the viscosity requirements of the forming process. Additionally, the conditioning system must deliver the glass at a controllable pull rate to ensure consistent product formation and reduce the likelihood of forming operation interruptions. It is difficult to achieve a completely uniform temperature over the forehearth exit plane, but it is possible to control the heat transfer processes at the boundary of the glass surface along the forehearth length to minimise temperature gradients in the exit profile. This is usually accomplished by a series of burners along the edges of the forehearth, the application of cooling along the forehearth centre and a control system designed to minimise exit plane temperature gradients. The careful design of the insulation package can also aid in minimising thermal nonuniformities.

The heat transfer processes in working ends and forehearths are very complex. Glass conditioning is characterised by banded internal radiation, natural convection and thermal coupling of combustion gas, refractory and glass. Very limited experimental methods are available capable of gathering data which would provide insight into the fundamental mechanisms of heat transfer between the various conditioning subsystems. The methods that do exist are often very expensive because of the high temperatures and corrosive environments the instrumentation must survive. Physical modelling

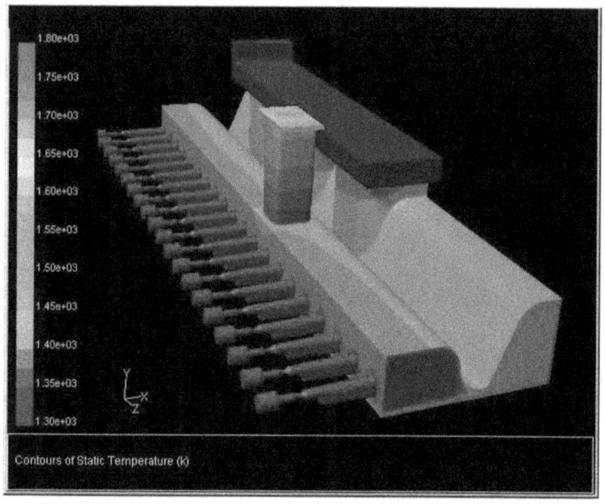


*Figure 1. Combustion space of cooling module [Colour available online]*

using dimensional analysis is also quite difficult. Developing small scale models of a complete forehearth or working ends is expensive and unless the convective and radiant heat transfer processes are also designed for similitude, does not circumvent the harsh environment which makes data acquisition difficult. Numerical simulation of the glass conditioning processes is an attractive line approach. Numerical techniques and computational power is now at the point where nearly all the important phenomena in forehearths and working ends can be accurately modelled. Once detailed mathematical simulations of the conditioning process are developed, determining the influence of process parameters on system performance becomes a relatively simple task.

The objective of this paper is to present a three-dimensional model for flow and heat transfer in working end/forehearth systems similar to those used in the continuous manufacture of glass containers (see Figure 1). In such systems it is important to know the temperature distribution, steady state and transient velocity profiles and pressure losses. The overall strategy for simulating the glass conditioning process consisted of coupling a highly turbulent reacting glass flow model to a model of the creeping glass flow. The

This paper was originally presented as an invited paper at the International Congress on Glass and has not been published elsewhere. This version has reworked figures to aid clarity.
DOI: 10.13036/17533546.62.5.Hyre

gas phase model included both the combustion space above the glass and the airflow associated with cooling the radiation plates located along the centreline of the forehearth. The simulation includes the effects of internal and external radiant effects, steady state and transient vortex rolls induced through natural convection and heat losses through the refractory and insulation surrounding the channels.

## Previous work

Numerical modelling has been used extensively in the glass industry to study flow and heat transfer in furnaces (Hallolin & Wajc;[1] Hoffman;[2] Murnane *et al*;[3] Wang & Zhou;[4] Studzinski *et al*;[5] *Glastech. Ber.*[6]). There has also been significant work in applying these techniques to forehearths (Choudhary;[7] Mercier & Deville;[8] Carling;[9] Lin[10]). The earliest work in this area has been reviewed in detail by McConnell,[11] with later reviews conducted by Carling and Lin. Nearly all early three-dimensional forehearth modelling decoupled the axial velocity components from the transverse components. The transverse velocities and longitudinal vortex rolls were neglected. The remaining two velocity components were then coupled via the continuity equation.

Choudhary[7] developed the first fully coupled three-dimensional forehearth model that included buoyancy driven transverse velocities. These models were used to investigate the effects of natural convection, throughput and radiation in the glass on velocity and temperature distribution throughout the glass delivery system. A radiation diffusion model was used to approximate internal radiation in the glass. With these models it was discovered that, although the natural convection velocities were much lower than the average velocity due to throughput, the presence of natural convection significantly altered the flow path lines in the system. Since Choudhary investigated glass associated with fibreglass production, it is anticipated that the transverse velocity magnitudes will be much larger for the dark green and amber glasses often used in the container glass industry. This is due to the shorter mean free path of radiation in the darker glasses resulting in larger temperature differences in the plane transverse to the direction of glass flow.

None of the models cited above included nongray internal radiation effects in the glass. In most numerical studies, the radiative transport was considered a short range process which can be considered as an isotropic diffusion of heat. This allows the definition of an 'effective' conductivity where

$$\lambda_{eff}=\lambda_{cond}+\lambda_{rad}$$

Radiative and conductive energy fluxes are then expressed by a common diffusive heat flux

$$\vec{q}_{diff}=-\lambda(T)\vec{\nabla}T$$

In addition to this simplification, while limited investigations on coupling the combustion space and glass in furnaces have been performed,[12] almost none of this work has been extended to include working ends and forehearths.

## Model formulation

In order to eliminate many of the assumptions and approximations built into current models, a three-dimensional glass conditioning model was constructed that included the effects of nongray internal radiation, buoyancy driven velocities and also coupled the combustion space and glass. Numerical modelling of such a system is complicated due to the radically different transport time scales in the combustion and glass domains. Fluid mixing and reaction is highly turbulent in the combustion space while the circulation flow and heat transfer in the highly viscous glass is laminar with buoyancy driven flow also playing a role. Because of the problem complexity, previous modelling approaches have often treated the domains separately, neglecting the coupling.

The model developed includes full coupling between the combustion space and glass flow. FLUENT is used to separately simulate the turbulent mixing, reaction and heat transfer in the gas phase and the laminar flow associated with the glass. Transport in the combustion space and glass flow is calculated iteratively, updating the appropriate coupling conditions. The combustion space uses the predicted local glass surface temperature distribution from the glass simulation and the glass relies on the local incident radiative and convective flux predicted by the combustion space model.

## Governing equations

### *Mass and momentum conservation*

In both the laminar glass sub-system and the turbulent combustion space subsystem, the flow was considered to be incompressible, steady and Newtonian. The equation of continuity and momentum are thus.

$$\frac{\partial u_i}{\partial x_i}=0 \tag{1}$$

and

$$\rho\frac{\partial}{\partial x_i}\left(u_i u_j\right)=-\frac{\partial p}{\partial x_i}+\frac{\partial \tau_{ij}}{\partial x_j}+\rho\beta\left(T-T_0\right)g_i \tag{2}$$

respectively. In the above equations, $\rho$ is the mass density, $u_i$ the velocity in the i-direction, $p$ the pressure (including the effects of gravitational head), $\beta$ the volumetric coefficient of expansion, $T$ and $T_0$ the glass temperature and its reference value and $g_i$ the gravitational body force. $\tau_{ij}$ is the viscous stress tensor given by

$$\tau_{ij}=\left[\left(\mu+\mu_t\right)\left(\frac{\partial u_i}{\partial x_j}+\frac{\partial u_j}{\partial x_i}\right)\right]+\frac{2}{3}\left(\mu+\mu_t\right)\frac{\partial u_l}{\partial x_l} \tag{3}$$

where $\mu_t$ is the turbulent viscosity (used in the combustion space only). For the turbulent combustion gas region, the terms representing the Reynolds stresses

$$\frac{\partial}{\partial x_j}\left(-\rho u_i' u_j'\right) \tag{4}$$

must also be added to the right hand side of Equation (2). In order to close the momentum equations, the Boussinesq approximation is used to model the Reynolds stresses and the renormalisation group method is used to represent the turbulent kinetic energy and its dissipation.

### Energy conservation

The energy equation is given by

$$\frac{\partial}{\partial x_i}\left(\rho u_i c_p T\right)=\frac{\partial}{\partial x_i}\left(k\frac{\partial T}{\partial x_i}+u_j\tau_{ij}\right)+\rho g_i \tag{5}$$

where $c_p$ is the specific heat and $k$ the phonon conductivity of the glass. For the combustion gas space, the conductivty is replaced by the effective thermal conductivity that includes the effects of turbulent viscosity.

In most studies the radiative transport is considered as an isotropic diffusion of heat. However, this approximation breaks down near the boundaries (forehearth walls) because the interface radiation interacts with the volume radiation of the semitransparent glass. An increase of temperature gradient near the wall results. This is often referred to as a temperature slip. For clear flint glass this effect can be significant in a forehearth. Since small differences in temperature gradients can lead to large differences in calculated viscosity, the Rosseland approximation was found to be unacceptable.

In this study, radiation effects were calculated through a separate radiation transfer equation:

$$\frac{dI(\vec{r},\vec{s})}{d\vec{s}}+\left(a+\sigma_s\right)I(\vec{r},\vec{s}) \tag{6}$$

$$=an^2\frac{\sigma T^4}{\pi}+\frac{\sigma_s}{4\pi}\int_0^{4\pi} I(\vec{r},\vec{s})\Phi(\vec{s}.\vec{s}')d\Omega$$

In the above equation, $\vec{\rho}$ is the position vector, $\vec{s}$ the direction vector, $\vec{s}$ the path length, $a$ the absorption coefficient, $n$ the refractive index, $\sigma_s$ the scattering coefficient, $\sigma$ the Stefan–Boltzmann constant, $I$ the total radiation intensity, $\Phi$ the phase function and $\Omega'$ the solid angle.

The discrete ordinates radiation model, included with FLUENT, was used to solve the radiative transfer equation for a finite number of discrete solid angles, each associated with a vector direction $\vec{s}$ fixed in the global Cartesian system. The discrete ordinates model solves as many transport equations as there are directions $\vec{s}$. The solution technique is the same as that used for the fluid flow and energy equations. Since the glass is nongray, the discrete ordinates implementation divides the radiation into the five wavelength bands discussed below. The radiation transport equation is integrated over each wavelength band within which the behaviour is assumed gray. The total intensity is then summed over the wavelength intervals. The advantage of using the discrete ordinates model is that it spans the entire range of optical thicknesses and allows for the accurate solution of radiation in semi-transparent media.

## Boundary conditions

A symmetry condition was assumed for the glass surface in the laminar flow model such that:

$$v=\frac{\partial w}{\partial z}=\frac{\partial u}{\partial z}-0 \tag{7}$$

The thermal boundary condition at the glass surface was assumed to be dominated by radiation. In this case:

$$k\left(\frac{\partial T}{\partial y}\right)_{y=H}=q_{rad_{net}} \tag{8}$$

The incident radiation is

$$q_{in}=\int_{\vec{s}.\vec{n}>0} I_{in}\vec{s}.\vec{n}\,d\Omega \tag{9}$$

and the net radiation leaving the surface is given by

$$q_{out}=\left(1-\varepsilon_a\right)q_{in}+n^2\varepsilon_a\sigma\left(T_a^4-T^4\right) \tag{10}$$

In the combustion space model, the glass surface was approximated as a solid wall. The radiant flux arriving from the glass was used as the thermal boundary condition for this wall. The radiant flux arriving at this wall from the combustion gases and roof was subsequently used in the laminar glass model.

### Forehearth/working end and combustion space walls

At the walls, the velocity components are set equal to zero and heat transfer is specified in terms of an overall effective heat transfer coefficient

$$-k\frac{\partial T}{\partial n}=U\left(T_a-T\right) \tag{11}$$

The overall effect heat transfer coefficient, $U$, was determined by constructing a separate FLUENT model of the glass/gas-wall-ambient interface and calculating the overall combined effects of conduction, convection and radiation. The values of $U$ on the walls were input as functions of spatial location.

The radiation transport equation boundary conditions were specified in the same manner as the boundary condition for the glass surface.

### Combustion space inlets

The burners on the forehearth and working end were assumed to operate near a stoichiometry of 1·0. In order to reduce computation time, the combustion gases were assumed to be at equilibrium upon entering the simulation. The chemical composition and equilibrium combustion temperature were calculated using Stanjan, a thermochemical based on minimising Gibbs free energy. The overall simulation can also run with the chemical kinetic calculations included in the combustion space model. This mode of operation will be used in the future for minimising pollutant formation. However, it was found that it did not have a significant impact on the velocity or thermal profiles in either the glass or the combustion space.

## Material properties

### Glass

Within the range of temperatures expected during the present study certain glass properties of the glass vary significantly and need to be taken into account in the numerical model. The viscosity was determined from the well known Fulcher[13] correlation

$$\log_{10} \eta = -A + \frac{B}{T - T_0} \tag{12}$$

where $T_0$ is the reference temperature, $A$ and $B$ the temperature coefficients of viscosity. These parameters were determined using a method developed by Lakatos.[14]

The specific heat of the glass was determined using a correlation developed by Ginther[15]

$$c_m = \frac{aT + c_0}{0{\cdot}00146T + 1} \tag{13}$$

where $a$ and $c_0$ are constants which are a function of glass composition. This method considers glasses as mutual solutions of oxides on the basis of random network theory.

The phonon conductivity of the glass was determined using the curve developed by McGraw[16] and was in the range 2·0 J/s m K for the conditions found in the feeder. An effective conductivity was not used on the models. Instead, the internal radiation was modeled using the discrete ordinates radiation model. For the temperatures involved in gob formation, the radiation of wavelength less than 0·8 μm is negligible (less than 0·1% of the black body radiation at 1150°C). In a manner similar to Gardon[17] the rest of the absorption spectrum was approximated by step functions with constant absorption coefficients for the bands 0·8–1·25, 1·25–2·125, 2·125–2·75, 2·75–4·5 and over 4·3 μm (where the glass is opaque). These bands correspond respectively to the effects of ferrous iron, an intermediate band, water and silica. The approximation is best for glasses with low iron content. The refractive index of the glass was taken as 1·5. Radiation to the surface of the refractory walls was assumed to be gray with an emissivity of 0·35.

The density of the glass was assumed constant at 2500 kg/m$^3$.

### Combustion and cooling gas

The combustion and radiation plate cooling gases were modelled as incompressible ideal gases. The molecular weight of the combustion gas was determined through the chemical equilibrium calculations. The transport properties and specific heats were assumed to be a function of temperature only.

### Refractory/insulation

All solid materials were assumed to have constant density and specific heat. The conductivity of each material was assumed to vary with temperature only.

## Initial model results

The conditioning system that was selected for analysis consisted of one working end with two straight forehearths. The two forehearths had different pull rates and different outlet temperatures. The low pull rate forehearth (100 usTPD) had a mass averaged outlet temperature of about 1390 K and the high pull rate forehearth (140 usTPD) had a mass averaged outlet temperature of about 1425 K. The working end was fed with glass through the throat at a temperature of 1630 K.

Figures 1 and 2 show the temperature contours of

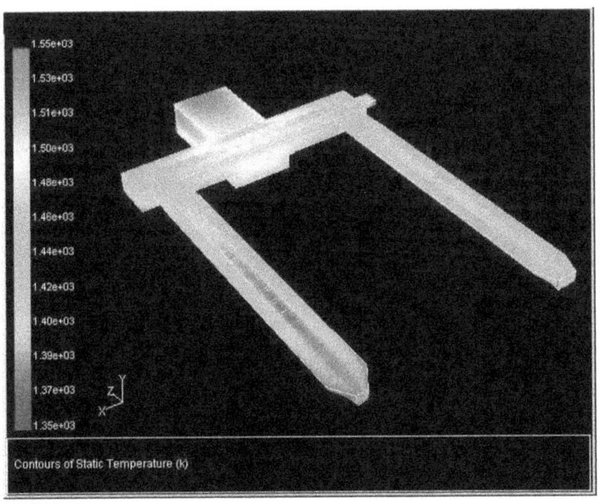


*Figure 2. Model of working end/forehearth glass [Colour available online]*

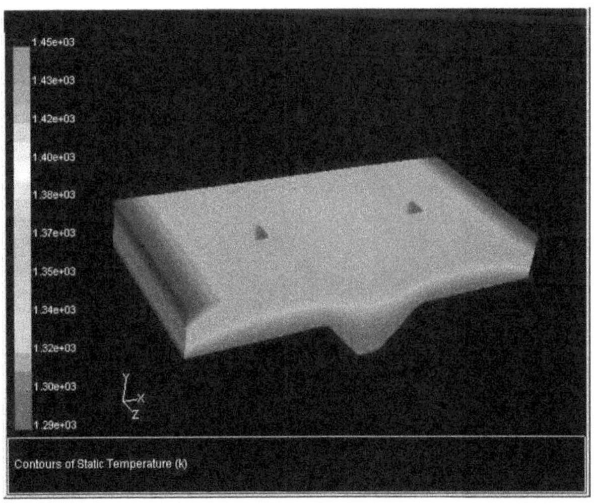


*Figure 3. Equalising section roof block [Colour available online]*

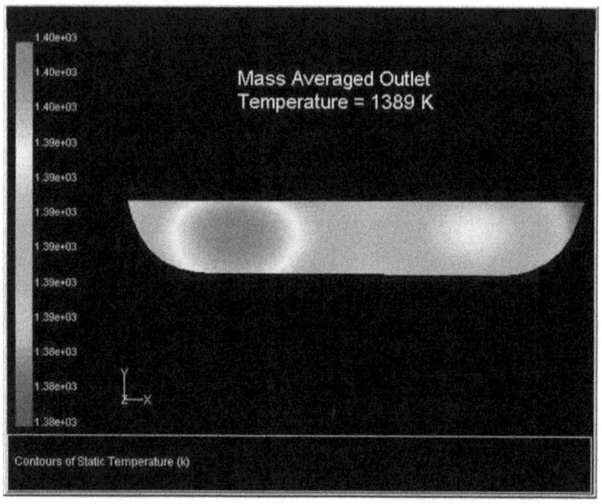


*Figure 4. Outlet plane temperature distribution: no natural convection, no internal radiation [Colour available online]*

the initial glass conditioning system models for the combustion space and the glass phase, respectively. This simulation did not include natural convection or internal radiation in the glass. The combustion space figure shows one cooling module. For the overall conditioning system simulation, several cooling modules were developed for the working end and forehearths. Each cooling module was iteratively linked to the glass phase model. Additionally, combustion space models were developed for the equalisation sections of the forehearths. These modules did not include radiation plates or radiation plate cooling and the roof structure consisted of only one lobe along the centreline of the forehearth (see Figure 3).

## Effects of natural convection on forehearth/working end performance

The effect of natural convection on the final outlet temperature distribution can be seen in Figures 4 and 5. These Figures represent the outlet plane temperature profiles of the low pull rate forehearth. The model associated with Figure 5 included buoyancy driven convection effects. This model resulted in a slightly lower mass averaged outlet temperatures. This is due to the cross stream mixing which results from the vortex rolls bringing hotter glass to the surface. The overall distribution of the temperature is very similar for the two cases. This is because the amount of convective transport is quite small (an order of magnitude less than the streamwise convection) and the overall mode of heat transport (radiation diffusion) has not changed.

## Effects of internal radiation on forehearth/working end performance

The effects of modelling the glass as a semi-transparent nongray medium can be seen by comparing Figures 5 and 6. The model associated with Figure 6 included both buoyancy driven convection cells and internal radiation. The primary effects of including internal radiation in the glass were the smearing of the temperature profile and higher temperature gradients at the forehearth/working end wall. The 'smearing' of the temperature profile is the direct result of the glass internally radiating to itself. The heat transport within the glass is no longer a diffusive process but now allows radiation from one particle of glass to all other particles of glass. This reduces the overall temperature gradients in the glass. The difference in temperature distribution near the walls of the forehearth are the result of the breakdown in the Rosseland (effective conductivity) model at solid boundaries. This is often referred to as tempearture slip.

## Effect of glass colour

If dark green glass was used instead of flint, the natural convection cells increased in intensity (see Table 1). This is due to the higher temperature gradients in the green glass which has a lower mean free path of radiation (lower effective conductivity). Therefore, it is more important to include these effects when modelling

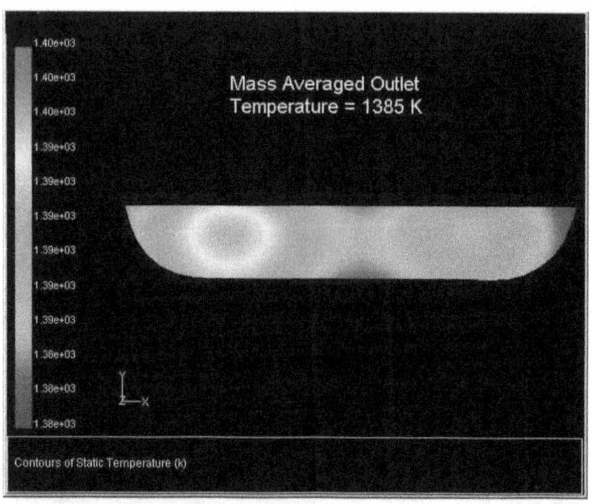


*Figure 5. Outlet plane temperature distribution:natural convection, no internal radiation [Colour available online]*

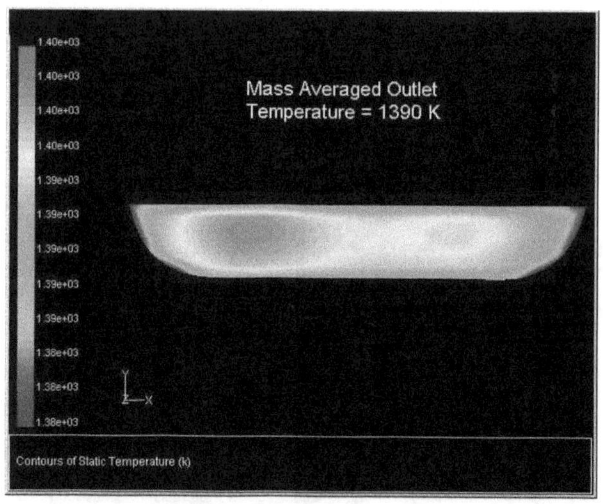


*Figure 6. Outlet plane temperature distribution: natural convection and radiation included [Colour available online]*

*Table 1. Magnitude of cross stream velocities in the working end: direct versus indirect cooling*

| | Flint glass | | Green glass | |
|---|---|---|---|---|
| | *Direct cooling* | *Indirect convective* | *Direct cooling cooling* | *Indirect convective cooling* |
| Max Y-velocity component (m/s) | 0·0008 | 0·0004 | 0·0012 | 0·0005 |
| Max Z-velocity component (m/s) | 0·0015 | 0·0006 | 0·002 | 0·0007 |
| Mass averaged Streamwise velocity (m/s) | 0·0022 | 0·0022 | 0·0022 | 0·0021 |
| Area averaged Streamwise velocity (m/s) | 0·001 | 0·001 | 0·0009 | 0·0009 |

glasses which exhibit larger resistances to heat flow. In contrast, the difference in using the internal radiation and the effective conductivity models in simulating dark green glass resulted in almost no difference in outlet plane temperature profiles. Since the inherent assumption made in the effective conductivity is that the radiation medium is optically thick, the performance is much better in glasses with a small radiation mean free path. As a result, it is probably not necessary to use the dicrete ordinates radiation model when evaluating dark, optically thick glasses.

## Using the models to evaluate two forehearth/working end cooling systems

As a first step it was decided to use the conditioning system model to evaluate two different cooling systems: (1) indirect convective cooling system (shown in Figure 1); and (2) direct radiative cooling system (tempearture contours shown in Figure 7).

The indirect convective cooling system makes use of radiation plates which are placed above the centreline of the glass. The radiation plates are cooled on the back side by blowing air through a small channel in the superstructure refractory above the plate. This reduces the temperature on the glass side of the plate increasing the radiative transfer. The plates and air run parallel to the glass flow in the working end and forehearths.

For purposes of analysis, the direct radiative cooling system has essentially the same roof structure as the indirect system except that the radiation plates and cooling channel have been removed. The glass is now able to radiate directly to the surroundings. This should result in larger heat removal rates than the indirect convective system due to the larger temperature differences between the glass and the surroundings as compared to the glass and radiation plate.

## Natural convection effects

One potential problem with using a direct radiative heat removal system is the development of longitudinal vortex rolls in the working end and forehearths. These vortices are driven by vertical density gradients in the glass (lower density at the bottom, higher at the top). Because the direct radiant system is transferring heat to the surroundings which are at a much lower temperature than the glass, relatively large density gradients can form leading to vortex rolls. This can result in conditioning system oscillations and makes controlling the working end and forehearth more difficult. These types of oscillations, often attributed to the furnace, have been seen in several existing glass container production facilities. Figures 8 and 9 show a comparison of the $y$ and $z$ velocity components of the vortices generated in the working end by using the direct and indirect cooling schemes. Table 1 summarises the magnitude of the velocity components normal to the streamwise direction.

As illustrated in both the Figures and Table 1 the direct radiant cooling configuration results in cross stream velocity components which are on the same order of magnitude as the streamwise velocity. Indirect cooling results in cross stream velocities which are more than a factor of two less than the direct cooling configuration. The difference is especially noticeable in green glass where the high absorption coefficients (low effective conductivity) creates larger temperature and density gradients than in flint glass.

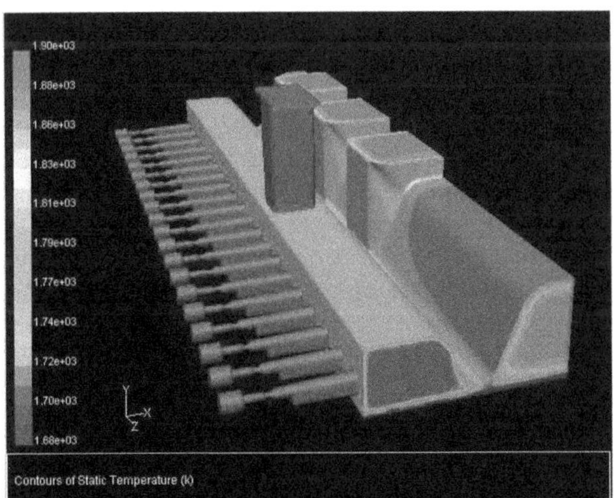


*Figure 7. Direct radiative cooling combustion space model—cooling module [Colour available online]*

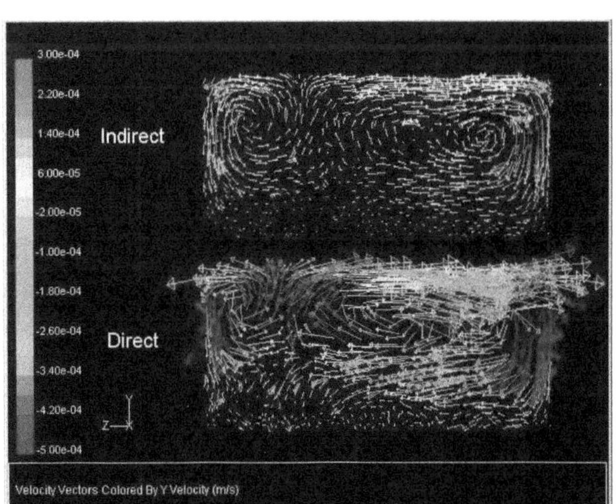


*Figure 8. Vortex cells in the working end: direct vs indirect radiation—green glass [Colour available online]*

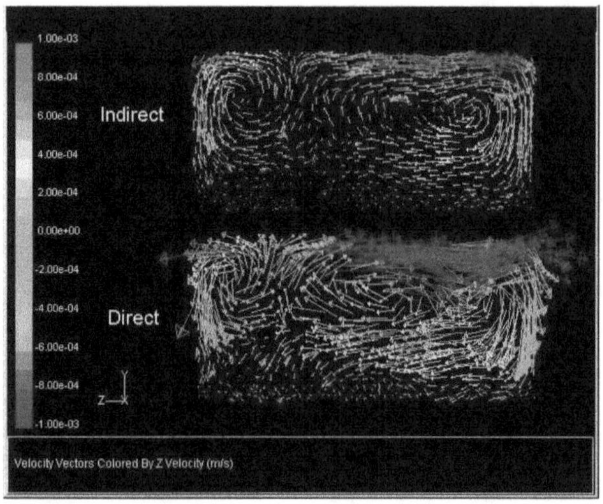


*Figure 9. Vortex cells in the working end direct vs indirect radiation—green glass [Colour available online]*

For green glass the predicted magnitude of the z velocity component is nearly as large as the streamwise component and about three times as large as the z velocity component when indirect cooling is used.

Vortex rolls can also be set up in the forehearths although not to the same extent as the working end. This is primarily because of the smaller characteristic lengths in the forhearth although the smaller rates of cooling also contribute to this phenomenon. Figure 10 shows the $y$ velocity components at the end of the first cooling module in the high pull rate forehearth. This is the most likely location for the onset of the convection cells. The cross stream velocities are of the order of $10^{-6}$ for the indirect cooling and on the order of $10^{-4}$ for the direct cooling. The cells in the indirect cooling case are so small that they would not produce any noticeable difference in the performance or operation of the forehearth. In the case of the direct cooling, however, the velocities are about 10% of the streamwise component. This will change the operation of the forehearth since the glass is no longer remaining in straight paths on the way to the feeder. The relationship between zonal cooling/heating and the forehearth outlet temperature profile will be difficult to establish.

## Effect of internal radiation on predicted flux removals from direct and indirect cooling

Table 2 shows a comparison of the predicted heat flux removal rates from the surface of the glass in

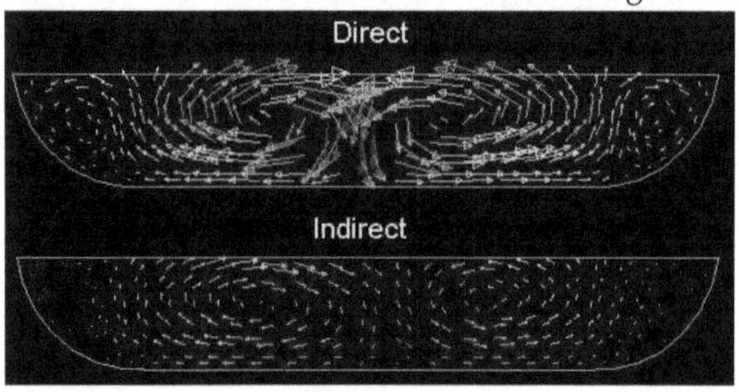


*Figure 10. Vortex rolls in the forehearth: direct vs indirect radiant cooling—green glass [Colour available online]*

**Table 2.** *Heat flux removal rates from working end—flint glass*

| | *Effective conductivity model* | | *DO radiation model* | |
|---|---|---|---|---|
| | *Direct radiant cooling* | *Indirect convective cooling* | *Direct radiant cooling* | *Indirect convective cooling* |
| Dimensionless heat removal rates | 0·125 | 0·060 | 0·075 | 0·043 |
| Theoretical max dimensionless removal rate | 0·284 | 0·071 | 0·284 | 0·071 |
| Dimensionless fractional $\Delta T$ per module | 0·39 | 0·18 | 0·23 | 0·15 |

the working end. In general the indirect cooling configuration removes about half the heat as compared to the direct radiant cooling configuration. When the semi-transparent, non-gray attributes of glass are taken into account by using the discrete ordinates radiation model, the heat removed by direct radiation to the surroundings is about 35% of the theoretical maximum and 60% of the value predicted by the radiation diffusion model. This confirms the results of others who have shown that using standard black/gray body analysis will result in incorrect predictions of heat removal rates from the glass surface.[16–21]

## Conclusions

A three-dimensional model for flow and heat transfer in working end/forehearth systems similar to those used in the continuous manufacture of glass containers has been constructed. In addition to modelling the glass, the simulation includes both the combustion space above the glass and the airflow associated with cooling the radiation plates located along the centreline of the forehearth. Internal and external radiant effects, steady state and transient vortex rolls induced through natural convection and heat losses through the refractory and insulation surrounding the channels are all included in the model.

A preliminary evaluation of using direct radiant heat transfer to the surroundings versus indirect radiation plate cooling of the working end and forehearths has been conducted. In general the direct radiant exchange with the surroundings increases the heat removal rate from the glass by a factor of two. This is much less than the theoretical difference in the two methods of heat removal. The reason for this difference lies in the temperature and wavelength dependency of the emissivity and the increased roof temperature along the centreline of the glass for the direct exchange system. Using a standard black or gray body radiation analysis does not include these effects and can lead to erroneous predictions of heat removal from the glass. The direct radiant exchange configuration also resulted in large vortex rolls in the working end and smaller rolls in the forehearth. These may cause operational problems since they are inherently transient in nature and cloud the connection between the forehearth outlet temperature profile and the conditioning zones. While not a part of this analysis, it should be mentioned that the direct radiant cooling configuration increased overall forehearth drawdown by about 1 cm.

In determining what type of cooling should be utilised in the forehearth and working ends, care must be taken to balance thermal, operational and control needs. Obviously the quickest way to remove heat is to completely remove the working end and forehearth roof but this is not possible because of glass channeling, devitrification and control issues. A balance must be struck between removing heat quickly and the operation and control difficulties that can be introduced through such phenomena as vortex rolls and increased drawdown. Numerical models are now at the point where these effects can be predicted and new designs developed to optimise glass conditioning thermal and control performance.

## References


1. Halloin, V. L. & Wajc, S. J. Physical and numerical simulations of creeping flow and heat transfer in forehearths. *Chem. Eng. Comm.*, 1996, **154**, 59–85.
2. Hoffman, O. R. Boussinesq approximation to compute the temperature and velocity distributions in glass melts. *Glastech. Ber.*, 1992, **65**, 239–45.
3. Murnane, R. A., Johnson, W. W. & Moreland, N. J. The analysis of glass melting processes using three-dimensional finite elements. *Int. J. Num. Methods Fluids*, 1988, **8**, 1491–511.
4. Wang, J. & Zhou, Z. Investigation into glass tank geometries by means of a mathematical model. *Glastech. Ber.*, 1992, **65**, 1–8.
5. Studzinski, J. Mathematical modelling of glass mass flow in a conventional glass tank furnace using isotope data. *Glastech. Ber.*, 1991, **64**, 169–78.
6. Papers presented at Session 1: Modeling of glass melting processes. *Proc. Second Int. Conf. on Advances in the fusion and processing of glass, Glastech. Ber.*, 1990, **63**(K), 1–102.
7. Choudhary, M. Mathematical modeling and flow and heat transfer phenomena in glass furnace channels and forehearths. *J. Am. Ceram. Soc.*, 1991, **74**, 3091–9.
8. Mercier, P. & Deville, M. Three-dimensional numerical study of glass flow in a forehearth. *Glastech. Ber.*, 1979, **52**, 131–40.
9. Carling, J. C. A reappraisal of the mathematical modeling of flow and heat transfer in glass tank forehearths. *Glass Technol.*, 1982, **23**, 201–22.
10. Lin, Y. S. Numerical simulation of unsteady, three-dimensional fluid flow and heat transfer ocurring in open channel forehearths. *PhD Thesis*, University of Toledo, OH. 1985.
11. McConnell, R. Modeling glass furnace, refiner and forehearth design and operation for energy efficiency and improved control. *PhD Thesis*, Purdue University, West Lafayette, IN. 1976.
12. Wang, L.-P., Wexler, A. S. & Zhou, Y. 2000 Statistical mechanical description and modelling of turbulent collision of inertial particles. *J. Fluid Mech.*, 2000, **415**, 117–153.
13. Fulcher, G. S. Analysis of recent measurements of the viscosity of glasses. *J. Am. Ceram. Soc.*, 1925, **8**, 339–55.
14. Lakatos, T. Viscosity-temperature relations in glasses composed of $SiO_2$-$Al_2O_3$-$Na_2O$-$K_2O$-$Li_2O$-CaO-MgO-BaO-ZnO-PbO-$B_2O_3$. *Glastek. Tidskr.*, 1976, **31**, 51–4.
15. Sharp, D. E. & Ginther, L. B. Effect of Composition and Temperature on the Specific Heat of Glass, *J. Am. Ceram. Soc.*, 1951, **34**(9), 260–271.
16. McGraw, D. A. Transfer of Heat in Glass During Forming, *J. Am. Ceram. Soc.*, 1961, **44**(7), 353–363.
17. Gardon, R. Calculation of temperature fistributions in glass plates undergoing heat treatment. *J. Am. Ceram. Soc.*, 1957, **41**, 200–9.
18. Glicksman, L. R. Massachusetts Institute of Technology. *Private communication*. 1999,
19. Roychowdhury, A. P. & Srinivasan, J. The modelling of radiation heat transfer in forehearths units in glass melting. *Warme-und Stoffubertrangung*, 1994, **30**, 71–5.
20. Belentepe, Y. Radiation Heat Transfer in an Absorbing Medium, as Applied to a Glass Forehearth. *AIChE Symp. Ser., Heat Transfer*, 1984, pp. 303–308.
21. Lee, K. H. & Viskanta, R. Comparison of the Diffusion Approximation and the Discrete Ordinates Method for the Investigation of Heat Transfer in Glass, *Glastech. Ber.*, 1999, **72**, 254–265.

Glass Technol.: Eur. J. Glass Sci. Technol. A, 2021, **62** (5), 170–174
Int. Congr. Glass, Volume 1. Invited Papers, Edinburgh, Scotland, 1–6 July 2001, 292–295

# Water in natural glasses

*V. Bykov & V. Anfilogov*
*Institute of Mineralogy, Miass, 456317, Russia*

Natural glasses usually are a product of the quenching of magmatic melts at the volcanic processes. Rapidly cooled lava results in volcanic glass units whose thickness depends upon chemical composition of the magma and its rate of cooling. Water is the most important volatile in magmas both in terms of its abundance and in its influence on physical and chemical melts properties. Water is present in nearly all terrestrial magmas and, for example, rhyolitic magmas usually contain ~4 wt% $H_2O$.[1] Water contents are decreased as magmas ascend because the solubility of water decreases with the decreasing of pressure. However, if the activity of water in the wall rock is greater than in the magma, water may by gained from surrounding wall rock.

Basaltic glasses are common on the sea floor. Basaltic magmas contain about 45 wt% $SiO_2$ and their low viscosity favours crystallisation. A glassy thin outer rim (rarely more than 1 cm thick) is formed by rapid quenching when hot magma contacts sea water. As a rule fresh glassy basalts contain up to 0·5 wt% $H_2O$. The small thickness of basaltic glasses, existing in contact with sea water, results in the disappearance of glasses in times as short as a few thousands to tens of thousands of years.[2] In contrast, highly polymerised rhyolitic magmas containing more than 70 wt% $SiO_2$ are extremely viscous and, even when erupted on dry land where the cooling rate is low, yield layers of obsidian up to tens of metres thick. Obsidians are non-vesicular rhyolitic glasses from lava flows and domes. Dissolved $H_2O$ concentrations in obsidians range up to 1 wt%.[3] It is known[4] that rhyolitic melts are able to preserve water to a larger degree than basaltic melts at the ascent of magma. Rhyolitic glasses with water contents more than 1 wt% are called perlites or pitchstones[5] These glasses are commonly judged to have undergone post eruptive hydration. Perlites are widely used at the production on a commercial scale of light poriferous materials which are produced by heating and degassing the glasses.[6]

Near infrared spectroscopy is the most powerful method for quantitative investigation of water species in glasses. Early infrared studies had indicated the presence of molecular water and hydroxyl groups as dissolving species in water bearing silicate glasses.[7–10] The technique has been refined and developed by Stolper and coworkers in a series of studies of water incorporation in a wide range of synthetic and natural melts and glasses of geological interest, including obsidians.[11–14] It was shown that water is dissolved as only hydroxyl groups at total water contents below about 1 wt%. For the higher water contents both hydroxyl groups and molecular water were found to be present in experimentally hydrous glasses.

## Experimental methods

Twelve volcanic water bearing natural rhyolitic glasses were studied by near infrared spectroscopy. Water contents ranged from 0·3 to 8·4 wt%. Composi-

This paper was originally presented as an invited paper at the International Congress on Glass and has not been published elsewhere. This version has reworked figures to aid clarity.
DOI: 10.13036/17533546.62.5.Bykov

*Table 1. Analyses and localities of glass samples*

| *Sample* | *$SiO_2$* | *$TiO_2$* | *$Al_2O_3$* | *$Fe_2O_3$* | *FeO* | *MgO* | *MnO* | *CaO* | *$Na_2O$* | *$K_2O$* | *$H_2O^+$* | *$H_2O^-$* | *Total* |
|---|---|---|---|---|---|---|---|---|---|---|---|---|---|
| PR-1 | 75·36 | 0·14 | 13·57 | 0·98 | – | 0·22 | 0·06 | 0·81 | 3·96 | 4·67 | 0·34 | 0·24 | 100·25 |
| O-77 | 74·17 | 0·13 | 14·14 | 0·31 | 0·47 | 0·09 | 0·05 | 0·69 | 4·12 | 4·01 | 0·42 | 0·05 | 99·06 |
| O-907 | 75·31 | 0·09 | 13·67 | 0·92 | 0·19 | 0·08 | 0·05 | 0·98 | 4·17 | 4·25 | 0·50 | 0·10 | 100·21 |
| O-399 | 76·03 | 0·29 | 12·63 | 0·59 | 0·25 | 0·10 | 0·06 | 0·77 | 3·99 | 4·26 | 1·27 | 0·10 | 99·77 |
| TSK-1 | 73·20 | 0·07 | 10·07 | 1·46 | 0·52 | – | 0·10 | 0·62 | 6·66 | 3·40 | 3·88 | 0·36 | 100·34 |
| P-907 | 73·26 | 0·10 | 13·24 | 0·74 | 0·12 | – | 0·05 | 0·84 | 3·47 | 4·45 | 4·20 | – | 100·47 |
| E-200 | 73·20 | 0·07 | 10·07 | 1·46 | 0·52 | – | 0·10 | 0·62 | 6·66 | 3·40 | 3·88 | 0·36 | 100·34 |
| DK-1 | 71·26 | 0·10 | 11·40 | 1·24 | 0·46 | 0·02 | 0·06 | 2·20 | 3·59 | 2·03 | 5·83 | 1·67 | 99·86 |
| MR-1025 | 69·06 | 0·28 | 13·72 | 1·00 | 0·35 | 0·25 | – | 0·85 | 3·99 | 3·95 | 6·17 | 1·37 | 100·00 |
| E-64 | 71·97 | 0·11 | 10·34 | 0·41 | – | 0·01 | 0·38 | 3·34 | 4·13 | 6·38 | 0·85 | 97·92 | |
| SN-1 | 75·36 | 0·14 | 13·57 | 0·98 | – | 0·22 | 0·06 | 0·81 | 3·96 | 4·67 | 8·08 | 1·60 | 99·45 |
| NG-1 | 71·1 | 0·05 | 11·07 | 0·67 | – | 0·15 | 0·03 | 1·09 | 4·00 | 1·60 | 8·37 | 1·53 | 99·96 |

*Localities of samples*: PR-1 obsidian - dome, Paravanskoe, Georgia; O-77, O-907, O-399 obsidians - flow, Ochotskoe, Chabarovskii region, Russia; P-907 perlite - flow, Ochotskoe, Chabarovskii region, Russia; TSK-1 perlite - Tashkeskenskoe, Taajikistan; E-200 perlite - dike, Magadanskii region, Russia; DK-1 perlite - dike, Magadan region, Russia; MR-1025 perlite - flow, Muchor-Talinskoe, Burijtia, Russia; E-64 perlite - dike, Darasoon, Zabaikalie, Russia; SN-1 perlite, Sosnovskoe, Zabaikalie, Russia; NG-1 perlite - dike, Negdankovskoe, Primorskii region, Russia tions and localities of the sample are listed in Table 1.

Transmission spectra were taken at wavenumbers of 4000–6000 $cm^{-1}$ on doubly polished plates of glass. Samples were prepared by grinding and polishing with alumina abrasive using kerosene as a lubricant and with diamond paste. A Specord-61NIR spectrophotometer was used to obtain spectra. Spectra were recorded in a linear absorbance mode versus wavenumber on chart paper. Spectra were simultaneously collected with computer interfaced to the spectrometer.

## Experimental results and discussions

Figure 1 shows typical near infrared spectra of obsidians and perlites. Absorption peaks are observed at 5230 $cm^{-1}$ (1·91 μm) and 4500 $cm^{-1}$ (2·22 μm). The background is due to the high energy tail of fundamental OH stretch peak and the iron absorption peak. The absorption at 5230 $cm^{-1}$ is produced by the combination stretching+bending mode of molecular water groups. The absorption at 4500 $cm^{-1}$ is attributed to the combination stretching+bending mode of Si(Al)–OH groups.[11,15] PR-1 samples, with 0·34 wt% total water, do not show the absorption at 5230 $cm^{-1}$ due to molecular water. However, O-399 obsidians, with 1·27 wt% total water display this peak. Peak at 5230 $cm^{-1}$ are the dominant in perlites spectra.

It is possible to determine the relative proportions of molecular water and hydroxyl groups in natural glasses by using peaks at 5230 and at 4500 $cm^{-1}$. The Beer–Lambert law is the basis for a quantitative relationship between water species concentration and the height of the absorption peak

$$C_{OH(H_2O)} = \frac{18A^{4500(5230)}}{d\rho\varepsilon_{(5230)}}$$

where C is the OH ($H_2O$) concentration expressed as a weight fraction, 18 the molecular weight of water, $A$ the height of the absorption peak, $d$ the thickness of the sample in cm, $\varepsilon$ the molar absorptivity in l/mol cm and $\rho$ the density of sample in g/l.

Once the molar absorptivities corresponding to these absorption peaks are determined then the concentrations of molecular water and hydroxyl groups can be determined quantitatively from the heights of these two peaks. The molar absorptivities for the molecular water band at 5230 $cm^{-1}$ ($\varepsilon_{5230}$=1·61) and for hydroxyl groups band at 4500 $cm^{-1}$ ($\varepsilon_{4500}$=1·73) were determined on the basis of investigation a wide range of experimentally hydrated rhyolitic glasses.[16]

Concentration of water species in the natural glasses in this study were determined on the base of equations

$$\frac{C_{OH}}{C_{H_2O}} = \frac{\varepsilon_{5230}}{\varepsilon_{4500}} \cdot \frac{A_{4500}}{A_{5230}}$$

$$C_{OH}+C_{H_2Omol}=C_{H_2O_{tot}}$$

It was suggested that relationship $\varepsilon_{5230}/\varepsilon_{4500}$ in natural glasses is the same as the one in experimentally hydrated rhyolitic glasses. The total water contents have been determined by the weight loss on ignition technique. Perlites are heterogeneous glasses to some degree, therefore, such a relative method is more adequate in this case than a method of absolute determination of concentrations of hydroxyl groups, of molecular water and of total water in the aluminosilicate glasses used by Stolper & coworkers.[11–13,17]

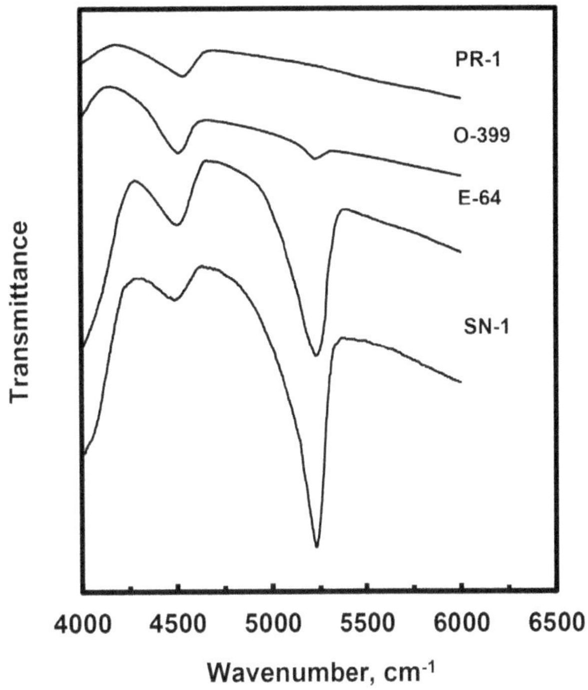


*Figure 1. Near infrared spectra of volcanic glasses*

Table 2 lists, for each sample, the measured values of total water content and water species concentration. The only hydroxyl groups are present in PR-1, O-77, O-907 samples with the total water concentration less than 0·5 wt%. A little amount of molecular water is detected in O-399 obsidians with 1·17 wt% $H_2O$. The total water contents ranged from 3·9 to 8·4 wt% in perlites and in the main water dissolves as molecular water although hydroxyl groups are present also. Spectrum of typical DK-1 perlite with 5·83 wt% total water has been obtained at liquid nitrogen temperatures in order to verify that the molecular

*Table 2. Concentrations of total water, hydroxyl groups, molecular water, primary and secondary water, degree of secondary hydration in the volcanic glasses*

| *Samples* | *$H_2O$ total (wt%)* | *OH (wt%)* | *$H_2O$ mol (wt%)* | *$H_2O$ prim (wt%)* | *$H_2O$ second (wt%)* | *n (%)* |
|---|---|---|---|---|---|---|
| PR-1 | 0·34 | 0·3 | 0 | 0·3 | 0 | 0 |
| O-77 | 0·42 | 0·4 | 0 | 0·4 | 0 | 0 |
| O-907 | 0·50 | 0·5 | 0 | 0·5 | 0 | 0 |
| O-399 | 1·27 | 1·0 | 0·3 | 1·3 | 0 | 0 |
| TSK-1 | 3·88 | 0·3 | 3·6 | 0·3 | 3·6 | 92 |
| P-907 | 4·20 | 1·1 | 3·1 | 1·6 | 2·6 | 62 |
| E-200 | 5·68 | 0·7 | 5·0 | 0·8 | 4·9 | 86 |
| DK-1 | 5·83 | 1·0 | 4·9 | 1·2 | 4·6 | 79 |
| MR-1025 | 6·17 | 1·4 | 4·8 | 2·8 | 3·4 | 55 |
| E-64 | 6·37 | 1·6 | 4·7 | 4·2 | 2·2 | 34 |
| SN-1 | 8·08 | 0·7 | 7·4 | 0·8 | 7·3 | 90 |
| NG-1 | 8·37 | 1·2 | 7·2 | 2·0 | 6·4 | 76 |

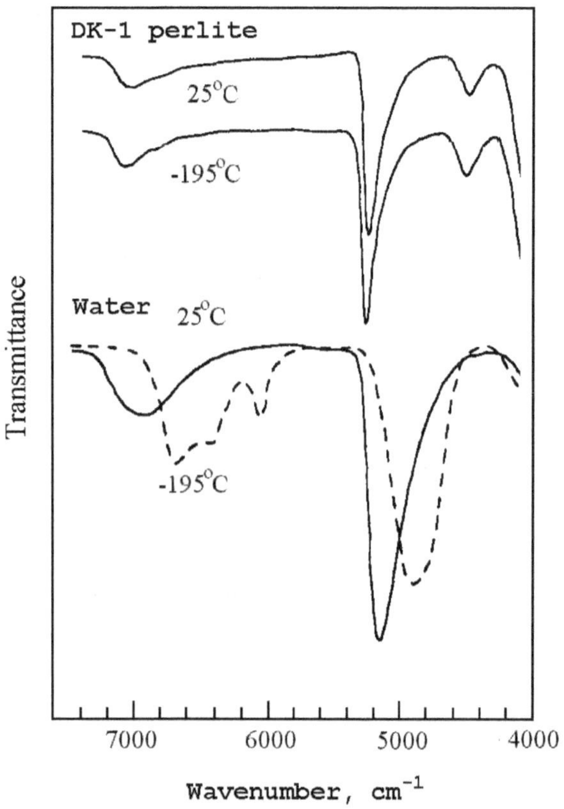


*Figure 2. Near infrared spectra of DK-1 sample and of water at room temperature and liquid nitrogen temperature. Spectra of water are from Ref. 12*

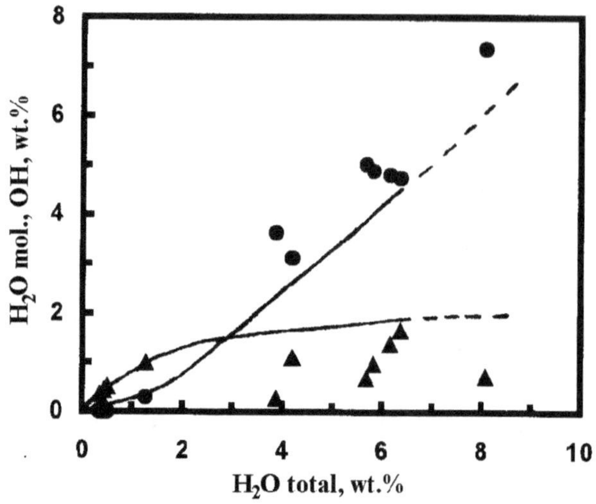


*Figure 3. Measured concentration of molecular water groups (circles) and of hydroxyl groups (triangles) versus total water content in the natural volcanic glasses. The curves show concentrations of these species in experimentally hydrated rhyolitic glasses formed by slow quenching from melts in the air*[17]

water detected spectroscopically is not simply submicroscopic fluid inclusion water. It should be noted that all samples of perlites were free of visible bubbles. Figure 2 shows that spectra of sample at room temperature and low temperature are identical. Ice bonds are not observed in the spectrum of glass at liquid nitrogen temperature. It means that molecular water is stucturally bound in perlites and not present in fluid inclusions as a separate phase.

Figure 3 shows the contents of molecular water and hydroxyl groups for rhyolitic glasses. The curves demonstrate concentrations of water species for experimentally hydrated glasses formed by slow quenching from melts in the air.[17] Hydroxyl groups are the dominant species of dissolved water at low total water contents. At a total water content of about 3 wt% equal amounts of water are present as hydroxyl groups and as molecular water. At higher total water contents it is dissolved mostly as molecular water and concentration of hydroxyl groups increases slightly.

It was proposed that the speciation of water in hydrated silicate glasses formed by quenching from melt reflects speciation in the melts.[11,13] However, water species were studied in aluminosilicate melts at high temperatures and pressure by *in situ* near infrared spectroscopy.[18,19] It was shown that the speciation of water strongly varies with temperature. Concentration of hydroxyl groups is significantly higher while the concentration of molecular water is lower in the melts than ones in the glasses at room temperature. Thus water species concentrations in the glasses reflect speciation of water at the glass transformation temperature. In natural glasses this speciation is 'frozen' at their formation and it characterises primary glasses formed at the cooling of magmatic melts.

Figure 3 shows measured concentrations of molecular water and hydroxyl groups in the studied natural glasses. There is a correlation between speciation of water in obsidians and the speciation in experimentally hydrated glasses formed by quenching from melts. It is in agreement with conventional assumptions that obsidians are primary glasses. Concentrations of molecular water are higher in all perlites than ones in experimentally hydrated glasses. It is proposed that these glasses have undergone post eruptive hydration. In this low temperature process water dissolves as molecular water without the formation of hydroxyl groups. Consequently concentration of hydroxyl groups characterises speciation of water at the formation of perlites from melts.

Process of dehydration hydration of E-200 perlite at a temperature of 200°C has been studied in order to confirm this assumption. Near infrared spectrum of sample after the heating in air at a temperature of 200°C for 2 h shows that the concentration of molecular water has been decreased on 2·8 wt% while the concentration of hydroxyl groups has not been changed (Figure. 4). Subsequent hydration of this glass at the temperature of 200°C and at the pressure 1000 bar restore concentration of molecular water. Contents of hydroxyl groups are not changed at the hydration as well as that at the dehydration. Early studies of low temperature dehydration hydration of perlites showed that about 3 wt% water can be dissolved at a temperature of 200°C and at the pressure 15·4 bar.[20] It was established that this process is reversible and the cycle of dehydration hydration can

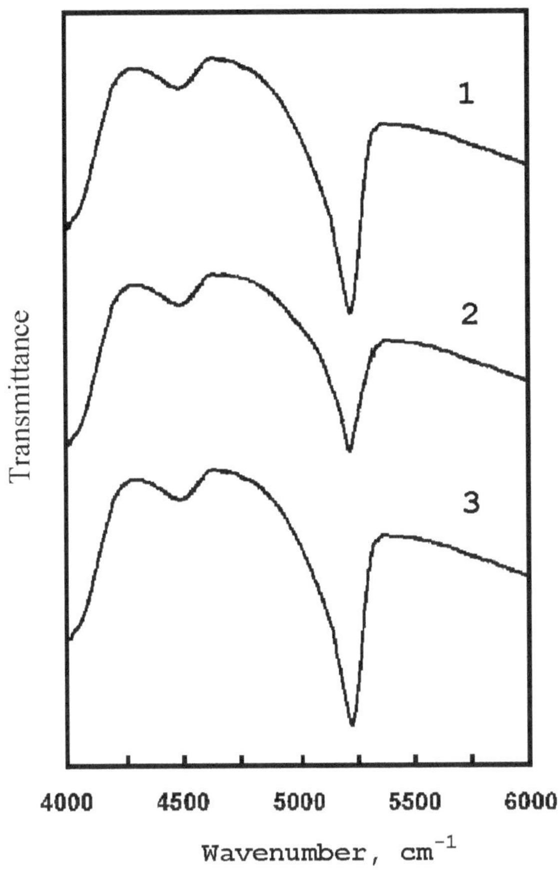


*Figure 4. Near infrared spectra of E-200 perlite: (1) initial perlite; (2) dehydrated perlite (200°C); (3) hydrated perlite (200°C, 1000 bar)*

be repeated many times without destruction of the glassesn. A NMR study of volcanic glasses including perlites has been carried out.[21] Three species of water have been established: hydroxyl groups, structurally bonded molecular water and movable molecular water. A movable water is incorporated in the smallest canals existing in perlitic structure of glasses and it is easily lost and gained by glasses at the processes dehydration and hydration.

Thus hydroxyl groups and structurally bonded molecular water are primary water dissolved in melts at the temperature higher than the glass transformation temperature. Secondary movable water was dissolved as molecular water in during post eruptive hydration of volcanic glasses. Concentration of hydroxyl groups is not changed at this low temperature process and it can be used to determine the amount of primary 'high temperature' water in volcanic water according to dependence of water species on total water content for experimentally hydrated rhyolitic glasses (Figure 3).

Table 2, for each sample lists the calculated values of concentrations primary structurally bonded water and secondary moveable water. Besides, the degree of the secondary hydration of glasses calculated as relationship of secondary water content to total water content are present. Figure 4 shows the histogram of the primary and the secondary water contents

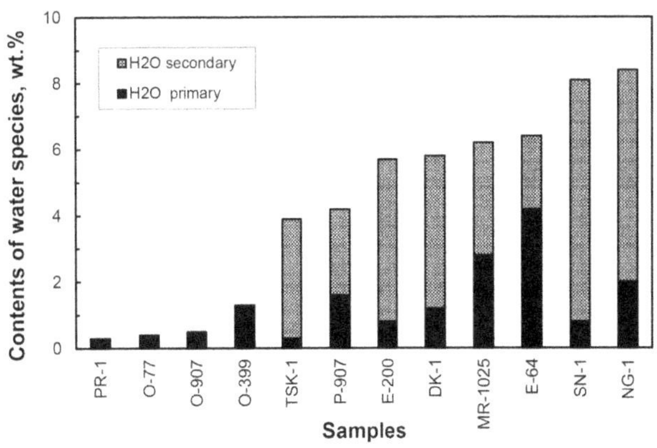


*Figure 5. Histogram of the primary and the secondary water contents for the volcanic glasses*

for volcanic glasses. Thus all studied perlites are secondary hydrated glasses in a varying degree. Secondary water contents range from 2·2 wt% (E-64 sample) to 7·3 wt% (SN-1 sample). Primary water contents are less than 1 wt% for TSK-1, E-200, CN-1 samples. Consequently, typical obsidians have undergone post eruptive hydration in this case. The degree of secondary hydration of obsidians mount to 92% in TSK-1 sample. Primary water contents are more than 2 wt% for NG-1, MR-1025, E-64 samples. Obviously these glasses are the product of quenching of magmatic melts with high water contents. Such glasses are called by the primary perlites. Doubtless these primary perlites have undergone secondary low temperature hydration during further evolution also. The degree of secondary hydration of perlites range from 34% (E-64 sample) to 76% (NG-1 sample).

Thus infrared spectroscopy can be used to determine quantitatively the concentrations of hydroxyl groups, of molecular water, of the primary and the secondary water in natural glasses. On this basis it is possible to draw some conclusions about genesis and evolution of volcanic glasses as well as about water contents in magmas. Besides it is impotant in order to elucidate glasses heat treatment conditions at the industrial production of poriferous materials from perlites.

## Conclusions

Hydroxyl groups are the dominant species of dissolved water in obsidians while water dissolves as both molecular water and hydroxyl groups in perlites. Concentrations of molecular water in the perlites are higher than ones in experimentally hydrated rhyolitic glasses formed by quenching from melts. It is due to post eruptive hydration of these glasses. Concentrations of hydroxyl groups are not changed at this low temperature process and reflect water speciation at the glasses formation from magmas. Concentrations of hydroxyl groups have been used to determine contents of the primary and secondary water in volcanic glasses.

## Acknowledgements

This research is supported by the RFFI grant no. 01-05-96426.

## References

1. Johnson, M. C., Anderson, A. T. & Rutherford, M. J. *Rev. Mineral.*, 1994, **30**, 281.
2. Yoder, H. S. & Tilley, C. E. *J. Petrology*, 1962, **3**, 342.
3. Freidaman, I. & Long, W. *J. Non-Cryst. Solids*, 1984, **67**, 127.
4. Khitarov, N. I. *Geokhimiya*, 1960, **7**, 586.
5. Epelbaum, M. B., Salova, T. P., Zavelskii, B. O. & Janev, I. *Water in volcanic glasses*. 1991. IEM Press, Chernogolovka.
6. Nasedkin, V. V. In *Perlites*. 1981. Edited by V.V. Nasedkin. Science Press, Moscow. Pp 17–42.
7. Keller, W. D. & Pichett, E. E. *Amer. Sci.*, 1954, **252**, 87.
8. Orlova, G. P. *Int. Geol. Rev.*, 1962, **6**, 254.
9. Ostrovskiy, I. A., Orlova, G. P. & Rudnitskaya, Y. S. *Dokl. Akad. Nauk SSR*, 1964, **157**, 149.
10. Bartholomew, R. F., Butler, B. L., Hoover, H. L. & Wu, C. K. *J. Am. Ceram. Soc.*, 1980, **9&10**, 481.
11. Stolper, E. M. *Geochim. Cosmochim. Acta*, 1982, **46**, 2609.
12. Stolper, E. M. *Contrib. Mineral. Petrol.*, 1982, **81**, 1.
13. Silver, L. A. & Stolper, E. M. *J. Geol.*, 1985, **93**, 161.
14. Zhang, Y., Stolper, E. M. & Wasserburg, G. J. *Geochem. Cosmochim. Acta*, 1991, **55**, 441.
15. McMillan, P. F. *Rev. Mineral.*, 1994, **30**, 131.
16. Newman, S., Stolper, E. M. & Epstein S. *Amer. Mineral.*, 1986, **71**, 1527.
17. Silver, L. A., Ihinger, P. D. & Stolper, E. *Contrib. Mineral. Petrol.*, 1990, **104**, 142.
18. Nowak M. & Behrens, H. *Geochim. Cosmochim. Acta.*, 1995, **59**, 3445.
19. Shen, A. & Keppler, H. *Amer. Mineral.*, 1995, **80**, 1335.
20. Nasedkin, V. V. *Water-bearing volcanic glasses, geneses and alteration*. 1963. Academic Press, Moscow.
21. Bondar, A. M. & Mikhailova, N. M. In *Perlites*. 1981. Edited by V. V. Nasedkin. Science Press, Moscow. Pp 153–65.

# International Year of Glass 2022

**The opening ceremony of the International Yeaar of Glass will take place at the Palace of Nations at the United Nations in Geneva, Switzerland on February 10–11 2022.**

The day will begin with an opening address from General Secretary United Nations António Guterres High Representative for the United Nations Alliance of Civilizations, M. A. Moratinos the Spanish Ambassador at UN Spanish Mission in New York along with Turkish and Egyptian UN representatives and Alicia Durán Glass and Teresa Medici.

Formal presentations start by placing glass in history; Ian Freestone, Professor of Archaeological Materials and Technologies, University College London, looks at "How it all began: The invention and re-invention of glass in the ancient world." He is followed by Dedo von Kerssenbrock-Krosigk, Director of the Glasmuseum Hentrich, Düsseldorf, speaking on "Glass: A History of Meaning."

James Carpenter, of James Carpenter Design Associates, New York, looks at the built environment with his talk "Light in the Public Realm." A theme continued by Emmanuelle Gouillart, Scientific Director, St. Gobain Research, Paris, with "Glass for Sustainable Construction."

Glass contributes to optics and photonic technologies, Kathleen A. Richardson, Professor of Optics, Materials Science and Engineering at CREOL/College of Optics and Photonics at the University of Central Florida, has a breadth of experience developing glasses for the communications and computing applications in her talk "Seeing the Future: Enabled by Infrared Glass Innovation." Naoki Sugimoto, General Manager, Materials Integration Laboratories & Technology, AGC, Yokohama, explains the role of glass in 5G communication technology; while Masashi Onishi, General Manager Optical Fiber and Cable at Sumitomo Electric Industries, Yokohama, describes the latest optical fibre technologies. Glass and quantum developments are at the very edge of delivering new processing power according to Lukas Novotny, Professor of Photonics, ETH Zurich.

Andy McConnell, a UK based journalist specialises in antique and vintage glassware, brings the vessel back into focus in his talk "Making glass visible." Andrea de Camargo, Professor of Physics, University of São Paulo, gives an insight into glass science in Brazil. The influence of glass on shaping our lives is the focus of the lecture by Courtney Calahoo of Genics and University of Alberta, Edmonton.

Frederik Kotz-Helmer of Glassomer, Freiburg, explains the role of fused silica glass in high-resolution 3D printing and injection moulding. Steve Jung of Mo-Si Corporation, Rolla, Missouri, reflects on the roles glass has played in healing. Leonid Glebov, Director of R&D, OptiGrate, Oviedo, Florida, explains the role of photo-thermo-refractive glasses in advanced laser applications. Samuel Poulain, CEO of Le Verre

*Palace of Nations at the United Nations Geneva is the venue of the opening conference of the International Year of Glass 2022*

Fluoré, Bruz, describes how fluoride glass technologies are meeting the XXIst century's challenges.

The first day of the opening ceremony finishes with a talk by Falko Langenhorst, Professor of Mineralogy, University of Jena, on glass in space.

Friday, February 11, marks the second day of the International Year of Glass Opening Ceremony with the first session covering sustainable development goals. The case of life science and energy is first expressed by Frank Heinricht, CEO, Schott, Mainz, with glass – an enabler in life science and pharma. Professor Julian Jones of Imperial College, London thene xplains the use of bioglass for regenerative medicine. Peng Shou, CEO, China Triumph International Engineering, Shanghai, looks at the building of solar power generation and it's potential across the globe. Ilkay Sökmen, Glass Technologies Director, Şişecam, Istanbul, describes how "Glass Creates Value."

A sustainability framework and the role of science and technology will be explained by Florian Kongoli of FLOGEN Technologies, Quebec. Vitaliano Torno, President Business Operations, Owens-Illinois & President of FEVE – the European Federation for Container Glass looks at the role of sustainable glass packaging in the circular economy, while Philippe Bastien, Chairman, Glass for Europe & Regional President, AGC Glass Europe, Belgium covers sustainability in the flat glass sector: a solid track record to transform an industry vision into reality.

Srikanth Sastry of the Jawaharlal Nehru Center for Advanced Scientific Research, Department of Science and Technology, Government of India will look at "Understanding glass formers and glasses: Survey of recent progress: India and beyond." Edgar Zanotto, Professor, Federal University of Sao Carlos covers "Education in Glass Art and Science - challenges of transdisciplinarity." Márcia Vilarigues, Professor, Universidade NOVA de Lisboa & Director of the Research Unit Glass and Ceramic for the Arts, Lisboa speaks on "Education in Glass Art and Science - challenges of transdisciplinarity." Corinne Claireaux, Academy Manager, CelSian, Eindhoven will cover "Empowering the industry through education and collaboration."

The age of glass proposed by Corning began the initiative for the International Year of Glass, Jeffrey Evenson the company CSO describes the strides being made in the Glass Age.

A panel discussion and closing presentation by Japanese artist Kimiake Higuchi on the nature of pâte de verre completes the formal proceedings of the opening ceremony.

The proceedings of the meeting will be webcast live and recordings will be available to view on online video platforms.

**Some Planned UK Events**

Contemporary Glass Society (CGS) will hold a series of monthly talks in London plus online.

Exhibition: Henry Moore Institute, Leeds with National Glass Centre and Corning Museum of Glass, from February 2022.

Illumination of Catcliffe Glasshouse Cone, Rotherham will happen in March.

Exhibition: Its all in the technique (CGS), at the National Glass Centre, Sunderland on 15 January–13 March.

Exhibition: The best of the best (CGS), Newcastle, February–March.

Exhibition: Craft in the Bay (CGS), Cardiff, April–May.

Exhibition: Glorious Glass, Leeds, April–July.

Exhibition: Joyful Reflections (CGS), Farnham, April–June.

Furnace Solutions (SGT), 8–9 June, venue to be confirmed.

Exhibition: The Best of the Best (CGS), Newcastle, June–July.

Exhibition: Earth/Sea/Sky, London, June–July

16th Conference on the Physics of Non-Crystalline Solids, Canterbury, 10–15 July.

Exhibition: Then, Now, The Future, Stourbridge, July–October.

UK Glass Biennale, Stourbridge, August.

International Festival of Glass, Stourbridge, 26–29 August.

Exhibition: A postcard from the Commonwealth (CGS), Stourbridge, August–September.

SGT History and Heritage, Cambridge, September.

Exhibition: The Best of the Best (CGS), Newcastle, September–October.

Exhibition: Bedazzled 25 glassmakers impress (CGS), York, September–October.

Exhibition: Razzle Dazzle FIVE X FIVE (CGS), York, September–October.

Exhibition: CGS at 25, London, September–October.

Glass Focus, British Glass Annual Conference & Awards, November.

Pyrex 100 year celebration, Sunderland.

# The Upcycled Glass Company of South Devon

Ian Hankey, Principal Technician at Fab Lab Plymouth in Plymouth College of Art, recently completed a successful campaign with Crowdfunder for the Upcycled Glass Project, working with Andy Bradford, a farmer from Brimpts Farm in Dartmeet, Dartmoor, and Richard Glass, an award winning designer and expert glassmaker. As a result, Ian and a number of colleagues are now launching the Upcycled Glass Company CIC.

The Upcycled Glass Project Crowdfunder successfully raised the initial funding of over £10,000 to build a new sustainable glassblowing workshop on Brimpts Farm where the team intend to implement a circular economy model and upcycle local waste glass (the majority of which cannot currently be recycled) into new glass suitable for use in handmade art and design products, utilising a mixture of new technologies, modern farming methods and historical glass mixing processes that go back to the 17th century.

*Ian Hankey (top left), Richard Glass (bottom) and Andy Bradford (top right)*

Devon Climate Challenge has provided another £5000 to the campaign, as part of the #BackTheFuture initiative, in which Devon County Council and Crowdfunder have teamed up to support projects who are tackling the climate emergency.

"We were delighted that the Upcycled Glass Company received a large match fund pledge from Crowdfunder's #BackTheFuture, to support turning its ideas into reality. This might not have happened without Crowdfunder and Plymouth College of Art's collaboration on the iMayflower project, leading to Ian's involvement through Fab Lab Plymouth. He quickly realised that crowdfunding would be a perfect way for him to raise pump priming money for his own creative business," said Dawn Bebe, Director of Crowdfunder.co.uk.

The concept for the Upcycled Glass Company brings together multiple strands of glass expert and glassmaker, Ian Hankey. After initially leaving school with no qualifications due to undiagnosed dyslexia, Ian has worked in the glass industry for over 40 years, from his first apprenticeship at Pilkington Glass and later managing the hot glass workshop at the Royal College of Art and managing Teign Valley Glass Studios, to his work to establish in 2014 and subsequently manage Fab Lab at Plymouth College of Art.

During this period, Ian studied for his BA(Hons) 3D Design at Buckinghamshire College, received a Commendation for his MA at the Royal College of Art, and while working as Hot Glass Instructor at the RCA, he also gained a Distinction on his PGCE in Teaching and Learning Art, Design and Communication at the London Institute. Ian has delivered papers about glass and methods and approaches of thinking and making at conferences hosted by the Society of Glass Technology in Cambridge, the V&A, Liverpool University, University for Creative Arts and Plymouth College of Art, and his research has been published by the Society, Columbia University and Plymouth College of Art.

In 2018, Ian stepped back from his role as Manager of Fab Lab Plymouth to a role as Principal Technician, in order to spend more time working in the glass department at Plymouth College of Art, and to pursue his personal work recreating a number of historical glass processes, particularly focusing on recipes and processes from the Renaissance. Alongside his work with glass, Ian has developed his own skills in the Fab Lab, learning specific and advanced CAD CAM processes.

Ian said: "Throughout my life, dyslexia and reflective rationality have shaped the path of my career, which is to say that I work with my hands and my heart. This direct approach has enabled me to solve problems at times when others could not. My tacit understanding of and approach to glass led me almost 20 years ago to help Imperial College and The V&A to develop a 'cure' for glass disease (the corrosion of historical Venetian glass), solving a problem that many material scientists and experts in the field could not.

"As a result of working with glass for most of my life, I've had incredible opportunities to develop new techniques, recreate old methods of glass blowing and think about how some of these skills can be combined to solve environmental problems that face us as a society. Adding my knowledge of glass to some of the collaborative opportunities that we create in Fab Lab Plymouth have led to so much innovation. I've spent over 20 years developing methods of reducing emissions and reducing the costs of building and running glass equipment, from creating a portable glass furnace that cuts emissions over traditional furnaces by up to 70% to developing more efficient solar panels and helping to develop small-scale $CO_2$ recapture systems.

"Whether we like it or not, we're living through a climate emergency. Everything I've seen and done in my life points to the fact that for every environmental problem that we solve, someone with money is going to come along and try to use that solution to create more money for themselves, undermining the positive change that could have benefited society. The only way to tackle societal problems of sustainability is to change the framework we're working in and find new ways to work together. The Upcycled Glass Company is my attempt to bring into the world one small part of the change that we need to live in a more sustainable way.

"Thanks to everybody who contributed to our successful Crowdfunder, we can start work building a small furnace in the grounds of Brimpts Farm on Dartmoor, which we hope to have finished by the end of 2021. Did you know that UK glassblowers pay up to £2500 a ton to import 'art' glass material from the EU? Worse than that, in the UK and EU, we recycle about 71% of container glass, but end-of-life building waste glass, such as window glass, is almost never recycled into new glass products, and mostly ends up in landfill.

"On Brimpts Farm in 2022, we plan to use a mix of new technologies, modern farming methods and historical glass mixing processes to use the furnace that we've built to develop a technique to refine the waste glass from nearby Teign Valley Glass Studios. By melting waste glass and using locally harvested bracken from Brimpts Farm (with the added bonus that in harvesting these plants we'll create wildflower meadows), we'll reduce emissions for road transport and shipping across the world, and demonstrate a way to reduce the landfill waste glass mountains, which are recognised globally as a major environmental problem.

"Additionally, through my work at Fab Lab Plymouth, I've provided technical support to Barnaby's Brewhouse, helping to develop a small-scale $CO_2$ recapture system. If I fit a similar system to the new glass furnace on Brimpts Farm, we'll be upcycling waste glass that companies currently pay to dispose of, and burning bracken from the local farmland to extract the chemicals that we need. If we power the furnace equipment with methane from the cows and farm slurry, at the end of the process, the only waste product that we create should be water.

"That's just the start of our sustainability plans, working as part of a circular local economy, where businesses work together for their own benefit, for the benefit of the community and the environment. Alongside acting as Managing Director of the Upcycled Glass Company CIC, I'm going to continue my work at Plymouth College of Art. Fab Lab Plymouth is based here, in an art school with over 160 years of history, allowing new technologies and traditional methods of making to come together, and this is where the solutions to so many problems in society begin to come together."

Ian is now preparing to launch the Upcycled Glass Company, with Brimpts Farm owner Andy Bradford, expert glassmaker Richard Glass, Dr Katie Shanks, Dr Sarah Fearn, Victoria Oakley ACR FIIC, and IT expert Jamie Kaye.

Dr Katie Shanks has expertise is in designing optics for concentrator photovoltaic applications and is currently working on the use of novel materials, such as plastics and biomimicry surface structures, for CPV optics, as well as building integrated CPV windows.

Dr Sarah Fearn is a Research Officer in the Department of Materials at Imperial College and manages the surface analysis lab. Her area of expertise is secondary ion mass spectrometry, ion beam analysis and materials science. It was during her first post-doctoral studies at Imperial in conjunction with the V&A, that Sarah was introduced to ancient glasses, their properties and possibilities, working with the help of Ian Hankey to monitor the changing composition of replica Venetian glass as it corroded in ambient air.

Victoria Oakley is a conservation professional with 40 years' experience working in museums, most recently as Head of Objects Conservation Section at the V&A Museum. Her practical expertise lies in the conservation of ceramics, glass, enamels and wax objects, with a particular interest in the chemical deterioration of historic vessel glass.

Together, as Upcycled Glass Company, this team of experts will collaborate to make the vision of a circular economy model for upcycling waste glass into a reality.

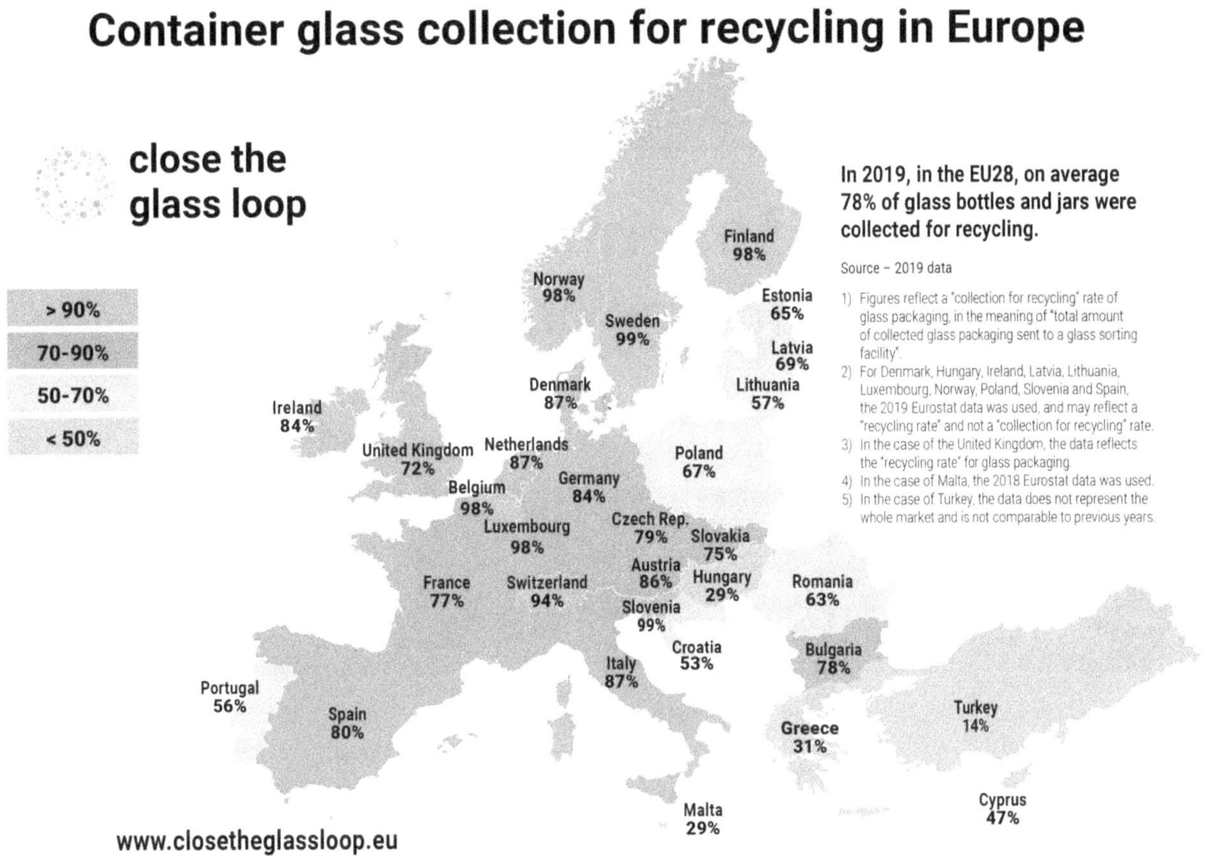


Latest value chain data published by **Close the Glass Loop** show that the EU28 average collection for recycling rate for glass packaging grew to the record rate of 78% in 2019. This represents a growth of 2% compared to 2018 which shows that the glass packaging value chain is delivering on its circular economy ambition

The multi-stakeholder stewardship partnership Close the Glass Loop aims to achieve a post-consumer glass container collection target of 90% by 2030, and to ensure that this is recycled into the containers. The vast majority of the 13.7 million tonnes collected go back to remelt new bottles and jars.

"The record high glass collection for recycling rate of 78% shows that our ambitious targets can be achieved with close cooperation between all players in the value chain, sharing best practices on innovative and tailored collection models, high quality sorting and recycling and communication campaigns for consumers" said Joachim Quoden, Managing Director of EXPRA (Extended Producer Responsibility Alliance), on behalf of Close the Glass Loop partners.

"Glass is a permanent material that can be endlessly recycled into new packaging. The more glass is selectively collected, the better its quality for more recycled glass that can be produced and incorporated in the production of new glass packaging" says Marine Ronquetti, Secretary General of FERVER (European Federation of Glass Recyclers), on behalf of Close the Glass Loop partners.

It is a priority for the glass collection & recycling value chain to ensure that the remaining 22% of glass packaging is collected and brought back into the loop. Close the Glass Loop unites glass producer, brand owners and fillers, the hospitality sector, glass treaters, Extended Producer Responsibility Schemes, collectors and municipalities under a multi-stakeholder European programme. It is supported by 11 national platforms to increase the quantity and quality of available recycled glass, so that people don't just recycle, but recycle more and better.

Municipalities was the focus of a Close the Glass Loop study, carried out by ACR+. The survey focused on two aspects:

- A description and comparison of the "framework conditions", e.g. every relevant policy, regulation, and economic instruments set in the seven targeted countries to enable and promote the recycling of glass packaging;
- An overview and comparison of how different local glass packaging waste collection systems operate across Europe, to highlight effective instruments as well as to better understand which contextual parameters seem to impact the generation of glass packaging waste and collection performances.

These two parallel analyses intend to allow a better understanding of the similarities and differences of glass packaging waste management across Europe. The study also identifies relevant framework conditions and local instruments that could contribute to high performances. By assessing the impact of different factors on local performances, it aims to define criteria that will make it easier to compare different schemes. https://closetheglassloop.eu/wp-content/uploads/2021/10/Municipalities-Study_FULL.pdf

# World's first all-oxide all-solid-state sodium (Na) ion secondary battery

Nippon Electric Glass has developed a new negative electrode material using a glass-ceramic for the all-solid-state Na-ion secondary battery, and became the first to successfully operate an all-oxide all-solid-state Na-ion secondary battery integrated with a glass-ceramic positive electrode and a solid electrolyte.

The proposed all-solid-state Na-ion secondary battery has an output voltage of 3 V, it possesses great application prospects and its performance is comparable to that of existing lithium-ion rechargeable batteries.

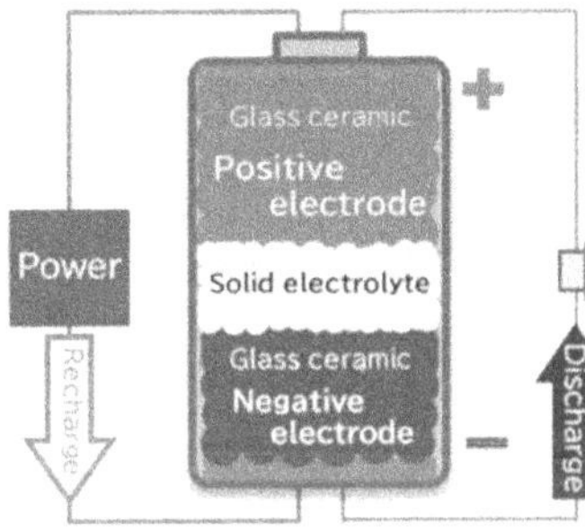


*All-solid-state sodium (Na) ion secondary battery structure*

The proposed all-solid-state Na-ion secondary battery utilises abundant metals, such as sodium and iron, and does not require any rare metals, such as lithium, cobalt and nickel, which are generally difficult to source for applications in conventional battery technologies. In addition, since the battery comprises a stable oxide material, it does not ignite nor generate harmful gas, even when pierced by a nail or knife.

Achieving a carbon-free society by commercialising the applications of all-oxide all-solid-state Na-ion secondary battery as it uses affordable resources and is very safe when operated.

High safety: The battery material consists entirely of inorganic oxides, thereby overcoming limitations such as ignition or generation of toxic substances during use and production.

Excellent battery performance: The battery can be operated at low temperatures owing to the softening fluidity of the glass used to integrate the positive and negative electrodes with the solid electrolyte and to improve the ionic conductivity. The solid electrolyte has enhanced cycle characteristics with minor deterioration due to the ion transfer. By developing high-potential active materials with simple structures, it is possible to manufacture batteries with high energy densities.

Abundant resources: The battery uses sodium that is present in abundance, there are no concerns related to supply compared to those associated with resources such as lithium and cobalt.

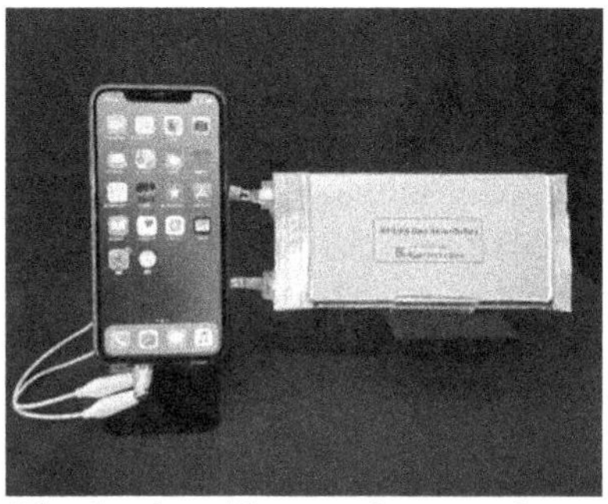

*The developed battery in operation*

# SGT Furnace Solutions training seminars

The usual Training Day held on the day before Furnace Solutions was cancelled due to the pandemic, instead the Society of Glass Technology have a series of five, one-off training seminars held online via Zoom on the second Wednesday of the month, starting in October. Times of presentations are 13.00 to 14.00 UK time, BST in October, GMT afterwards. The speakers and topics are presented below.

Wednesday 13th October 2021
Thermal balance and furnace screening: a periodic audit to reduce emissions and increase energy efficiency presented by Walter Battaglia of Stazione Sperimentale del Vetro (SSV).

Wednesday 10th November 2021
Forehearth Audits – is your glass conditioning equipment at its best? Presented by Alan Stephens of Fives Stein.

Wednesday 8th December 2021
Forehearth Colour Technology: Today's performance needs and new challenge presented by Pauline Darbouret of Ferro.

Wednesday 12th January 2022
NOx emission and possibilities to reduce presented by Matthias Hagen of Luft- und Thermotechnik Bayreuth (LTB).

Wednesday 9th February 2022
Challenges of Tin Bath Blocks in the "Float Process" and how to test them? Presented by Jan Theron of Lucideon.

www.sgt.org
Recordings can be seen on the SGT YouTube channel:
https://www.youtube.com/channel/UCe2tywIcr11dor_0cdrT5Vg

---

# Furnace Solutions
# &
# Training Day

The Melting Technical Committee (MTC) of the Society of Glass Technology (SGT) invite you to mark your diaries for the 2022 Furnace Solutions on 9th June and Training Day on the 8th June.

Lucideon has converted their conference room to office space so the venue for the return of the person-to-person event is still to be confirmed.

Further details (including the Geoff Evans Dinner and hotel accommodation) will be announced on the website
http://www.furnacesolutions.co.uk/

# NEWS

Ed Miliband, the Shadow Secretary of State for Business, Energy and Industrial Strategy* visited **British Glass** member **Ardagh Glass** Packaging's Doncaster plant to discuss the impacts of escalating energy prices on the glass industry. The MP for Doncaster North spoke with representatives from both organisations on the ongoing issues and the lasting impact the sector is facing as a result of the prolonged energy price crisis.

British Glass also provided the minister with questions to be raised in parliament to address the lack of action from government in responding to the crisis.

The Rt Hon Mr Miliband (centre of picture) said: "The Government's failure to have a plan that secures our energy supply has left our country dangerously exposed.

"The silence from ministers on support for energy intensive firms speaks volumes about what little value the Conservatives place in the future of these proud industries."

Technical Director at British Glass, Dr Nick Kirk (second on the left), said: "We are very grateful to the Shadow Secretary for taking the time to visit one of our member's factories and to discuss the very real impact that continued high energy prices are having on our sector.

"Glass manufacturers and other energy intensive industries have been dealing with the burden of extortionately high prices for far too long and with no action, support or intervention from government in sight, our industry continues to suffer and face the real possibility of impacting the food & drink and construction supply chains.

"As the representative body for the glass industry we will continue to support for our members on this matter and we hope that following today's meeting, government will address our concerns and the pressing need for action to resolve the crisis."

*Reshuffled a few days later to Shadow Secretary of State for Climate Change and Net Zero.



Electricity generator **EDF** and environmental services group **Veolia** have announced the creation of **Waste2Glass**, a 50/50 joint venture that will develop a new sector based on Veolia's GeoMelt vitrification technology.

Vitrification has long been considered by international nuclear authorities as a suitable choice for high level nuclear waste because of its durability, which is measured in hundreds of thousands of

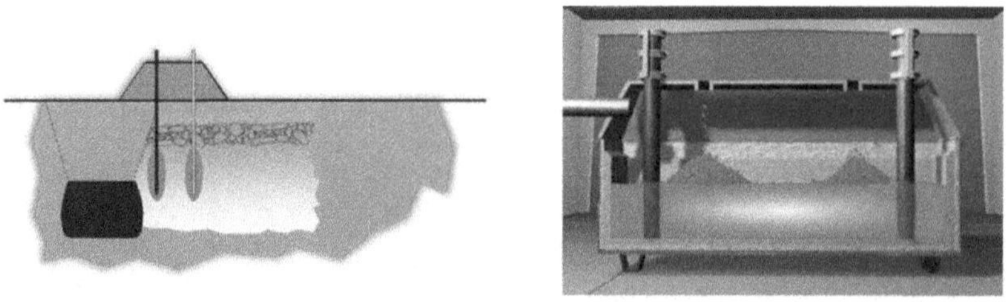

*Waste2Glass GeoMelt technology offers in situ melting (left) and in container melting (right)*

years. GeoMelt produces an ultra-stable glass that is about ten times stronger than concrete and more durable than granite or marble.

Due to its technical nature and cost, vitrification has until now been reserved for highly radioactive waste. Thanks to the complementary know-how of the two partners, Waste2Glass will be able to take up the challenge of the industrial deployment of the GeoMelt technology, which will make it possible to broaden and simplify the use of the vitrification process for a wider range of waste types.

Antoine Frérot, Chairman and CEO of Veolia said: "Our business and our Purpose as a world leader of the ecological transformation is to offer innovative technologies and solutions for the management of complex waste such as hazardous and radioactive waste. I am delighted that we have taken this further step in our collaboration with EDF with the creation of Waste2Glass. It will enable a real change of scale through the industrialisation of GeoMelt, which will make it possible to treat radioactive waste more safely and more economically, with a reduction in storage volumes."

Jean Bernard Lévy, Chairman and CEO of EDF, said "After Graphitec, the creation of Waste2Glass illustrates not only the quality of the cooperation between our two companies, but also the EDF Group's commitment to the development and industrial application of truly innovative solutions for the treatment of radioactive waste, a key issue for bringing nuclear power into the mainstream of sustainable development and helping to build a carbon-free future."

With an expansion in Durham, North Carolina, **Corning**'s high-volume manufacturing facility can produce up to 500 million glass vials, which can store nearly 4.3 billion doses annually and aid the progress of public health.

Corning has helped combat COVID-19, ramping up production of its damage-resistant Valor® Glass vials and enabling delivery of over 3 billion doses of COVID-19 vaccines in the past 20 months. Along with Valor Glass, the Durham facility will manufacture Corning's new Velocity® Vials, Corning Pharmaceutical Technologies' recently announced innovative technology that directly responds to pharmaceutical supply-chain challenges. The coated borosilicate vials are engineered to improve drug quality and increase pharmaceutical filling line efficiency, compared to conventional borosilicate vials.

Approximately $261 million of federal funding from the Biomedical Advanced Research and Development Authority (BARDA) allowed Corning to expand production at Durham and two other Life Sciences facilities. BARDA, a US Department of Health and Human Services office, works with industry partners to secure the development of the necessary vaccines, drugs, therapies, and diagnostic tools for public health medical emergencies.

The workforce at Durham currently handles the facility's annual capacity of 185 million vials (or approximately 1.6 billion doses), meeting BARDA's contract terms within 15 months – three months ahead of schedule. The facility

Glass packaging company, **Croxsons**, has made changes to its leadership structure as part of their succession plans. Tim Croxson, currently the company's COO, has taken up the position of CEO with immediate effect, in place of his father, James Croxson, who steps aside to concentrate more on developing Croxsons' international concerns. The change of leadership comes after a period of extraordinary growth for the family firm, who celebrate their remarkable 150th anniversary year in 2022.

In taking the business forward and increasing its market share, both James and Tim have enjoyed a fruitful partnership together over the past 20 years.

Global sales, has enabled Croxsons to extend its reach and strengthened its brand – a development that was rewarded with a Queen's Award in 2019 for exceptional progress being made in international growth and exports. Regularly featuring in the Sunday Times Fast Track 200, Croxson has offices in the UK, Australia, New Zealand. USA and Hong Kong.

"Having been in the business for over 50 years, I have seen both the packaging industry and my own company change extensively in that time," said James Croxson, 70. "The company has grown from being wine barrel suppliers and bottle washers in 1872 to now a Global packaging supplier, providing over 50 different countries with millions of bottles, jars, caps and corks. As we approach 150 years of being a family owned business, we are proud of our company, its heritage, its five generations, its great team and all its achievements.

"As is the norm in a fast growing SME, I have done every role and have built a great team over the past 50 years, who have been a joy to direct and lead. My main focuses going forward are international sales, continuing to build our international sales teams and giving the next generation some inevitable pearls of wisdom. And like Tim, I look forward to the next chapter in Croxsons' illustrious history."

Tim Croxson added: "Moving into the CEO seat at Croxsons is really a natural progression for both James and I. Our combined strengths and the team we have built around us, have helped put the company into an incredibly strong position. Alongside the board, we've both worked hard to future-proof our offering and be clear on our purpose, thereby ensuring we continue to be relevant for the next 150 years. I look forward immensely to continuing that self-belief, creating value for our customers, and maintaining our uniqueness."

---

has over 224,000 square feet of manufacturing space, including eight converting lines, three IOX tanks, two wash lines, one drying oven, and three coating lines, along with supporting equipment and spare parts.

**Hypermemo**, a photonics high-growth company in Joensuu, Finland, has received an investment of €250,000 from the EIC Fund managed by the European Innovation Council operating under the European Commission. The investment agreement also includes an option for additional €4.5 million equity funding. It will enable the company to transfer to industrial production, new recruitments and move to wider operating and production facilities.

The company has successfully developed, patented and commercialised an innovative glass cutting and finishing method using a pulsed carbon dioxide laser, which enables excellent accuracy and speed. It is a considerably more energy efficient and ecological alternative to existing solutions.

Globally, industrial glass processing is a market of around €1 billion and growing at an annual rate of 4.45%. International growth has been part of Hypermemo's plans from the beginning, as its customers are large global glass manufacturers. The company strives to conclude the first commercial contracts at the beginning of 2022.

"If everything goes as planned, we will make the first deliveries to customers at the end of 2022. With the help of new funding, we will move to industrial scale production and recruit more people to support the growth. The aim is to strengthen all areas from product design to production and sales to marketing," explains Erik Raita, COO of Hypermemo.

**Foamit Group**, part of the Finnish group **Partnera Oyj**, has acquired Norwegian foam glass manufacturing company **Glasopor**. This strengthens Foamit Group's position as the leading foam glass manufacturer in the Nordic region and one of the largest players in the industry globally.

Foam glass is a very environmentally friendly and lightweight building material made from recycled glass, the demand for the material is supported by an increased demand for products in the circular economy.

"Foamit Group aims to expand and further develop its business activities around glass recycling and foam glass manufacturing. The acquisition of Glasopor gives us access to new markets as well as increased capacity for the production of foam glass, as well as new expertise in the use of foam glass in various construction projects. The supply of raw materials is also improving. The acquisition of Glasopor supports our growth strategy and enables us to participate in ever larger projects, "says Kalle Härkönen, CEO of Foamit.

Glasopor was founded in 2002 and has two foam glass factories in Norway, one in Skjåk and one in Fredrikstad. The company has 34 employees.

Dental reinforcement brand **Hi-Fiber** of Milan, Italy has established a partnership agreement with Lisbon based **Malo Clinic**, a company known for its oral rehabilitation and dental aesthetics. The partnership includes clinical trial cases using Hi-Fiber's metal-free dental reinforcement structures designed and 3D printed from advanced materials and the company's continuous fibre manufacturing (CFM) technology.

CFM, which 3D prints continuous fibre thermoset composites, comes from parent company **Moi Composites**. The manufacturing system incorporates a single strand of continuous fibre composite material, proprietary software and robotics to smartly design, 3D print and rapidly cure

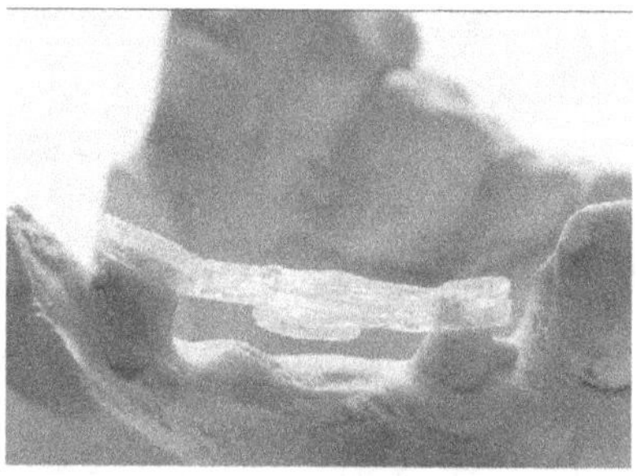

*3D printed from a single strand of transparent thermoset fibreglass, Hi-Fiber reinforces the tiniest of spaces, ensuring accuracy, efficiency, and maximum performance of continuous pre-tensioned fibres. The resulting reinforcement bar is the closest possible representation to natural bone behaviour with higher strength levels*

the reinforcement precisely where needed, eliminating waste, saving time and resources.

Hi-Fiber has had five years of analysis, refinements and clinical cases, becoming a widely adopted solution in Italy for the reinforcement of fixed and removable prosthesis, overdentures, maryland bridges and orthodontic splints. Similarly, **Moi Dental**, a division of Moi Composites, has entered into partnerships with leading companies in the sector such as **VITA Zahnfabrik**, **ETK/Lyra** and **3Shape**.

**Ardagh Group** has agreed to acquire **Consol Holdings**, the leading producer of glass packaging on the African continent, for an equity value of ZAR10.1 billion ($635 million).

Headquartered in Johannesburg, and founded in 1946, Consol is the market leader in South Africa, operating four well-invested glass production facilities. It serves a broad range of leading international, regional and domestic customers, principally in the beer, wine, spirits, food and non-alcoholic beverage sectors.

In the year to June 30, 2021, Consol reported consolidated revenues of ZAR9.0 billion ($566 million). South Africa represented approximately 90% of revenues, with the balance represented by smaller production facilities in Kenya, Nigeria and Ethiopia. Glass consumption in Consol's markets is projected to continue to increase, driven by long-term trends, including population growth, rising income levels and shifts to premium one-way sustainable glass packaging.

Ardagh expects to finance the acquisition through a combination of its own cash resources and the assumption of ZAR5.7 billion ($358 million) existing net debt at Consol. Completion of the acquisition is subject to certain conditions, including regulatory approvals that are expected to be obtained in the second quarter of 2022.

Ardagh Chairman, Paul Coulson, commented: "We are very pleased to expand our European and North America presence in glass packaging into Africa with the acquisition of Consol, a high-quality business, led by an excellent management team. Consol is a market leader in the region, with great relationships across a diversified domestic and multinational customer base. Virtually all of Consol's multinational customers

# Glass Focus Awards 2021

Businesses from across the glass supply chain came together in Manchester to celebrate the many achievements of the industry over the last 12 months. The 2021 Glass Focus Awards organised by British Glass returned to a live event after a video conference led presentation in 2020.

For the second year in a row Encirc were crowned Company of the Year and also took home the Innovative Solution and Strengthening business through people awards while Saint-Gobain Glass won three awards including Rising Star, Health and Safety action and Sustainable Practice alongside Morley Glass & Glazing. Other winners on the night included Allied Glass, The Armstrong Partnership and The Glass Company.

British Glass CEO Dave Dalton said: "It's no secret that our industry has faced, and is still going through, some difficult times over the last year which makes it all the more special to be able to welcome everybody back to the Glass Focus Awards to celebrate what we have all achieved together in person.

"We have seen some truly remarkable entries this year, from the incredible innovations that allow our industry to continually evolve and the sustainable actions that drive us forward to our goal of reaching net zero carbon emissions, to the beautiful designs that adorn our buildings and bottles, the unrivalled talent of our workforce and everything in between.

"It's been a pleasure to welcome everybody back to the awards; we are all extremely proud of the achievements of our industry over the last year and we look forward to building on these in the coming year and beyond."

The full list of winners from the Glass Focus Awards 2021 are:

Design of the Year – container (sponsored by Rockfield): Clean T by Allied Glass

Design of the Year – flat (sponsored by *Glass Worldwide*): Ubiquitous Energy by The Glass Company

Innovative solution (sponsored by Glass Technology Services): The world's most sustainable glass bottle by Encirc

Health and safety action (sponsored by Arco): Onsite defibrillators by Saint-Gobain Glass

Strengthening business through people (sponsored by Glass Futures): Ignite programme by Encirc

Rising Star (sponsored by the Worshipful Company of Glass Sellers): Will Ruane at Saint-Gobain Glass

Marketing Impact (sponsored by Friends of Glass): Encirc: Big 5 by The Armstrong Partnership

Sustainable Practice (sponsored by *Glass International*): GreenVision Fund by Saint-Gobain Glass and Morley Glass & Glazing

British Glass Company of the Year: Encirc

---

are also customers of Ardagh. We look forward to welcoming Consol to the Ardagh family and to investing in the long-term growth of the African market, driven by consumer trends and rising sustainability awareness."

Bruce MacRobert, Chairman of Consol, remarked that: "The Consol team has built a great business, with an established reputation for delivering quality products to a growing customer base. Ardagh's long-term presence in, and commitment to, glass packaging, makes it the ideal owner to continue this progress."

**Johnson Matthey** has announced the sale of its Advanced Glass Technologies (AGT) business, a provider of specialist glass enamels and precious metal pastes supplied mainly to the automotive sector. AGT will be sold to **Fenzi Holdings**, a manufacturer and supplier of materials for flat glass

**Eviosys**, **Grown.Bio** and **Verescence** have come together to innovate on a 100% recyclable and reusable cosmetic gift pack. This collaboration is responding to a market of continued demand for customisation, as consumers look for tailored and unique products with the planet in mind. The limited-edition gift set will include 100% infinitely recyclable and reusable glass bottles, carbon negative mushroom packaging inserts, and a reusable 100% infinitely recyclable metal box.

Grown.Bio's mushroom packaging, made only from mycelium, an agricultural by-product offers the opportunity to be used for other purposes, and if not, can be composted.

Eviosys, has adapted their original product to work well with the mushroom insert and Verescence's glass products. Made of recycled metal it can be kept and reused for decades, or be recycled.

Verescence's new perfumery range CARA, offers a standard screw neck, allowing the consumer to refill the bottle at home or in-store. All standard bottles from La Collection Verescence have been produced with 20% cullet from household waste collection. The Verre Infini 20 formulation combines 20% cullet and a glass tint adapted to the aesthetic requirements of luxury brands. Already produced in France and Spain, this glass composition will be gradually deployed in the United States in the first quarter of 2022, then in South Korea in 2023.

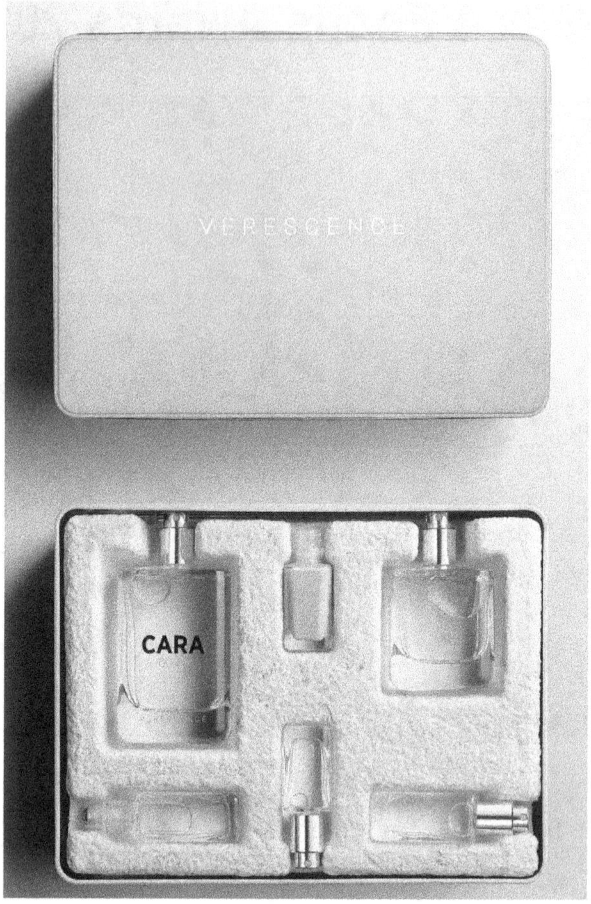


---

processing, for £178 million.

The transaction is subject to works council consultation in the Netherlands and the satisfaction of customary completion conditions.

In the year ended 31st March 2021, AGT generated sales excluding precious metals of £66 million, with an operating profit of £17 million.

Fenzi Group's Chief Executive Officer, Alessandro Fenzi said, "We are delighted to welcome the AGT to the Fenzi Group. AGT leadership in glass enamels for the automotive and industrial sectors fits perfectly with our strategy to be the largest and most complete partner for the glass industry worldwide. This partnership brings new innovative technologies to our global customer base. We believe that AGT's growth will be enhanced under the Fenzi Group."

The Turkish international companies **Şişecam** and **Ciner Group** are putting a joint US$4 billion in the US soda ash industry. With the planned investment, Şişecam will become one of the World's largest soda ash producers.

Şişecam, a global player in the chemicals and glass industries, and its partner Ciner Group, one of the largest players in the soda ash industry, leap forward from the strategic step they took in 2019 in the field of natural soda ash, to a new phase. The companies are expanding the scope of their cooperation and changing the structure of their partnership.

Şişecam and Ciner Group have been development JV partners in the Pacific natural soda ash development project since 2019. As part of the new investment plan, Şişecam will become a 60% shareholder by acquiring an additional share in the Pacific Project. In addition, Şişecam will also purchase a 60% percent share in the Atlantic Project and Ciner Resources, the existing natural soda ash production facility of Ciner Group in Wyoming. The Pacific and Atlantic development projects with a targetted combined production capacity of ~5 million tonne/year is expected to start production during 2025 and is planned to gradually reach full capacity by 2027.

Prof. Ahmet Kırman, Chairman of Şişecam remarked that the investment, which is the largest investment made in the US by Turkish companies ever, is also one of the largest investments made abroad from Turkey. "With these investments, Şişecam's controlled soda ash production capacity will exceed 10 million metric tonnes, compared to the existing capacity of 2.5 million metric tonnes today. The plan also includes the establishment of the largest soda ash production facility globally in a single location."

The strong demand for soda ash is expected to grow in the coming years. The market is expected to reach 60 million metric tonnes globally by the end of 2021. When compared to synthetic soda ash production, producing natural soda ash provides a decrease of approximately 50% both in carbon intensity and water consumption. "This ecologically friendly natural soda ash production is perfectly aligned with Şişecam's long term sustainability-oriented perspective" Prof. Kırman said.

**Bormioli Pharma** has become a member of **Glass Futures**, the UK-based not-for-profit collaboration of universities, companies and research institutions that aims to bring about a technological and economic revolution of great impact on the industry. The pilot facility, The Global Centre of Excellence for Innovation is being built in St Helens, UK.

The agreement will give Bormioli privileged access to innovative technologies for glass production, strengthening the company's position and reducing impact on the environment.

**Paşabahçe**, the Turkish glassware brand, has expanded its 100% Recycled Aware Collection, which contributes to the circular economy, with new products. A greener forest, a bluer sea and a cleaner future is possible. Therefore, Paşabahçe breathes new life into glasses that have completed their life cycle and invites to use recycled glassware for a sustainable living. With a lower carbon footprint and reduced industrial waste output, the Aware Collection combines sustainability with aesthetics and modern style in its wide range of products.

Inspired by the turquoise colour of the Bosphorus, the Aware Collection brings together sustainability, aesthetics, and modern style. Launched with four products in 2020, the Aware Collection now includes 25 products consisting of glass, mug, stemware, jar, carafe that appeal to consumers who are environmentally conscious and aware.

Paşabahçe developed the Aware Collection based on its vision of leaving a better world for future generations in line with Şişecam's Care for Next sustainability approach. In 2011, Şişecam launched the "Glass and Glass Again" initiative, one of Turkey's most comprehensive sustainability and social responsibility efforts. To date, more than 1.3 million tons of glass waste have been recycled under this initiative.

**Beatson Clark** is investing £1 million in expanding its on-site recycling plant in South Yorkshire and installing new, more efficient equipment. The company partnered with **REDWAVE** to complete a turnkey plant expansion which begun operation.

Throughput at the recycling plant is expected to increase by an average of 50% over the year, and the project has created 12 full-time jobs which were previously contracted.

As part of the investment Beatson Clark has installed additional optical sorters, vibratory feeder conveyors, new screening machines and a new JCB telehandler.

Additional processing stages have been introduced using new vibrating screens and dedicated optical sorters, improving quality and reducing waste.

The new equipment and the additional inspection stages have reduced the false rejects of glass which were removed with the CSP (ceramics, stones and pottery) by over 50%, which in turn has increased the plant efficiency. This makes the process more sustainable and means less waste glass going to aggregate for re-use rather than being truly recycled within the furnaces.

"The addition of new optical sorters makes the process more precise and reduces the amount of good glass which gets rejected by the machines and sent to waste, thereby increasing the amount of quality cullet that can be used in the furnace," said Patrick Pagdin, Recycling Plant Manager at Beatson Clark, "We are now seeing an increased output of both white flint glass cullet and gramber (green and amber glass cullet) thanks to easier automated sorting and improvements in the colour sorting process."

Beatson Clark has also installed a 300 $m^2$ canopy to cover infeed material and protect it from inclement weather conditions; this will improve processing during the winter months as the material will remain relatively dry.

Glass comes into Beatson Clark's recycling plant from the hospitality sector, which is beginning to recover from the pandemic, and from local authority kerbside collections of domestic glass.

The post-consumer recycled content rates for containers manufactured by Beatson Clark are 30% for white flint and 45% for amber glass. The recycling plant used to generate on average 50% of this cullet; following these investments it now generates over 75% of the cullet.

The **Heye** PlantPilot is an intelligent information management system for the glass manufacturing plant. Operators and managers have to keep the overview of the actual production tendency in the plant: Where are the losses along the line? How do we retrieve events occurred during the production? How do we improve the efficiency of the plant? All these key questions can be answered by the help of the PlantPilot.

To boost efficiency, improve quality and finally remain competitive it is crucial to stay state-of-the-art not only with regard to reliable equipment but even when it comes to Industry 4.0 requirements of a "smart glass plant".

As part of the maintenance agreement, the software is regularly updated and expanded with new functions and can also be quickly revised and adapted to latest production requirements. Hence, the functionality of the software can be guaranteed over a long period of time.

**Saint-Gobain** subsidiary **CertainTeed** is investing $400 million to expand production capacity at four key manufacturing locations in the United States. The combined investments will reduce waste by more than 50% and cut $CO_2$ emissions by more than 20%.

"This is an investment in our company's future, and also in our future ability to ensure security of supply for our customers," said Mark Rayfield, CEO of Saint-Gobain North America and CertainTeed. "We will continue to look for opportunities to strategically grow and expand our business, all while striving to become the world's leader in light and sustainable construction."

At the company's insulation plant in Chowchilla, California, Saint-Gobain will invest $32 million in a new furnace, which will allow the plant to expand its production by 13% at that location and will begin coming online in 2022.

At the company's plant in Athens, Georgia, Saint-Gobain will invest $30 million to start up an additional production line that will increase the plant's capacity of loose-fill insulation. This expanded production is expected to begin in the fourth quarter of 2023 and will lead to the creation

**Şişecam** has announced plans to invest in a new furnace with a 155,000 tonne annual capacity at its Eskişehir glass packaging production facility. This new investment is in response to the growing potential of glass packaging arising from the rapidly expanding demand for hygienic solutions across the world. The new €84 million glass packaging furnace is planned to be commissioned at the beginning of 2023. Aimed at boosting the company's global glass packaging production capacity to 3.2 million tonne/year, this major investment will provide input to numerous sectors.

Şişecam Chairman of the Board of Prof. Dr. Ahmet Kırman elaborated on the new investment will realize in Eskişehir: "Şişecam has decided to invest in a new furnace in Eskişehir, where it currently has four glass packaging furnaces in operation. This new investment will allow us to meet growing glass packaging demand in Turkey resulting from heightened health awareness while also implementing our export strategies. Şişecam, currently running glass packaging production operations in Turkey, Russia, Ukraine, and Georgia, also announced in June that it would invest in a greenfield facility with a 330,000-ton annual capacity in Hungary."

Together with its joint venture partner **AGC Flat Glass North America**, SCHOTT has entered into definitive agreements to divest its shares in **SCHOTT Gemtron Group** to **SSW Advanced Technologies**, a portfolio company of **Trive Capital**. Gemtron specialises in the production and sale of flat glass products mainly for the home appliance industry in North America.

*The Vetri Speciali Black Carafe has been selected for the ADI Design Index 2021 Award. Characterized by sinuous lines, defined reliefs and its unmistakable inclined mouth cut, the Black Carafe has been designed and built with a black glass colour with 100% filtering power. A high quality product, with an elegant finish that has maximum protection from light.*

**Vetropack Italia** has set the foundations for a new glassworks to be built in Boffalora sopra Ticino.

In a context where ESG (environmental, social and corporate governance) matters are becoming increasingly relevant, the company is renewing its commitment to integrate sustainability throughout the production chain. Circular economy, social and environmental sustainability are at the heart of **Vetropack Group**'s sustainability strategy. The construction of the new production site is in fact the result of Vetropack Italia's desire to innovate through a structure that complies with its own sustainability objectives outlined in the Group's Strategy 2030, which falls within the European Union's development plans aimed at increasing the percentage of recycled glass, a secondary raw material, used in production.

With this in mind, the company has selected the industrial site of Boffalora sopra Ticino, located at a strategic junction of the area's road network, which will allow the recovery of a partially used, but not yet abandoned, area. This unique project is part of the plan to improve the efficiency and quality of production processes and will enable the company to meet the ever-increasing demand for its products and respond to the increasingly stringent requirements of its customers.

Johann Reiter, CEO of Vetropack Holding, said: "The new industrial plant meets the main innovations and sustainability specifications and will be the flagship of our production sites, allowing us to achieve production efficiency and sustainability in line with our 2030 Strategy."

Sergio Antoci, General Manager of Vetropack Italia, added: "Sustainable development is both a local and a global challenge, for a quality economic recovery, health, safety and social cohesion. After a long search, we selected the land in Boffalora sopra Ticino because it meets our needs on several levels: it is in fact an industrial site strategically located at an important transport hub, which will lead to improved logistics and ensure the preservation of jobs in the region and the recovery of an industrial site, which fully embraces the Regional Directive governing land consumption."

**Encirc** is set to lead the UK glass container industry's journey to net zero following the Government's decision to bring hydrogen to North West England via the **HyNet** project.

The decision to select the transformational low carbon and hydrogen energy scheme for 'track one' status follows months of support from Encirc and other stakeholders. Of the six regional bids, HyNet and the Teesside-based East Coast Clusters schemes were chosen, meaning these projects are in line for a slice of the Government's £1 billion fund and can start decarbonising as early as 2025.

The HyNet scheme will unlock a low carbon economy and put the region at the forefront of the UK's drive to net zero. Encirc plans to be one of the first organisations to make use of the hydrogen to melt glass instead of less sustainable fuels. The company is already part of Net Zero North West, an industry-led cluster investing in industrial decarbonisation and clean growth projects in the region.

The Government's decision will support Encirc in its journey to net zero operations, enabling the further expansion of its Elton facility and creating more than 200 new jobs by 2026.

Adrian Curry, Managing Director at Encirc, commented: "The HyNet project will play a key role in our ambition to decarbonise. This green light is great news for the environment, for the North West region, for industry, and for jobs.

"We have been pushing for Government to select the HyNet North West project due to the transformational impact we see it having on our region. We're excited to lead the way as one of the first in the industry to be on this hydrogen pathway, as we take steps to get to net zero."

Aston Fuller, General Manager at **Glass Futures** said: "Glass Futures is working closely with glass manufacturers across the world to make their operations sustainable. Low carbon fuels, such as hydrogen will play an essential role in helping our industry switch away from fossil-based fuels to enable the UK to meet it's 2050 zero-carbon targets. It's fantastic to see the North West access the necessary investment to develop a hydrogen infrastructure."

While ambitious, HyNet is low-cost and meets the major challenges of reducing $CO_2$ emissions from industry, providing fuel for transport and heating for homes and businesses. The project is set to reduce carbon dioxide emissions by 10 million tonnes a year by 2030.

**Vidrala Group** have renewed their renewable enrgy agreement with **ACCIONA Energía**. The renewal emphasises the company's commitment to reducing the environmental impact of every plant and constant pressure to improve energy efficiency. The agreement implies the supply of 343 GWh of additional green energy over the next two years.

In addition to the commitment to use green energy, the adoption of ACCIONA Energía Greenchain technology entails a significant improvement in efficiency at the Portuguese plants as it displays real time information about clean energy generation. The technological platform Greenchain will allow Vidrala to track real time consumption of renewable energy and check at the same time the amount of $CO_2$ emissions avoided as well as the associated equivalences such as, for example, the number of planted trees or cars retired from the roads. This additional and innovative function showcases our contribution to minimising the impact of its activity on the climate change and how it embraces a more sustainable future.

During the last two and a half years The company has reduced $CO_2$ emissions by 125,000 tons, equivalent to removing 30,000 vehicles from circulation or planting 90,000 trees. Over the next two years, the two plants in Marinha Grande expect to reduce emissions by around 100,000 tons of $CO_2$.

Jorge Escudero, Energy Category Manager at Vidrala considers the renewal of the agreement with ACCIONA Energía a, "natural step after the good results achieved since Vidrala first started its relationship with ACCIONA Energía. Since 2018 we have registered a continuous reduction of our carbon print. However the path to carbon neutrality and full sustainability is long and challenging. Greenchain solution allows us to verify the origin of the electric energy we consume. We consider this as another step aimed at combining Vidrala's full productivity with social responsibility and concern for the community where it is integrated".

The **Energy Transition Initiative-Center for Global Commons** (ETI-CGC), an industry-academia collaboration platform in which **AGC** participates has been launched.

ETI-CGC is an industry–academia collaboration platform conceived by the **University of Tokyo**'s Center for Global Commons (CGC) with participation from leading Japanese companies in materials, energy, transportation, and infrastructure sectors that support its philosophy. By using the University of Tokyo's platform to promote inter-industry and inter-company collaboration, the ETI-CGC participants will consider and propose paths and policies that will enable both Japan as a whole and its industries to achieve net zero carbon emissions by 2050.

The AGC Group has set the initiative of "promoting sustainability management" as one of the key strategies in its medium-term management plan, AGC plus-2023. With the goal of achieving net zero carbon by 2050, the AGC Group is working to reduce greenhouse gas emissions in its own production

McGregor, Texas, and the greater Waco area," said Knauf Insulation North America CEO Matt Parrish. "At Knauf Insulation, we are excited to expand our North American manufacturing operations to McGregor, Texas. As part of a global, family-owned business, we believe the timing is right, and McGregor is exactly the type of community where we desire to expand and partner for the future."

"I am personally thrilled to be on the ground in McGregor leading this project," said Knauf SVP Strategic Projects, Kevin McHugh. "Watching this incredible facility rise from the Texas soil and become one of our industry's leading production facilities worldwide will be an incomparable experience. Hiring, training, and empowering people from Central Texas to carry out Knauf's vision and the mission of this facility will positively impact Central Texas for decades to come."

In cooperation with the City of McGregor, McGregor Economic Development Corporation, Greater Waco Chamber of Commerce, McGregor Independent School District as well as partners and community members across McLennan County, the facility will directly employ 166 Knauf Associates who will be hired locally. Additionally, construction of this facility will create numerous jobs in the skilled trades as well as opportunities for local businesses and service providers.

**Saint-Gobain Glass** has completed the major re-build and installation of our new state-of-the-art float glass furnace at Eggborough, East Yorkshire.

The former furnace had reached the end of its design life and the replacement needed an enlarged capacity to reduce energy usage and boost manufacturing efficiency and production capacity.

Eggborough resumed production in August 2021. This upgrade will provide greater automation and enhanced control systems, with the use of the very latest equipment and integrated digital technologies.

It is a major milestone in the

The traditional Czech brand Pilsner Urquell is launching a new returnable bottle, thereby strengthening its commitment to sustainability. The elegant bottle with striking embossing is produced by **Vetropack Moravia Glass**.

Glass is equally suitable for one-way and returnable bottles. A returnable bottle can be reused more than 40 times. In light of the debate surrounding the environmental impact of packaging, the demand for returnable glass bottles is growing. Vetropack strives to produce its packaging as ecologically as possible. For example, the use of recycled glass and returnable containers is being systematically increased in order to conserve natural resources as much as possible.

The **Plzeňský Prazdroj** brewery is taking another important step towards sustainability: a new returnable bottle for the world-famous Pilsen beer with its striking embossing that does away with the aluminium foil at the neck of the bottle, while the plastic labels are replaced by recyclable paper ones.

The reliefs on the new bottle are very striking. The words "The Original Pilsner" have been added to the front, emphasising the pre-eminence of this beer in the Pilsner category. The back of the bottle features a revised relief of the gate of the traditional brewery and the inscription "Pilsner Urquell". The raised elements are not just decorative, they represent tradition, quality and the product's local roots. Thanks to the reliefs, the bottle also sits very well in the hand, which the brewers are sure to appreciate.

---

processes and developing and selling high-performance energy-saving glass and other products that contribute to achieving net zero carbon worldwide.

AGC joined ETI-CGC because it agreed with the following principles: "The success or failure of decarbonization will determine Japan's future in the world", "We need a scientific and integrated pathway to decarbonization for all of Japan", and "We aim to contribute to global carbon neutrality by creating a model that is useful for countries with diverse regional circumstances".

The AGC Group supplies various materials, such as architectural, automotive, and display glass; chemicals; and ceramics in about 30 countries and regions. AGC intends to contribute to the activities of the ETI-CGC based on the various knowledge it has gained through its businesses to date.

**Knauf Insulation North America** is expanding capacity with the contruction of a 600,000 square feet state-of-the-art facility at the Industrial Park in McGregor, Texas. Knauf Insulation is a leading global manufacturer of fibreglass insulation, delivering thermal and acoustical solutions for residential, commercial, industrial and OEM applications.

"This is a great day for our employees, customers, vendors, shareholders, the citizens of history of the business as £30 million was put into this project, the largest industrial investment in the UK by Saint-Gobain since the Eggborough plant was built in 2000.

The project has been several years in development and planning to ensure customers received the same level of supply during the construction. This involved stock build-up in advance and imports of glass from sister plants in Europe which helped fulfil orders.

The new facility is designed to manufacture high-performance products developed in the UK to meet the specific needs of the UK window industry. Better reliability, higher yield and a better product is the result of the upgrade. There has never been a greater need to protect the planet and invent a more sustainable future, the new design will result in a 20% energy reduction.

Increased supply of sustainable carbon-saving glass to the UK is really important for future homes and buildings as they move to become more energy efficient.

Steven Severs, Managing Director at Saint-Gobain Glass, described his overriding emotion as "pride" upon completing this mammoth project safely, on time and on budget. He said: "It's a major step forward in Saint-Gobain's net-zero journey and delivering our global strategic purpose to Make the World a Better Home."

***NSG Group*** *has collaborated with* ***Cohda Design*** *to create new glass applications that deliver power to devices without the need for wires. Cohda combines NSG TEC™ electrically conductive glass with its patented Power-Tap®wireless power technology. P-Tap technology allows electricity or data to be transferred wirelessly through a transparent conductive coating to power any device. This allows glass to act as a wire, transferring power out through contact alone. Electricity typically flows into the glass via a connection at the perimeter of the glass surface. This creates the positive and negative charge, which flows invisibly through individual layers within the conductive glass' lamination and into any connected device placed on apertures, or contact points, on its surface.*

*The collaboration has a vast array of applications for designers, architects and engineers to locate powered devices within transparent structures. Early applications have included retail, where in-store display stands can wirelessly charge electronic devices such as mobile phones and cameras; and in automotive, to power head-up display screens that are embedded in a vehicle's windscreen.*

*For example, in art galleries or museums embedded LED lights can illuminate a glass display cabinet without aesthetically disruptive wires. In kitchens of the future, P-Tap technology can be installed in a clear glass counter, with power concentrated at various spots. This can provide induction heating for pans, power kitchen appliances such as kettles and toasters, and even charge your phone wirelessly, all on the same surface of glass.*

**Triveni Renewables** new 240 tpd manufacturing plant in India will feature an effective and environmentally friendly production line. The company commissioned **EME**, **SORG** and **Grenzebach** to construct the facility.

EME will provide Triveni with a batch plant and cullet return system, while SORG is responsible for the melting furnace. Grenzebach will deliver the annealing lehr and the cutting line.

The joint work will allow Triveni Glass, through its subsidiary company Triveni Renewables, to take advantage of India's growing demand for high quality solar glass.

Triveni's decision to commission a SORG melting furnace is based on SORG's extensive experience in the solar and figured glass industry. It also highlights the benefits of the long-term partnership between EME, SORG and Grenzebach to provide glassmakers worldwide with the best equipment for flat glass production.

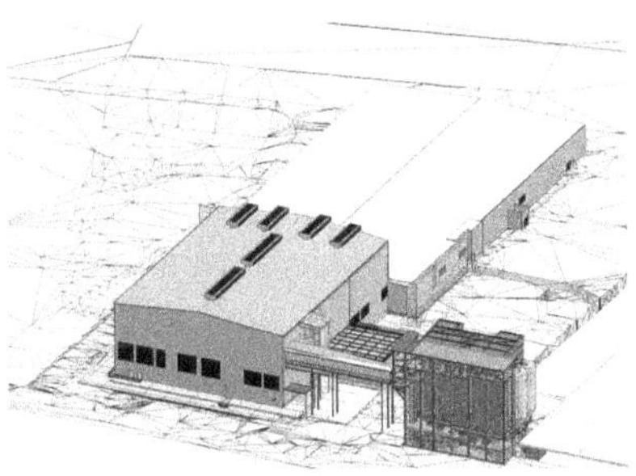

**Noelle+von Campe** processes up to 550 tons of glass to make approximately 1.8 million glass containers daily. The family-run company is located in Boffzen, Germany, and supplies more than 35 countries.

A completely new furnace with a maximum of 420 tons daily is proposed for the site in addition to the existing melting furnace with 320 tons melting capacity. The batch and cullet would need to supply two furnaces instead of one furnace. **CM. Project. Ing** was responsible for the complete coordination of the construction activities including detailed planning of the buildings and supply media.

**Zippe Industrieanlagen** received the order for an extension of the existing batch house and a completely new cullet return system. Cullet addition is 50%. The ZIPPE scope of supply includes the engineering, the equipment, the silos, as well as the complete mechanical and electrical installation including the supervision of the works.

Corning, NY

Properties · Composition · Applications · Structure

# Borate X – O – Phosphate III

**July 17–21, 2023**

## Revised Conference Dates

The 10th International Conference on
**BORATE GLASSES, CRYSTALS AND MELTS**
**July 17–19, 2023**

This marks the 10th in the series of conferences on borate glasses, crystals and melts, and will return to New York, USA close to where the first Borate Conference was held (Alfred, 1977). The most recent meeting took place in Oxford, UK (2017). The Borate Conference is a forum for the discussion of current scientific research on borate and B-containing materials. The conference will be dedicated to Dr. E.I. (Stratos) Kamitsos from the National Hellenic Research Foundation (Athens), to honor his lifetime of work in the field of borate glasses.

The 3rd International Conference on
**PHOSPHATE GLASSES, CRYSTALS AND MELTS**
**July 19–21, 2023**

This conference continues the tradition established with the first two conferences held in Pardubice, Czech Republic (2014) and Oxford, UK (2017). It will cover all aspects of phosphates and P-containing materials, including P-O, P-S, P-Se and P-O-N networks. Topics of interest will cover basic science, applications and technologies. The Phosphate Materials conference will be dedicated to Professor Ladislav Koudelka, University of Pardubice, Czech Republic, recognizing his career in phosphate and borophosphate glasses.

## Conference Themes

This joint meeting, to be held at the Corning Radisson Hotel, is comprised of two conferences that will run sequentially. On the third day of the meeting there will be a combined day dedicated to borophosphate glasses followed by a conference dinner to mark the close the Borate Conference and the start of the Phosphate Conference. Topics will encompass anything related to borate and phosphate materials, including:

- **Energy**
- **Healthcare**
- **Optical and data storage materials**
- **Property-structure relationships**
- **Fundamentals of the glassy state**
- **Novel glasses, crystals and melts**
- **Technology and Industry**

## Conference Chair

Dr. Randall Youngman
Corning Incorporated
youngmanre@corning.com

## Local Organizing Committee

R. Youngman (chair), J. Wu (Corning), L. Ma (Corning), L. Huang (RPI), A. Goel (Rutgers), D. Möncke (Alfred), A. Clare (Alfred)

***www.boratephosphate2020.org***

*__Dassault Systèmes__, __Ardagh Group__ and __EXXERGY__ will begin 'virtual twin' trials on __Diageo__'s iconic Johnnie Walker bottle in January 2022 to research and develop a coating that will enable the glass bottle to be lightweighted without compromising its strength and shape.*

*A virtual twin is a real-time digital representation of a product or process that is used to model, visualise and predict new innovations before any physical trials take place. The trial will research and develop an external coating for the glass bottle to reduce the micro-cracks in the glass surface. This will allow the glass to be much lighter while maintaining its strength. The bottle will still maintain its 100% recyclability and, if successful in the virtual trial, glass bottle testing from summer 2022.*

*Ardagh's work with EXXERGY will develop the coatings for Diageo. Dassault are providing contract research services using its BIOVIA applications to create a nanoscale virtual twin of the coatings, simulate their interaction with the surface of the glass and test their efficacy.*

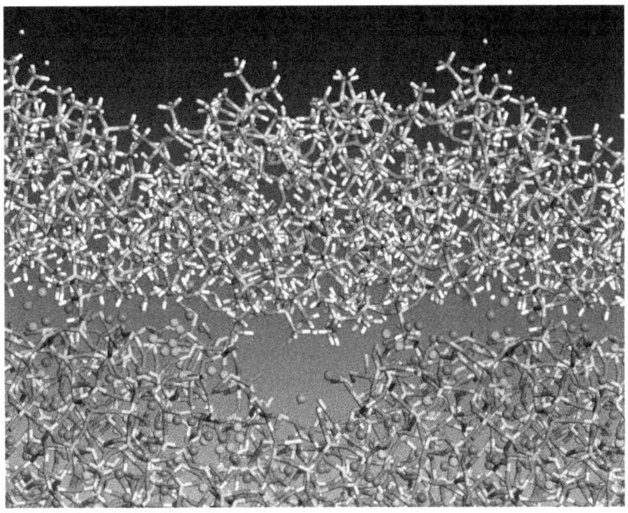

*Structural model of the coating on top of the glass, blunting or repairing any micro-cracks*

*John Sadlier, Chief Sustainability Officer at Ardagh Group, commented: "Ardagh Group has pioneered the lightweighting of glass and as a leading global supplier of infinitely recyclable, sustainable metal and glass packaging, we have a responsibility to respond to the sustainability challenges we all face. Together with our customers and supply chain partners, we are eager to explore the potential of digitalisation to drive new and innovative lightweighting solutions."*

*Lightweighting is one of the solutions for reducing the glass sector's $CO_2$ impact and carbon emissions will be reduced in both the manufacturing of the bottle and transportation of the finished goods. Should this collaboration project be successful, it will be a major breakthrough for virtual twin technology in supporting the sustainability ambitions of the glass industry and enabling the decarbonisation of the sector as a whole.*

*"We're excited to be part of this pioneering trial that embraces the principles of our Diageo Sustainable Solutions platform which encourages innovation and collaboration in developing scientific solutions to the impact of climate change. We've made a commitment in our Society 2030 strategy to have net zero carbon emissions across all operations by 2050 or earlier, and should this trial be successful, it's a great step towards us achieving that", said Lucy Fishwick, Head of Procurement – Grain to Glass Sustainability, Diageo.*

### SCHOTT receives funding to develop climate-friendly glass production

In the last quarter of 2020, **SCHOTT** announced its intention to become climate-neutral by 2030. A year later and the company has been awarded two grants totalling €4.5 million from the German Minister of the Environment Svenja Schulze to support the development of climate-friendly glass melting processes. The funding comes from the "Decarbonization in Industry" grant programme of the German **Federal Ministry for the Environment**.

The government subsidies will be used for two development projects that aim to replace the fossil fuel natural gas with electricity from renewable energy sources in heating the company's glass melting units. With the new technologies, SCHOTT plans to make a substantial reduction in its carbon emissions.

"Technological transformation is our greatest challenge on the path toward climate neutrality. It requires us to fundamentally rethink glass production and make some groundbreaking innovations. In particular, we are focusing on electrification through green electricity and hydrogen technology. The government research funding will help support our efforts in this area. That is why we are very grateful to have been awarded these grants," explained Dr Frank Heinricht, Chairman of the Board of Management at SCHOTT.

The first development project focuses on the melting process for pharmaceutical glass. Here, the research funding amounts to around €784,000. The second project involves the melting process for specialty glass used in technical applications, this grant amounts to €3.72 million. Specialty glass is melted at temperatures of up to 1700°C. Because glass melting is a highly complex process, developing new technologies in this area is associated with big challenges.

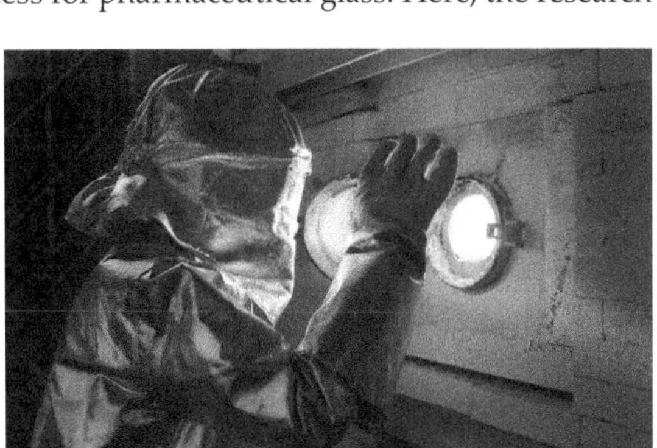

SCHOTT aims to achieve the 2030 goal of climate neutrality through a plan across four fields. In addition to technological change, these include continued improvement in energy efficiency, switching to 100% green electricity, and compensating for any unavoidable residual emissions.

# Stoelzle Glass Group wins over by innovative decoration techniques

**Stoelzle Glass Group** has used recent packaging awards events to demonstrate its expertise in the manufacture and decoration of outstanding high-quality glass containers. The company presented innovations, which set a clear focus on eco-friendly aspects and recycling, both key topics in the Group's common sustainability strategy at all sites.

An ink made from old tyres: For quite a long time the Stoelzle innovation teams together with external experts have been working on the development of new, eco-friendly inks. An innovative, environmentally-friendly black printing ink has been developed from rubber waste, exclusively for Stoelzle. The Recovered Carbon Black, which is used as a basis of new inks, derives a 100% from end-of-life tyres. Printed bottles obtain a deep black silk-screen effect, which is unlikely to be distinguished from decoration with common black inks. The recycled ink saves carbon and contributes to reduce rubber waste at the same time. The recycled ink decor is available in dark black as well as in shades of grey and is available in all three decoration sites of the Stoelzle Group, in France, Poland and the UK.

Embossed Lettering: Normally, embossing is done directly via the mould during production of the bottles. Stoelzle also offers a cold end embossing, which can be applied to the bottles like a label that remains invisible. The secret of this decorative technique is the perfect cutting of the embossed features. From looking closely or even touching, no one would guess that such a decoration might be a labelling! The embossing can be done on flint bottles as well as on already coated ones. There is almost no limitation in terms of colours: gold, silver, multicolour, gradient, whatever suits best. In addition, embossed bottles can still be recycled.

Next generation of screen-printing: The Stoelzle decoration teams have set themselves the goal to further develop the screen-printing techniques in order to allow utmost fine, distinctive, detailed, and almost photo-realistic artworks. The screen-printing is still based on classic inks, but on a much higher level of quality and broad choice of colours.

Ecofriendly powdercoating: the Quali Glass coat 2.0 technique, Stoelzle's special powder coating process, has won the "coup de coeur" at the Formes de Luxe Awards for being both ecofriendly and appealing. The innovative, eco-friendly powder coating is completely solvent free and helps save up to 77% $CO_2$ emissions compared to standard lacquering.

**Stoelzle Masnières Parfumerie** is very proud of their first collaboration with the **Editions de Parfums Frédéric Malle** teams. Perfume Gun, a unique collection of home fragrance guns was the first bottle treated with the innovative powder coating technique Quali Glass Coat. Applied to amber bottles, commonly used in the pharmaceutical industry, the 40 μm powder coating gives a perfectly smooth surface with a beautiful high-end matt finish.

Tigital: Another kind of powder based decoration is the innovative "Tigital" technology, which can achieve photo-realistic artworks whilst saving 85% $CO_2$ emission compared to classic screen printing. This technique suits very small lots and allows a high level of flexibility.

# CONFERENCE DIARY

**8 December 2021**
Furnace Solutions training seminar: Forehearth Colour Technology: Today's performance needs and new challenge presented by Pauline Darbouret of Ferro.
*Web www.sgt.org*

**12–17 December (new dates)**
14th Pacific Rim Conference on Ceramic and Glass Technology (PACRIM 14) including Glass & Optical Materials Division 2021 Annual Meeting (GOMD 2021), Vancouver, BC, Canada.
*Andrea Ross, The American Ceramic Society, 550 Polaris Parkway, Suite 510, Westerville, OH 43082, USA. Email aross@ceramics.org Web www.ceramics.org*

**2022 UN International Year of Glass**
**5–7 January**
Recent advances on the glass problem, Lausanne, Switzerland.
*Elisabeth Agoritsas, EPFL CECAM, Avenue de Forel 2, BCH 3103, 1015 Lausanne, Switzerland. Email helpdesk@cecam.org Web www.cecam.org/workshop-details/48*

**12 January**
Furnace Solutions training seminar: NOx emission and possibilities to reduce presented by Matthias Hagen of Luft- und Thermotechnik Bayreuth (LTB).
*Web www.sgt.org*

**22–27 January**
Photonics West, San Francisco, USA.
*SPIE, PO Box 10, Bellingham, WA 98227-0010, USA. Email MeetingInfo@SPIE.org Web spie.org*

**9 February**
Furnace Solutions training seminar: Challenges of Tin Bath Blocks in the "Float Process" and how to test them? Presented by Jan Theron of Lucideon.
*Web www.sgt.org*

**10–12 February**
Opening Ceremony of the International Year of Glass 2022, Geneva, Switzerland.
*Web www.IYoG2022.org*

**16–18 February 2022**
Glass Performance Days, Tampere, Finland.
*Brown Onduso, GPD, Vehmaistenkatu 5, 33730 Tampere, Finland Email brown.onduso@gpd.fi Web gpd.fi/*

**17–18 February**
Glassman Asia, Seoul, Korea.
*Ken Clark, Quartz Business Media, Quartz House, 20 Clarendon Road, Redhill, Surrey RH1 1QX, UK Email ken-clark@quartzltd.com Web www.glassmanevents.com*

**24 February**
Simulating glasses (webinar)
*Sara Bonella, EPFL CECAM, Avenue de Forel 2, BCH 3103, 1015 Lausanne, Switzerland. Email helpdesk@cecam.org Web www.cecam.org/workshop-details/1105*

**3–5 March**
glasspex INDIA 2022 plus GlassPro India, Mumbai, India.
*Vivek Bohra, Senior Project Manage Email bohrav@md-india.com Web www.glasspex.com*

**6–10 March**
Optical Fiber Communications Conference and Exhibition (OFC), San Diego, California USA.
*The Optical Society, USA. Email custserv@osa.org Web www.ofcconference.org*

**8–10 March**
JEC World, Paris, France.
*Web www.jec-world.events/*

**9–10 March**
Sustainable Industrial Manufacturing (SIM), Brussels, Belgium.
*Nicola Davison, Quartz Business Media, Quartz House, 20 Clarendon Rd, Redhill, Surrey, RH1 1QX, UK. Email nicoladavison@quartzltd.com Web sustainableindustrial-manufacturing.com*

**15–18 March**
Unified International Technical Conference on Refractories (UNITECR 2022) 17th Biennial Worldwide Congress on Refractories, Chicago, USA.
*The American Ceramic Society, 550 Polaris Parkway, Suite 510, Westerville, OH 43082, USA. Email aross@ceramics.org Web www.ceramics.org*

**18–20 March**
2022 International Flameworking Conference, Salem, NJ, USA.
*Salem Community College, 460 Hollywood Avenue, Carneys Point, NJ 08069, USA. Email info@salemcc.edu*

**2–3 April**
Deco '22, The Society of Glass and Ceramic Decorating Products annual conference, Columbus, OH, USA.
*Myra Warne, Society of Glass and Ceramic Decorating Products, PO Box 2489, Zanesville, OH 43702, USA. Email info@sgcd.org Web www.sgcd.org*

**3–6 April**
TMS World Congress on Artificial Intelligence in Materials and Manufacturing (AIM 2022), Pittsburgh, Pennsylvania, USA.
*TMS Meeting Services, 5700 Corporate Drive Suite 750, Pittsburgh, PA 15237, USA. Web www.tms.org/AIM2022*

**3–7 April**
Photonics Europe 2022, Strasbourg, France.
*SPIE Europe, 2 Alexandra Gate, Ffordd Pengam, Cardiff, CF24 2SA Wales. Email info@spieeurope.org Web spie.org*

**5–6 April**
GlassPrint 2022, Düsseldorf, Germany.
*Email sales@glassworldwide.co.uk Web www.glassprint.org*

**13–16 April**
China Glass 2022, Shanghai, China.
*Web www.chinaglass-expo.com*

**18–20 April**
Int. Conf. on Glass Science and Technology from the Golden Era of the Pharaohs to a Sustainable Future, Egypt.

**19–22 April**
13th European Conference on Industrial Furnaces and Boilers, Algarve, Portugal.
*Raquel Reis, Cenertec, Rua Gago Coutinho, 185-187, 4435-034 Rio Tinto, Portugal. Email raquel.reis@cenertec.pt Web infub.pt*

**11–12 May**
Glassman Latin America, Monterrey, Mexico.
*Ken Clark, Quartz Business Media, Quartz House, 20 Clarendon Road, Redhill, Surrey RH1 1QX, UK Email ken-clark@quartzltd.com Web www.glassmanevents.com*

**18–20 May**
REFRA PRAGUE 21, 21st Conference on modern refractory materials and key achievements in high temperature technologies, Prague, Czech Republic.
*Ing. Jan K. Luxa, Silikátová společnost České republiky, Novotného lávka 5, 110 01 Praha 1, Czech Republic. Email hluxa@seznam.cz Web www.silikaty.cz*

**18–21 May**
GAS 2022, Glass Art Society conference, Tacoma, WA, USA.
*Glass Art Society, 2208 NW Market Street, Suite 200, Seattle, Washington 98107, USA. Email info@glassart.org Web www.glassart.org*

**22–26 May**
2022 Glass and Optical Materials Division Annual Meeting (GOMD2022), Baltimore, USA.
*Andrea Ross, The American Ceramic Society, 550 Polaris Parkway, Suite 510, Westerville, OH 43082, USA. Email aross@ceramics.org Web www.ceramics.org*

**6–9 June**
Mir Stekla 2022, Moscow, Russia.
*Ekaterina Razumeitseva, Project Head, EXPOCENTRE, Moscow, Krasnopresnenskaya nab., 14, 123100, Russia. Email re@expocentr.ru Web www.mirstekla-expo.ru/en/*

**8–9 June**
Furnace Solutions and Training Day, venue TBC, UK.
*Christine Brown, Society of Glass Technology, 9 Churchill Way, Chapeltown, Sheffield S35 2PY, UK. Email christine@sgt.org Web www.sgt.org*

**16–17 June**
XXXIV ATIV Int. Conf., Where Glass Science, Art and Technology Meet Together, Parma, Italy.
*ATIV Secretariat, MV Congressi spa Via Marchesi 26 d 43126 Parma, Italy. ativ@ativ-online.it Web www.ativ.eu*

**21–24 June**
Techtextil 2022, Frankfurt, Germany.
*Email techtextil@messefrankfurt.com Web techtextil.messe-frankfurt.com*

**23–24 June**
Challenging Glass Conference 8: International Conference on the Architectural and Structural Application of Glass, Ghent, Belgium.
*Email cgc8@challengingglass.com Web www.challenging-glass.com*

**29 June–3 July**
Glassman South America, Sao Paulo, Brazil.
*Ken Clark, Quartz Business Media, Quartz House, 20 Clarendon Road, Redhill, Surrey RH1 1QX, UK Email ken-clark@quartzltd.com Web www.glassmanevents.com*

**3–8 July**
26th International Congress on Glass, Berlin, Germany. Hosted by the DGG in their centenary year.
*Annette Doms, Deutsche Glastechnische Gesellschaft, Siemenstrasse 45, D-63071 Offenbach, Germany. Email dgg@hvg-dgg.de Web www.hvg-dgg.de*

**4–8 July**
Ninth International Conference on Optical, Optoelectronic and Photonic Materials and Applications (ICOOPMA) and EuroDIM: 14th Europhysical Conference on Defects in Insulating Materials, Ghent, Belgium.
*Web icoopma-eurodim.ugent.be*

**6–8 July**
World Glasstech-MEXICO / Glass Technology Expo Mexico, Guadalajara, Mexico.
*Glasstech Mexico, Av. Santa Fe 170, 01210 Lomas de Santa Fe, Mexico. Email info@glasstechmexico.com Web www.glasstechmexico.com*

**10–15 July**
Physics of Non-Crystalline Solids, now also including Society of Glass Technology Annual Meeting Canterbury, UK.
*Christine Brown, Society of Glass Technology, 9 Churchill Way, Chapeltown, Sheffield S35 2PY, UK. Email christine@sgt.org Web www.sgt.org*

**10–14 July**
Ninth Int. Congress on Ceramics (ICC9), Krakow, Poland.
*European Ceramic Society AISBL, Av. Gouverneur Cornez 4, B-7000 Mons, Belgium. Email ecers@bcrc.be Web ecers.org/*

**18–21 July**
65th ASGS Symposium, Sarasota, Florida, USA.
*Email philip.legge@gmail.com*

**20–28 August**
ICOM Prague 2022, Prague, Czech Republic.
*Secretariat, ICOM Czech Republic, Moravian Museum, Zelný trh 6, 659 37 Brno, Czech Republic. Email lehman-nova@icom-czechia.cz Web prague2022.icom.museum*

**24–29 July**
Sol-Gel 2021, International Sol-Gel Society Conference, Lyon, France.
*Université Claude Bernard Lyon 1, Cellule Congrès - SOLGEL 2021, 43, bd du 11 Novembre 1918, 69622 Villeurbanne, France. Email solgel2021@univ-lyon1.fr Web www.solgel2021.fr*

**27–30 August**
International Festival of Glass, Stourbridge, UK.
*Ruskin Glass Centre, Wollaston Road, Amblecote, Stourbridge DY8 4HF, UK. Email ifg@rmlt.org.uk Web www.ifg.org.uk*

**15 September**
Going Global: neue Herausforderungen an die Provenienz-forschung, Graduate Workshop, Romont, Switzerland.
*Vitrocentre Romont Au Château PO Box 150, CH-1680 Romont, Switzerland. Email sarah.tabbal@vitrocentre.ch Web www.vitrocentre.ch*

**20–23 September**
Glastec 2022, Düsseldorf, Germany.
*Anke Seltmann, glastec, D-40474 Düsseldorf, Am Staad (Stockumer Höfe), Germany. Email seltmanna@messe-duesseldorf.de Web www.glasstec-online.com/*

**2–7 October**
Tenth International Conference on Multiscale Materials Modeling: including Multiscale Modeling of Glasses and Structurally Disordered Materials, Baltimore, Maryland, USA.
*MMM 10 Conference, 3400 North Charles Street, Department of Mechanical Engineering, Johns Hopkins University, Baltimore, MD 21286, USA. Email mmm10@jhu.edu Web mmm10.jhu.edu*

**7 October**
Celebrating the Birth of English and Irish Crystal Drinking Glass, 1640–1702, London, UK.
*Colin Brain, Association for the History of Glass. Email cbrain@interalpha.co.uk Web www.historyofglass.org.uk*

**9–13 October**
ACerS 124th Annual Meeting with Materials Science & Technology 2022, Pittsburgh, USA.
*The American Ceramic Society, 550 Polaris Parkway, Suite 510, Westerville, OH 43082, USA. Email customerservice@ceramics.org Web www.ceramics.org*

**11–13 October**
GlassBuild America 2022, Las Vegas, USA.
*National Glass Association, 1945 Old Gallows Road, Suite 750, Vienna, VA 22182, USA. Web www.glassbuildamerica.com/*

**17–22 October**
TCM-TOEO 2022 Conference: 8th International Symposium on Conductive Materials & 12th International Symposium on Transparent Oxide and Related Materials for Electronics and Optics, Heraklion, Crete.
*George Kiriakidis, Nik. Plastira 100, 70013, Heraklion, Crete, Greece. Email info@mitos.com.gr Web tcm2021.org/*

**18–20 October**
GlassBuild America, Las Vegas, Nevada, USA.
*National Glass Association NGA, 8200 Greensboro Drive, Suite 302, VA 22102 McLean, USA. Email attend@glass.org Web www.glass.org*

**19–21 October**
XXXIV ATIV Int. Conf., Where Glass Science, Art and Technology Meet Together, Parma, Italy.
*ATIV Secretariat, MV Congressi spa Via Marchesi 26 d 43126 Parma, Italy. ativ@ativ-online.it Web www.ativ.eu*

**26–28 October**
Glasstech Asia 2022, Singapore.
*Web www.glasstechasia.com.sg/*

**5 November**
International Conference on Glass in Museums: Research, Context, Communication, Frauenfeld, Switzerland.
*Email sarah.keller@vitrocentre.ch Web historisches-museum.tg.ch and vitrocentre.ch*

**9–11 November**
Recent Advances in Glass and Ceramics Conservation 2022. Sixth ICOM Interim Meeting of the Glass and Ceramics Working Group, Lisbon, Portugal and online.
*Unidade Investigação VICARTE – Vidro e Cerâmica para as Artes, NOVA School of Science and Technology, Hangar III – Campus da Caparica, 2829-516 Caparica, Portugal. Email vicarte.secretariado@fct.unl.pt Web eventos.fct.unl.pt/icomcc_gc_2022*

**2023**

**3–5 February**
Future Past, Ausglass 2023, Melbourne, Australia.
*Ausglass, GPO Box 101, Canberra ACT 2601, Australia. Email secretary@ausglass.org Web ausglass.org*

**5–9 March**
Optical Fiber Communications Conference and Exhibition (OFC), San Diego, California USA.
*The Optical Society, USA. Email custserv@osa.org Web www.ofcconference.org*

**4–9 June**
2023 Glass and Optical Materials Division Annual Meeting (GOMD 2023), New Orleans, USA.
*The American Ceramic Society, 550 Polaris Parkway, Suite 510, Westerville, OH 43082, USA. Email customerservice@ceramics.org Web www.ceramics.org*

**2–7 July**
18th European Ceramics Society Conference, Lyon, France.

**10–14 July**
Nonlinear Optics, Honolulu, Hawaii, USA.
*The Optical Society, 2010 Massachusetts Ave, NW, Washington, DC 20036 USA. Email custserv@osa.org Web www.osa.org*

**17–21 July**
Tenth International Conference on Borate Glasses, Crystals and Melts and Third International Conference on Phosphate Materials, Corning, USA.
*Dr Randall Youngman, Corning Incorporated Email youngmanre@corning.com Web boratephosphate2020.org/*

**~~August~~ (moving to 2024)**
NCM15, Structure of Non-Crystalline Materials 15, combined with the SGT Annual Meeting, UK.
*Christine Brown, Society of Glass Technology, 9 Churchill Way, Chapeltown, Sheffield S35 2PY, UK. Email christine@sgt.org Web www.sgt.org*

**31 October–2 November**
GlassBuild America, Atlanta, Georgia, USA.
*National Glass Association NGA, 8200 Greensboro Drive, Suite 302, VA 22102 McLean, USA. Email attend@glass.org Web www.glass.org*

**2024**

**June**
Tenth International Conference on Optical, Optoelectronic and Photonic Materials and Applications (ICOOPMA), Pardubice, Czech Republic.
*Center of Materials and Nanotechnology and Department of General and Inorganic Chemistry, Faculty of Chemical Technology, University of Pardubice, Cs. Legii sq. 565, 532 10 Pardubice, Czech Republic. Email info@icoopma.com Web www.icoopma.com*

**2–6 July**
XVIII ECerS Conference, Lyon, France.
*European Ceramic Society AISBL, Av. Gouverneur Cornez 4, B-7000 Mons, Belgium. Email ecers@bcrc.be Web ecers.org/*

**14–19 July**
Tenth Int. Congress on Ceramics (ICC10), Montreal, Canada.
*The American Ceramic Society, 550 Polaris Parkway, Suite 510, Westerville, OH 43082, USA. Email customerservice@ceramics.org Web www.ceramics.org*

**20–23 August**
Materials Challenges in Alternative & Renewable Energy 2023 combined with the Sixth Annual Energy Harvesting Society Meeting (EHS 2023), Bellevue, Washington, USA.
*The American Ceramic Society, 550 Polaris Parkway, Suite 510, Westerville, OH 43082, USA. Email customerservice@ceramics.org Web www.ceramics.org*

Glass Technol.: Eur. J. Glass Sci. Technol. A, 2021, 62 (6), 198–204

# Four testing and evaluation methods of the spreading effects on E-glass woven fabrics

*Xi Wang,*[a] *Tianyong Zheng*[a,1] *& Xiangchun Ning*[b]

[a] *School of Textile Science and Engineering of Tiangong University, Tianjin, China 300387*
[b] *Henan Guangyuan new material Co., Ltd., Anyang, China 456500*

Manuscript received 11 January 2021
Revision received 26 October 2021
Manuscript accepted 18 November 2021

*The testing and evaluation methods of the spreading effects on E-glass woven fabrics are studied. Multiple improved methods are proposed based on the existing methods, which are summarised into four types, including the surface method, the section method, the air permeability method and the wettability method. These four testing and evaluation methods are used to separately test several different types of E-glass woven fabrics. The surface method shows that the porosity of glass fabrics dropped by 35–58% after spreading and the section method shows that the height of yarn amplitude dropped by 5–11%. The air permeability method shows that the air permeability of glass fabrics dropped by 25–37% and the wettability method shows that the contact angle between diluted epoxy and glass fabrics dropped by 12–27%. According to different fabric types and various application requirements, the spreading effects on E-glass woven fabrics can be comprehensively evaluated using various methods, which can help E-glass and CCL companies to better evaluate the quality of spread fabrics.*

## 1. Introduction

Electronic woven glass fabrics, abbreviated as E-glass fabrics, are interwoven with warp and weft composed of multiple continuous glass filaments. They are fibre reinforced materials of copper clad laminate (CCL).[(1)] The rapid development of the electronics industry has put higher requirements on E-glass fabrics and spawned the use of spreading technology used on E-glass fabrics. Currently, the mainstream spreading technology typically uses high-pressure jets with deionised water as the main medium to perform physical processing on grey fabrics to obtain spread fabrics that have undergone stretching and thinning treatment.

The intuitive performance of E-glass fabrics after spreading is that the yarns are opened and flattened, the width increases, and the pores between yarns becomes smaller. The surface area of warp and weft increases while the protrusions at the interwoven points decrease. The fabric surface becomes smoother and softer. This has many positive effects. Firstly, the contact area between spread fabric and resins is larger, the penetration speed of resins is faster, the penetration time is shorter and the penetration is more uniform, causing spread fabrics to have better resin wettability, improving the peeling strength of laminates.[(2)] Meanwhile, the surface of spread fabrics is smoother and the laminates become smoother, which is conducive to drilling.[(3,4)] The inner wall of the hole is smooth and the hole position accuracy is higher. Defects such as different hole wall smoothness and different conduction quality conditions resulting from drilling between the pores of two yarns could be avoided.[(5,6)] The workability and dimensional stability of laminates have been improved. Additionally, the internal stress of yarns is released to a certain extent due to spreading, the thermal expansion and contraction degree of laminates is relatively reduced. Fibres and resins in laminates are relatively uniform and tightly combined, which can effectively improve the thermal stress of the laminate.[(7)]

As spreading technology for E-glass fabrics becomes gradually more mature, the spreading effect becomes one of the key factors in determining the quality of spread fabrics, which directly affects the grade and price of products. However, the present evaluation methods for this spreading effect are still limited. The tested objects are not direct enough. There are still no intuitive, accurate, and efficient evaluation methods. In this paper, the testing and evaluation methods of spreading effects are summarised into four types, including the surface method, the section method, the air permeability method, and the wettability method. Firstly, the surface method of directly observing the front of the fabric with an optical microscope is introduced. In addition, an improved semi-automatic porosity measurement method using a laser scanning confocal microscope is also introduced. The second method is the section

[1]Corresponding author. Emailzty_zzti@126.com
DOI: 10.13036/17533546.62.6.001

method, including observing the yarn's cross section and measuring the yarn amplitude. The third method is the air permeability method, measuring the air permeability of the whole piece of fabric with a fabric air permeability tester. The fourth method is the wettability method, including the evaluation method to determine the impregnating time of resin and the evaluation method to measure the contact angle between the diluted epoxy and the fabric. These four testing and evaluation methods were used to test and compare several different types of E-glass fabrics.

## 2. Surface method

The most intuitive way to evaluate the spreading effect is to observe and measure the surface changes of E-glass woven fabrics. Generally, a microscope is used to directly observe the front of the fabric sample, test the width and interval of warp and weft, and then calculate the area of the interweaving points of warp and weft, the pore areas, as well as the porosity.[8] The spreading effect of E-glass woven fabrics can be evaluated by comparing the changes in this data before and after spreading, as shown in Figure 1.

Sixty pieces samples of 2116, 1080, 106 glass fabrics before and after spreading were taken to test the width and interval data of the warp and weft. Then the average values were obtained. The average interweaving point area, the average pore area, and the porosity was also calculated. The statistical data is shown in Table 1.

Porosity refers to the ratio of the pore area to the total area.[9] The schematic diagram of the porosity calculation of woven fabrics is shown in Figure 2.

The specific calculation method is shown in Equation (1):

$$K = \frac{100X_k Y_k}{(X_j + X_k)(Y_w + Y_k)} \quad (1)$$

where $X_k$ is warp interval, $Y_k$ is weft interval, $X_j$ is warp width, $Y_w$ is weft width, $K$ is porosity.

By comparing the data in Table 1, the spread fabric warp width of 2116, 1080, 106 glass fabrics increased by 3·8%, 32·4%, and 12·9%, respectively. The weft width increased by 6·4%, 27·5% and 16·3%, respectively. The porosity decreased by 35·2%, 57·6% and 36·4%, respectively.

The above-mentioned testing method can measure the changes in the width of yarns and the porosity of fabrics. It is one of the commonly used methods at present, but it depends on a large number of manual measurements to ensure accuracy. The efficiency of the test is therefore limited, and image processing technology can solve such problems.[10] We propose an semi-automatic porosity measurement method based on the principle of image segmentation, which can improve the efficiency and accuracy. Firstly, for this method, a laser scanning image for image segmentation is required. Using the Keyence VK-X110 laser scanning confocal microscope to scan the surface of the fabric sample with a red semiconductor laser and obtain the laser scanning image, as shown in Figure 3.

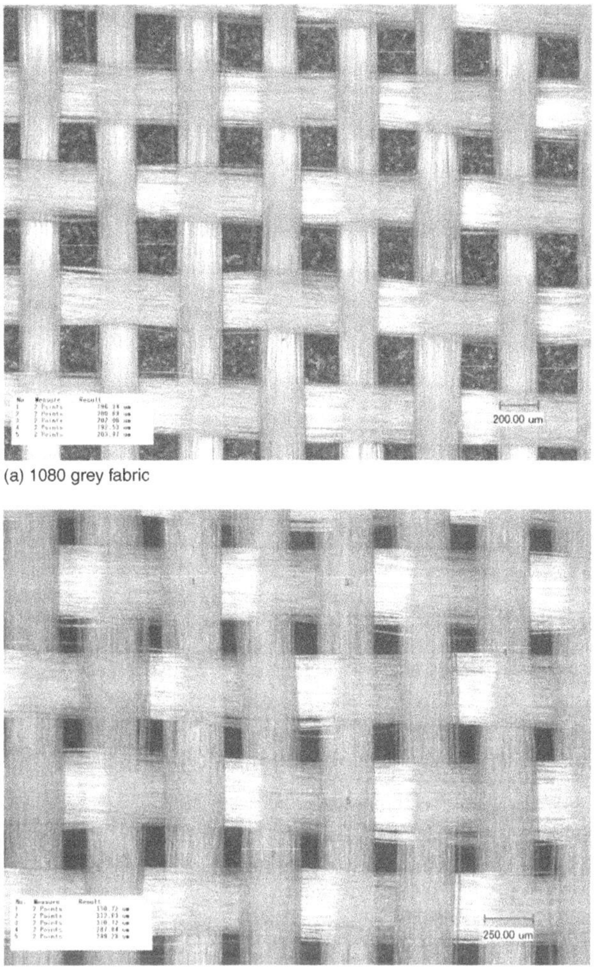

(a) 1080 grey fabric

(b) 1080 spread fabric

*Figure 1. Surface test of E-glass woven fabrics [Colour available online]*

Then the laser scanning image was imported into Image-Pro Plus image analysis software. The image segmentation tool was used to segment the image by

*Table 1. Statistical data of surface test of different E-glass fabrics*

| Fabric type | Warp width (μm) | Warp interval (μm) | Weft width (μm) | Weft interval (μm) | Interweaving point area (μm²) | Pore area (μm²) | Porosity (%) |
|---|---|---|---|---|---|---|---|
| 2116 Grey | 329·5 | 96·4 | 344·5 | 92·8 | 113512·8 | 8945·9 | 4·80 |
| 2116 Spread | 341·9 | 82·2 | 366·4 | 70·0 | 125272·2 | 5754·0 | 3·11 |
| 1080 Grey | 232·6 | 186·5 | 286·4 | 227·1 | 66616·6 | 42354·2 | 19·68 |
| 1080 Spread | 308·0 | 127·4 | 365·1 | 145·5 | 112450·8 | 18536·7 | 8·34 |
| 106 Grey | 223·6 | 217·6 | 271·0 | 185·5 | 60595·6 | 40364·8 | 20·04 |
| 106 Spread | 252·4 | 192·4 | 315·1 | 131·6 | 79531·2 | 25319·8 | 12·74 |

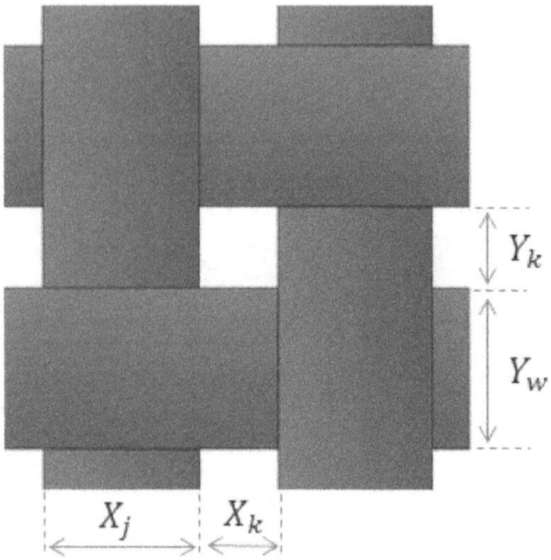


*Figure 2. Schematic diagram of porosity calculation of woven fabrics*

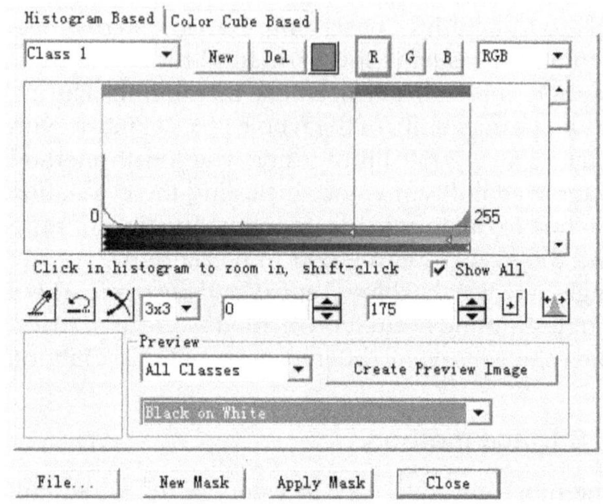


*Figure 4. Image segmentation by setting RGB value [Colour available online]*

setting the RGB value. The histogram of RGB statistical values provided by the software makes it easy to select the RGB value range and is not affected by the operator perception. In most cases, the upper limit of *R* value needs to be adjusted to within 200 for the laser scanning image of E-glass woven fabric such as Figure 3. The upper limit of *G* value should avoid the statistical range of most *R* values in the histogram, which is usually slightly lower than 240. The *B* value basically does not need to be adjusted and remains unchanged from 0–255. The image segmentation tool is shown in Figure 4.

The obtained pore image of E-glass woven fabric is shown in Figure 5, where the black pixels represent pore between the yarns and the white pixels represent the area covered by the yarns.

It can be seen from Figure 5 that the pore represented by the black pixels are not ideal rectangles, but are irregular and discontinuous. After comparison with the laser scanning image, it is not difficult to find that these irregular and discontinuous pores exactly match the pore situation on the laser scanning image, including the uneven yarn width. The actual situation such as the separation between certain fibres and the main body of the yarns can be reflected either. Therefore, this black and white image can truly reflect the actual situation of the pores on the fabric surface, which is more accurate than the traditional method of measuring the width and length of yarns and calculating the pores approximate to rectangles.

Then the automatic area measurement tool of the Image-Pro Plus software was used, which can easily count the total area of the black pixels without measuring the area of each pores, as shown in Figure 6.

The area of the black pixels refers to the total pore area on the sample. The porosity can be obtained by counting the percentage of the black pixels area to the total pixels area. Table 2 shows the porosity data of 1080 and 2116 glass fabric before and after spreading, which were measured using the image segmentation method.

Based on the porosity data from Table 2, it can be further calculated that the porosity of 2116 and 1080 E-glass fabrics decreased by 36·2% and 56·8%, respectively, after spreading. This is similar to the

*Figure 3. The laser scanning image of 1080 glass fabric [Colour available online]*

*Figure 5. Fabric pore image after image segmentation, see Figure 3 to the left*

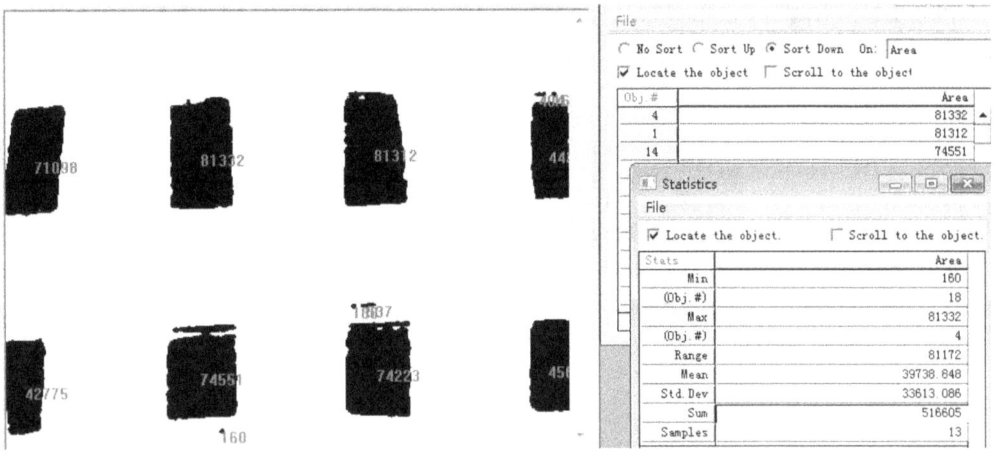


*Figure 6. Automatic measurement of total pore area [Colour available online]*

*Table 2. Statistical data of porosity measurement by using the image segmentation method*

| *Fabric type* | *Total image area (μm²)* | *Total pore area (μm²)* | *Porosity (%)* |
|---|---|---|---|
| 2116 Grey | 3413333·32 | 146090·67 | 4·28 |
| 2116 Spread | 3413333·32 | 93184·00 | 2·73 |
| 1080 Grey | 3413333·32 | 600746·66 | 17·60 |
| 1080 Spread | 3413333·32 | 259754·67 | 7·61 |

porosity changes calculated by traditional manual measurement methods. This semi-automatic measurement method can save lots of manual operations and avoid errors caused by manual measurement. Additionally, the real shape of the pores on the fabric surface and the pixel data can also improve the measurement accuracy.

## 3. Section method

It is quite intuitive to evaluate the spreading effect from the surface of the fabric sample. Relatively intuitive results can also be obtained by observing and measuring a sample's section. The section of the fabric sample can be divided into two types, including the cross section and the longitudinal section of the yarns.

The common practice is to observe the fibre stratification of the yarn's cross section. By taking a slice of the fabric sample's cross section and using a microscope to observe it, or use a laser scanning confocal microscope to directly scan the fabric sample's cross section, researchers can get an image of the yarn's cross section from the fabric sample. The fibre stratification of the yarn's cross section was observed, and the spreading effect was evaluated by comparing the changes in the number of fibre layers, as shown in Figure 7. For example, the weft of grey fabrics have 6–8 layers of fibres, the weft of spread fabrics can have as few as three layers of fibres, and high-quality spread fabrics have only two layers of fibres.

We propose an evaluation method for measuring the height of yarn amplitude from the yarn's longitudinal section image. In the traditional textile industry, the wave curve formed by bent yarn in the fabric is called the yarn amplitude, and the altitude difference of the yarn centre line is called the height of yarn amplitude.[11] Considering that the yarn of E-glass fabrics is thin and the yarn amplitude is relatively flat, to avoid measurement errors caused by inaccurate situations of the yarn's centre line, the method for measuring the altitude difference between the peak and the valley was used as the statistical data of the height of yarn amplitude, as shown in Figure 8.

After obtaining the yarn's longitudinal section image of the fabric sample, the perpendicular distance measurements from a line tool of the Image-Pro Plus software was used. The height of yarn amplitude can be obtained by taking the connecting line of the two valleys in the yarn waveform as the reference line, and then selecting the peak of the yarn waveform as the measurement point. 75 fabric samples of 2116, 1080, 106 before and after spreading were taken to test the heights of warp and weft amplitude from the longitudinal section. The test results are shown in Table 3.

It can be seen from Table 3 that the thinner the fabric, the smaller the height of yarn amplitude. The

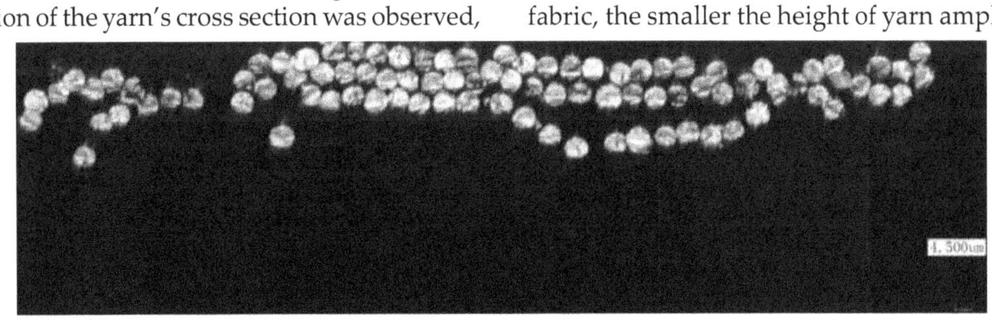

*Figure 7. Cross section image of 106 spread fabric's weft (D900 yarn)*

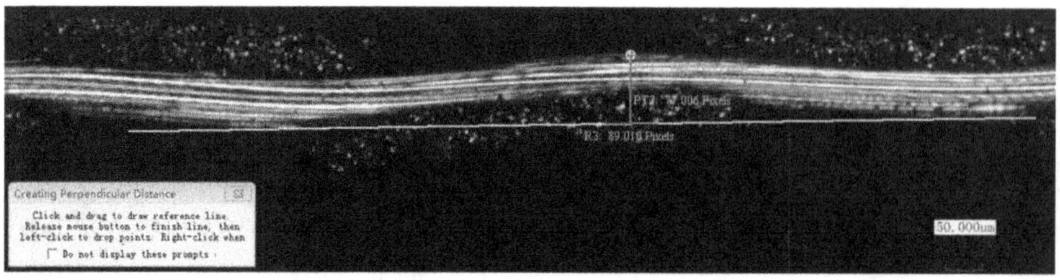

*Figure 8. Height of yarn amplitude test in E-glass woven fabric [Colour available online]*

*Table 3. Statistical data of the heights of yarn amplitude*

| *Fabric type* | *Yarn type* | *Average height of yarn amplitude of grey fabrics (μm)* | *Average height of yarn amplitude of spread fabrics (μm)* | *Decrease ratio of height of yarn amplitude (%)* |
|---|---|---|---|---|
| 2116 | Warp | 109·1 | 101·0 | 7·4 |
| | Weft | 103·1 | 92·0 | 10·8 |
| 1080 | Warp | 83·6 | 77·1 | 7·8 |
| | Weft | 82·6 | 74·9 | 9·3 |
| 106 | Warp | 66·1 | 62·9 | 4·8 |
| | Weft | 49·6 | 45·3 | 8·7 |

heights of warp amplitude were generally larger than the heights of weft amplitude. The heights of warp and weft amplitude of spread fabrics were generally smaller than grey fabrics, while the heights of weft amplitude decreased more, indicating that the spreading effects for the weft were more effective.

## 4. Air permeability method

The air permeability method tests the whole piece of fabric by using a fabric air permeability tester, which evaluates the spreading effect through the air permeability change before and after spreading.[12,13] When the sample area and pressure difference are specified, the air flow passing vertically through the sample within a certain time can be tested and the rate of air flow passing through the sample, i.e. air permeability, can be calculated. The sample shall be clamped on the circular test bench, and the sample shall not be close to the selvedge and free of wrinkles. Meanwhile, the sample shall have sufficient tension to make the sample flat without deformation. The suction ventilator is started to let the air pass through the fabric sample, and then the pressure drops and gradually approaches the specified value by adjusting the flow. After 1 min or when the air flow reaches stability, the pressure difference on both sides of the fabric sample is recorded. Under the same conditions, different parts of the same sample shall be tested at least 10 times. After the air permeability of the 106, 1080, 2116, and 7628 E-glass woven fabrics were tested, the statistical data were obtained, as shown in Table 4.

It can be calculated from Table 4 that after spreading the air permeability of fabric 106 dropped by 25·2%, the air permeability of fabric 1080 dropped by 36·4%, the air permeability of fabric 2116 dropped by 36·8% and the air permeability of fabric 7628 dropped by 28·7%.

The air permeability test can reflect the air permeability of the whole sample with low cost and a convenient operation. It is an indirect method used to characterise the spreading effect on fabrics. However, this method can only provide measurement data for the fabric as a whole, which cannot reflect the microscopic situation of spreading. It is also not as intuitive as the surface method or the section method.

## 5. Wettability method

The three evaluation methods mentioned above all quantitatively describe the physical characteristics of E-glass fabrics before and after spreading. However, one of the most important meanings of spreading is to better permeate resins into fabrics. These three evaluation methods have no direct relationship to the resin wettability of fabrics. The spreading effect is not equivalent to the change in these physical quantities themselves.

Among the existing evaluation methods, there is an evaluation method for resin wettability of glass fabrics, which sets the impregnation time of resins, namely, the time from the resin injection to the disappearance of the reference line as the reference basis for comparison of fabrics before and after spreading. However, the definition of the impregnation time in this method depends on the subjective judgment of the tester, and it is not accurate sometimes. Moreover, it can only provide a time data, which is not intuitive enough. In this paper, we propose a method to test the contact angle to evaluate the spreading effect of E-glass woven fabrics.

The Dataphysics OCA20 contact angle measuring system was used to measure the contact angle of E-glass fabrics, and DEGBA epoxy resin for FR-4 copper clad laminate was used. Firstly, the epoxy

*Table 4. Statistical data of air permeability test of E-glass fabrics*

| *Fabric type* | *Average nozzle pressure difference (Pa)* | *Average air permeability ($mm^3/s$)* | *Sample pressure difference (Pa)* | *Sample area ($cm^2$)* |
|---|---|---|---|---|
| 106 Grey | 2203·1 | 3060·0 | 200 | 20 |
| 106 Spread | 1239·8 | 2287·4 | 200 | 20 |
| 1080 Grey | 2826·4 | 2180·0 | 200 | 20 |
| 1080 Spread | 1154·6 | 1386·2 | 200 | 20 |
| 2116 Grey | 2497·7 | 818·6 | 200 | 50 |
| 2116 Spread | 835·4 | 517·2 | 200 | 50 |
| 7628 Grey | 2220·6 | 77·7 | 200 | 50 |
| 7628 Spread | 1154·3 | 55·4 | 200 | 50 |

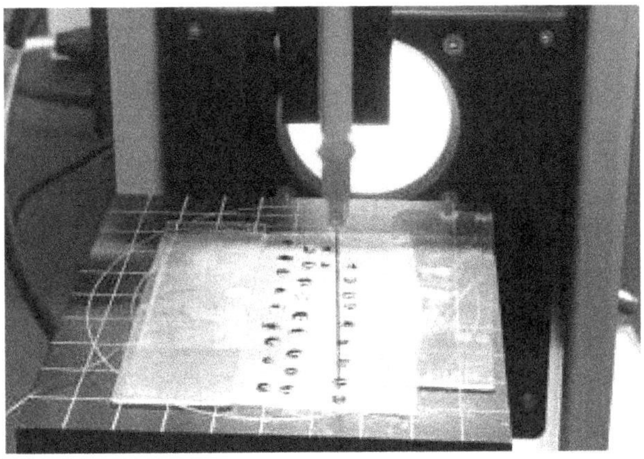

*Figure 9. Diagram of the contact angle test [Colour available online]*

resin with acetone was diluted at a volume ratio of 1:5 and then the mixed solution was injected into a syringe with a 0·5 mm diameter needle. Meanwhile, a sample of glass fabrics was cut within 10×10 cm and placed horizontally on the sample table. There was a background light source and a camera on both sides of the sample table, as shown in Figure 9.

The light brightness, the focal length of the camera, and the position of the needle were adjusted to make the image captured by the camera clear with significant black-white contrast. Then about 0·2 ml of the resin mixture was slowly injected from the needle, lightly touching the fabric surface as a droplet. Then the image was captured after the droplet fell on the fabric surface to obtain a clear static image. After the droplet stayed on the fabric surface for a period of time, it would gradually infiltrate and the contact angle would also change accordingly. Therefore, the image should be captured within 2–5 s after the droplet falls on the fabric surface.

Then, the droplet was integrated through image processing. At the contact points on the left and right sides of the droplet with the fabric surface, the angles between the tangent of the droplet surface and the baseline of the fabric surface were the left and right contact angles, as shown in Figure 10.

The spreading effect can be evaluated by comparing the changes in the contact angles before and after the spreading, as well as the resin wettability of E-glass fabrics. The contact angle tests of fabrics 1080, 2116, and 7628 glass fabrics before and after spreading showed that the contact angle was reduced by 26·5%, 16·7%, and 11·8%, respectively. The smaller the contact angle, the more fully spread the glass fabrics, the better the resin wettability and the better spreading effect.[14] The wettability method is a more effective and more direct method for evaluating the spreading effect of E-glass fabrics, which can be used as a useful supplement for other evaluation methods. In addition, the size used in the processing of E-glass woven fabrics has an influence on the contact angle. The relationship between size content and contact

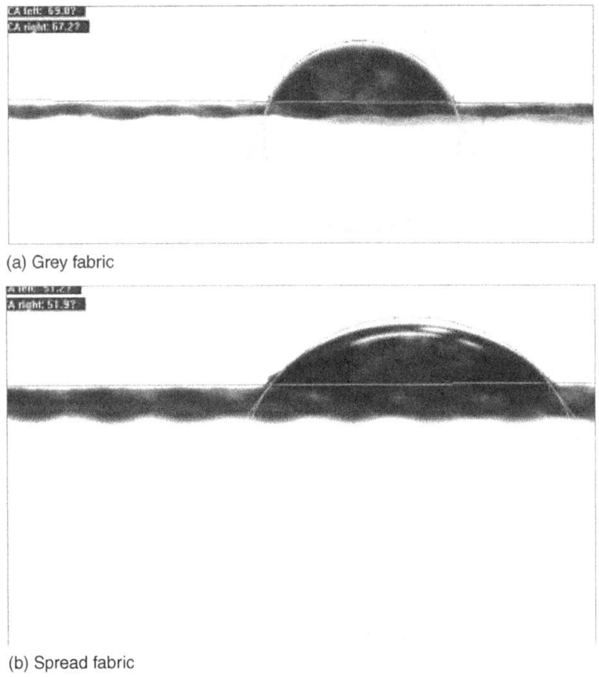

(a) Grey fabric

(b) Spread fabric

*Figure 10. Contact angle between diluted epoxy and fabrics [Colour available online]*

angle could be studied by comparing the contact angle between diluted epoxy and E-glass woven fabric with different size content.

## 6. Discussions

### *6.1. Selection of evaluation methods*

Each evaluation method can provide quantitative test data. The impact ratios of spreading effect on glass fabrics can be calculated by comparing the data changes before and after spreading. However, it cannot be mechanically assumed that these ratios should be exactly the same or even close, because various methods test different physical quantities, which are not comparable. Each evaluation method reflects the spreading effect from a specific perspective, so one or multiple methods could be used to comprehensively evaluate the spreading effect based on the actual situation of different fabric types and different application requirements.

### *6.2. Difference between warp and weft spreading*

Although it is not possible to directly compare difference ratios of various physical quantities under different methods, it can be seen from all evaluation methods that the spreading effect on the warp is generally not as good as that of the weft. Since the warp of woven fabrics is sizing at the same time as the warping during the preparation before weaving, the integrity of yarns is greatly improved. Moreover, during the weaving process, the warp is subjected to great tensions, making the spreading effect not as effective on the weft. Therefore, the spreading effect on warp is the key to the evaluation of the spreading effect on E-glass fabrics.

### 6.3. Thickness measurement of E-glass fabrics

The section method should easily be considered as the measurement of fabric thickness, but the actual situation is that traditional fabric thickness measurement method are not suitable for E-glass fabrics.[15,16] Since the thickness of E-glass fabrics is often only a few tens of microns, while the minimum accuracy of traditional fabric thickness tester is generally above 10 μm.[17] Additionally, the friction coefficient of glass yarns is much smaller than that of cotton yarns. The pressure weight of the fabric thickness tester will cause deformations on glass fabrics, further affecting the measurement accuracy. Therefore, the fabric thickness tester is not reliable on E-glass fabrics.

The laser confocal displacement sensor for non-contact thickness measurement is also not suitable. Although its accuracy is high enough to meet the thickness measurement requirements of E-glass fabrics, but glass fabrics are different with CCL. Glass fabrics that have not been coated with resins have pore, which will cause large errors in the laser confocal displacement sensor measurement. The same problem also exists with the measurement of surface roughness. The measurement of surface roughness is valuable for CCL, but it is not feasible for glass fabrics for the same reason. Due to these pore in glass fabrics, the calculations for surface roughness of glass fabrics have no practical significance.

## 7. Conclusion

In this paper, four evaluation methods of the spreading effect on E-glass fabrics were discussed. Each of the four evaluation methods for the spreading effect of E-glass fabrics has its advantages and disadvantages. In the surface method, the traditional surface test is intuitive and simple but requires a lot of manual measurement and a long time, while the semi-automatic porosity measurement based on image segmentation can improve the efficiency and accuracy. The measurement efficiency of modern optical microscope has been greatly improved. On a microscope that could perform analysis within 60 s, the surface method would be preferred instead of laser confocal microscope plus off-line processing. The section method is also quite intuitive, but it is tedious and time-consuming to make fabric sample slices and obtain section images. The measurement of the height of yarn amplitude is still done by manually processing the images. The air permeability method is simple, fast and low-cost, but it is only a macroscopic representation, which can only indirectly reflect the spreading effect, and is not as intuitive as other methods. The wettability method can reflect the resin wettability of glass fabrics directly and effectively. The measurement of contact angles is intuitive, but it is still necessary to measure the contact angle manually, which is not the most convenient method. Additionally, there are still more characterisation methods and technical means worth trying and researching, laying a foundation for further improving the quality of spread fabrics.

## References


1. Zhang, Z.Y., Yuan, L., Liang, G. Z. & Gu, A. J. Fabrication and origin of flame retarding glass fibre/bismaleimide resin composites with high thermal stability, good mechanical properties, and a low dielectric constant and loss for high frequency copper clad laminates. *Rsc Advances.*, 2016, **6** (24), 19638–46.
2. Opening eyes on fibre weave and CAF. IPC APEX Expo 2009, Las Vegas, NV, United States, 31 Mar.–2 Apr. 2009. Edited by R. Dudek, J. Kuhn & P. Goldman, IPC - Association Connecting Electronics Industries.
3. Aoyama E, Nobe H & Hirogaki T. Drilled hole damage of small diameter drilling in printed wiring board. *J. Mater. Proc. Technol.*, 2001, **118** (1–3), 436–41.
4. Evaluation and improvement of micro-drilled-hole wall quality in printed wiring boards made of glass fibre-reinforced plastics. 2009 ASME InterPack Conference, IPACK2009, San Francisco, CA, United States, 19–23 July 2009. Edited by K. Ogawa, H. Nakagawa & M. Matsudaira, American Society of Mechanical Engineers (ASME), 2010.
5. Fibre weave effect: practical impact analysis and mitigation strategies. DesignCon 2007, Santa Clara, CA, United States, 29 Jan.–1 Feb. 2007. Edited by J. Loyer, R. Kunze & X. Ye, International Engineering Consortium, 2007.
6. Through-hole drilling condition of FR-4 printed wiring boards based on preventing ion migration. Int. Electronic Packaging Technical Conference and Exhibition. 6–11 July 2003; Haui, HI, United states. Edited by E. Aoyama, T. Ikuta, T. Hirogaki, K. Yamauchi, H. Nobe, T. Katayama *et. al.*, American Society of Mechanical Engineers.
7. Li, H., Eng, D., Tang, C. & Westbrook, P. Low dielectric glass fibre development – new printed circuit board base materials. *Glass Technol.: Eur. J. Glass Sci. Technol. A*, 2013, **54** (2), 81–5.
8. Zupin, Z., Hladnik, A. & Dimitrovski, K. Prediction of one-layer woven fabrics air permeability using porosity parameters. *Textile Res. J.*, 2012, **82** (2), 117–28.
9. Dubrovski, P. D. & Brezocnik, M. Using genetic programming to predict the macroporosity of woven cotton fabrics. *Textile Res. J.*, 2002, **72** (3), 187–94.
10. Jensen, M., Keding, R., Fjendbo, S., Poschwatta, H. H. & Yue, Y. Z. Effect of bubbles on the characterisation of striae in glasses. *Glass Technol.: Eur. J. Glass Sci. Technol. A*, 2010., **51** (4), 147–52.
11. Zheng, T. Y., Cai, Y., Jing S. J. & Zhang, X. J. Study on non-devastating measurement and reconstruction of the three-dimensional geometric structure of woven fabrics. *Textile Res. J.*, 2011, 81 (10),1027–38.
12. *Textiles. Determination of the permeability of fabrics to air. Textiles - Determination de la permeabilite a I air des etoffes.* ISO 9237: 1995. 1995, 1–15.
13. Ince, M. E. & Yildirim, H. Air permeability and bursting strength of weft-knitted fabrics from glass yarn. Part II: knit architecture effect. *J. Textile Inst.*, 2019, **110** (7), 1072–84.
14. Tu, Y. Y., Zou, H. L., Lin, S. D. & Hu, J. W. Understanding the mechanism for building woven fabrics with wettability ranging from superhydrophobic to superamphiphobic via an aqueous process. *React. Funct. Polymers*, 2017, **119**, 75–81.
15. *Textiles. Determination of thickness of textiles and textile products.* Textiles Determination de lepaisseur des textiles et produits textiles (ISO 5084:1996). 1997, 1–7.
16. Fontaine, S., Durand, B. & Freyburger, J. M. Fabric thickness dynamic measurement during a classic uniaxial tensile test. *Exp. Mech.*, 2002, **42** (1), 84–92.
17. Musa, A. B. H., Malengier, B., Vasile, S., Van Langenhove, L. & De Raeve, A. Analysis and comparison of thickness and bending measurements from fabric touch tester (FTT) and standard methods. *Autex Res. J.*, 2018, **18** (1), 51–60.

www.ingramcontent.com/pod-product-compliance
Lightning Source LLC
LaVergne TN
LVHW060212210926
848481LV00008B/2702